01/12/05

Mammalian Social Learning
Comparative and Ecological Perspectives

Social learning commonly refers to the social transfer of information and skill among individuals. It encompasses a wide range of behaviours that include where and how to obtain food, how to interact with members of one's own social group, and how to identify and respond appropriately to predators. The behaviour of experienced individuals provides natural sources of information, by which inexperienced individuals may learn about the opportunities and hazards of their environment, and develop and modify their own behaviour as a result. A wide diversity of species is discussed in this book, some of which have never been discussed in this context before, and particular reference is made to their natural life strategies. Social learning in humans is also considered by comparison with other mammals, especially in their technological and craft traditions. Moreover, for the first time discussion is included of the social learning abilities of prehistoric hominids.

HILARY BOX is a Senior Lecturer in the Department of Psychology at the University of Reading, and has previously been President of the Primate Society of Great Britain and a Vice-President of the International Primatological Society.

KATHLEEN GIBSON is Professor and Chair of Basic Sciences and Adjunct Professor of Neurobiology and Anatomy at the University of Texas, Houston, and Adjunct Professor of Anthropology at Rice University. She has previously been Chair of the Section of Biological Anthropology and a member of the Executive Board of the American Anthropological Association.

Symposia of the Zoological Society of London

The series *Symposia of the Zoological Society of London* was originally established in 1960 and published the invited contributions to international meetings held by the Society to explore a wide variety of zoological topics.

The series has been published by Cambridge University Press since 1997, and it is now evolving to focus particularly on conservation biology, while continuing to include volumes on other zoological topics. With the addition of specially commissioned contributions to augment those arising from the Society's meetings, an integrated, comprehensive and authoritative treatment of the subject is ensured.

Symposia of the Zoological Society of London 72

Mammalian Social Learning:

Comparative and Ecological Perspectives

Edited by

Hilary O. Box and Kathleen R. Gibson

CAMBRIDGE
UNIVERSITY PRESS

PUBLISHED BY THE PRESS SYNDICATE OF THE UNIVERSITY OF CAMBRIDGE
The Pitt Building, Trumpington Street, Cambridge CB2 1RP, United Kingdom

CAMBRIDGE UNIVERSITY PRESS
The Edinburgh Building, Cambridge CB2 2RU, UK   www.cup.cam.ac.uk
40 West 20th Street, New York, NY 10011–4211, USA   www.cup.org
10 Stamford Road, Oakleigh, Melbourne 3166, Australia
Ruìz de Alarcón 13, 28014 Madrid, Spain

First published 1999

Printed in the United Kingdom at the University Press, Cambridge

Typeset in Minion 10/13pt [VN]

A catalogue record for this book is available from the British Library

Library of Congress Cataloguing in Publication data

Mammalian social learning : comparative and ecological perspectives/edited by Hilary O. Box
and Kathleen R. Gibson.
    p. cm. – (Symposia of the Zoological Society of London)
    Based on a conference held in London, November, 1996.
    ISBN 0 521 63263 3 (hb)
    1. Social behavior in animals – Congresses. 2. Learning in animals – Congresses. 3.
Psychology, Comparative – Congresses. 4. Mammals – Ecology – Congresses. I. Box, Hilary O.
(Hilary Oldfield), 1935– . II. Gibson, Kathleen Rita. III. Series.
QL775.M35 1999
599. 156 – dc21   98–41965 CIP

ISBN 0 521 63263 3 hardback

Contents

Contributors

Ágnes Bilkó
Department of Ethology
Eötvös Loránd University
Budapest
Hungary

Janette Wenrick Boughman
Department of Zoology
University of Maryland
College Park
MD 20742, USA

James R. Boran
Zoology Department
Cambridge University
Cambridge, UK
Current address
Seawatch Foundation
70 Stratford Street
Oxford
OX4 ISW, UK

Hilary O. Box
Department of Psychology
University of Reading
3 Early Gate
Whiteknights
Reading RG6 6AL, UK

Donald M. Broom
Department of Clinical
Veterinary Medicine
University of Cambridge
Madingley Road
Cambridge CB3 OES, UK

Richard W. Byrne
Scottish Primate Research Group
Department of Psychology
University of St Andrews
St Andrews, Fife KY16 9JU
Scotland, UK

David B. Croft
School of Biological Science
University of New South Wales
Sydney
New South Wales
Australia 2052

Chris G. Faulkes
Conservation Genetics Group
Institute of Zoology
Zoological Society of London
Regent's Park
London NW1 4RY, UK

Kathleen R. Gibson
Department of Basic Sciences
University of Texas Houston
Dental Branch
PO Box 60028
Houston, TX 77225, USA

Barrie K. Gilbert
Department of Fisheries and
Wildlife Ecology Center
Utah State University
5210 Old Main Building
Logan, UT 84322–5210, USA

Sara L. Heimlich
Hatfield Marine Science Center
Oregon State University
Newport, Oregon, USA

Karen Higginbottom
School of Applied Science
Griffith University
PMB 50
Gold Coast Mail Centre
Queensland 9726
Australia

Robyn Hudson
Institut für Medizinische
Psychologie
Goethestr. 31
D-80336
München
Germany

Vincent M. Janik
Woods Hole Oceanographic
Institution
Department of Biology
Redfield, MS
34 Woods Hole, MA 02543, USA

Barbara J. King
Department of Anthropology
College of William and Mary
Williamsburg, VA 23187–8795,
USA

Andrew C. Kitchener
Department of Geology and
Zoology
National Museums of Scotland
Royal Museum
Chambers Street
Edinburgh EH1 1JF
Scotland, UK

David R. Klein
Institute of Arctic Biology
University of Alaska
Fairbanks AK 99775, USA

Kevin N. Laland
Sub-Department of Animal
Behaviour
University of Cambridge
Madingley
Cambridge CB3 8AA, UK

Phyllis C. Lee
Department of Biological
Anthropology
University of Cambridge
Downing Street
Cambridge CB2 3DZ, UK

Steven Mithen
Department of Archaeology
University of Reading
Whiteknights
Reading RG6 6AA, UK

Cynthia J. Moss
Amboseli Elephant Research
Project
African Wildlife Foundation
PO Box 48177
Nairobi, Kenya,
East Africa

Jan A. J. Nel
Department of Zoology
University of Stellenbosch
Private Bag X 1.
Matieland 7602
South Africa

Thelma Rowell
Professor Emeritus
Department of Integrative
Biology
University of California at
Berkeley
California 94720, USA
and
West Chapel House
Chapel-le-Dale
Ingleton via Carnforth LA6 3JG,
UK

Benoist Schaal
Laboratoire de Comportement
Animal
CNRS ura 1291, Inra
Station de Physiologie de la
Reproduction
Nouzilly
France

Stephen J. Shennan
Institute of Archaeology
University College London
31–34 Gordon Square
London WC1E 0PY, UK

Richard M. Sibly
School of Animal and Microbial
Sciences
University of Reading
PO Box 228
Reading RG6 6AJ, UK

James Steele
Department of Archaeology
University of Southampton
Highfield, Southampton SO17
1BJ, UK

Gerald S. Wilkinson
Department of Zoology
1210 Zoology/Psychology Bdg
University of Maryland
College Park, MD 20742–4415,
USA

Preface

This volume is based upon a conference that we organised in London in November, 1996, as a tripartite enterprise of the Zoological Society of London, The Mammal Society and The Primate Society of Great Britain under the title 'Social Learning in Mammals'. The impetus for the meeting came from a symposium organised by one of us (HOB) together with Dorothy Fragazsy (USA) and Elisabetta Visalberghi (Italy) on social learning in primates at the XIVth Congress of the International Primatological Society in 1992. The preparation of my own paper (HOB) further encouraged my interest in wider comparative and ecological perspectives. A proposal was put to and accepted by the Zoological Society of London.

A very pleasant dinner at a lovely, but what turned out to be a somewhat dubious, hotel near Gatwick Airport led to an invitation to KRG to join the project. Her interests in the evolution and maturation of primate and human brains, cognition and language complement my own. Much of the academic programme for the meeting was arranged when KRG came to the UK as a visiting scholar at Wolfson College, Oxford from January to July, 1996.

Although there are many excellent examples of social learning among animals other than mammals, notably in birds, our brief with the Zoological Society of London and with The Mammal Society did restrict us to mammals. This was no hardship, however. We both work with mammals, and the issues in which we are both particularly interested are readily discussed with reference to mammalian taxa. Mammals live in a wide diversity of habitats, have much diversity in their social systems, and eat a wide range of food. The behaviour and/or products of experienced individuals that have relevant information about the opportunities and dangers of their environments provide natural sources of information by which inexperienced individuals may develop and modify their behaviour. The acquisition and integration of such information, and the biobehavioural propensities that facilitate it, are part of the biological complex of individual development, in many cases throughout life. They are part of the adaptive complex of responsiveness to the environment that has consequences for survival and fitness.

The term 'social learning' has commonly been used to refer to the social transfer of information and skill among individuals – most usually of the same species. It encompasses a wide range of functional behaviour that includes learning to identify and respond appropriately to predators, learning what to eat, where to find food, and how to process it, and learning how to interact with other members of one's social group and how to choose mates. Social influences, learning mechanisms, and the biobehavioural responsiveness that facilitate them vary with reference to the phylogenetic, ecological, sensorimotor and cognitive adaptations of different taxa. For example, social influences involve the transfer of information, as from parents to their offspring (vertical transmission), as from adults to young animals other than their own offspring (oblique transmission), and among age-mates (horizontal transmission) – see Laland, Richerson and Boyd (1996).

Mechanisms of social learning vary extensively (see Zentall 1996, Heyes 1996). These include olfactory information that is acquired *in utero*, during nursing and from faecal pellets. They also involve cognitive mechanisms that include imitation and language. Individuals of different species may also vary significantly in their opportunities for, and use of, socially mediated skills and information; this has much to do with their biobehavioural propensities to respond to their environments.

Furthermore, information that we now have clearly shows that socially mediated behaviour is ecologically relevant in a very wide diversity of animal taxa, and that such behaviour is not necessarily more important to groups such as the simian primates, which have long been regarded as having a 'special position' in terms of mental abilities, than to many other animal groups (Box 1994, Laland, Richerson and Boyd 1993). An important contemporary perspective is that we no longer consider simplistic phylogenetic associations about mechanisms and functions of social learning. A growing appreciation of this perspective emphasises the importance of a functional approach to consider the scope of the phenomena – to study comparatively in what taxa and under what conditions social learning occurs and is relevant to the life strategies of individuals of different groups. Hence, it is important to study the occurrence and functions of social learning based upon specific knowledge of the biology of different species.

Historically, however, scientific interest in social learning has been markedly biased in both theoretical considerations and in the range of species that has been studied, especially in nature. Consequently, it is not surprising that social learning has never been adequately represented in the literature in behavioural biology and that the whole area is still undersubscribed, despite a recent upsurge of interest. The relative lack of interest in this area is an important omission in our research and in our teaching, where much can be done to stimulate interest. Despite intense interest in some groups such as monkeys and apes, and some rodents, we have little or no information for the majority of species. In fact, biologists and especially field biologists have rarely considered their animals from social learning perspectives.

Many of the contributors to the volume are field workers. Moreover, their contributions cover a wide diversity of taxa that vary greatly in ecological strategies, social systems, and in mental and sensorimotor skills. Thereby we fulfil another of our main aims, namely, to increase the database of comparative information. We do this in two ways. First, we provide new information on species that had not previously been discussed or had been very little discussed in this domain. Second, we provide additional information for animals that are much more familiar in a social learning context.

The mammals include our own species, which relies so heavily on social learning for the development of behavioural strategies, and our simian primate relatives, about which so much controversy has been generated in social learning contexts. Another of our main aims is to place humans firmly within a comparative mammalian framework by delineating the ways in which human cognition, social structure and development facilitate human social learning propensities, especially the learning and transmission of technological and craft traditions. The volume also charts new ground by including, for the first time, discussions of the social learning abilities of prehistoric hominids, whose brain size and technological capacities were intermediate between those of modern humans and great apes. These discussions of modern human and ancient hominid social learning capacities, and of the biological traits that facilitate them, not only help explain the human condition, they also provide new insights into possible contexts of social learning in other mammals.

Another of our major aims is to introduce additional perspectives for ways in which we conceptualise social learning, and a section of this volume is concerned with a variety of relevant material. For example, the transfer and acquisition of information between individuals involves ways in which individuals may transform that information in accordance with their own perceptual, mental and behavioural repertoire (see King, chapter 2). Moreover, interactions among biological systems within individuals will influence their opportunities to obtain and use socially mediated information (see Box, chapter 3).

Moreover, sound experiments that use well-controlled conditions in captivity have always been a major concern among social learning theorists. Such experiments may also have profound implications for interpretations of social learning among wild animals. The examples given in this volume fulfil both these conditions. The acquisition of dietary preferences from olfactory cues provided at different stages of preweaning development in young rabbits (see Hudson *et al.*, chapter 8) and the transmission chain experiments with rats reported by Laland (chapter 10) are exemplary cases. Hence, we have a number of aims in presenting this volume. We also give separate introductory comments to each of the six parts into which the volume is organised.

In recent years there have been substantive theoretical and methodological advances since earlier exemplary and influential discussions (see Zentall and Galef 1988). *Social Learning in Animals: The Roots of Culture*, edited by Heyes and Galef in 1996 bears excellent testimony to such developments. The overarching aim of this volume is to provide additional information for a wide diversity of species and a variety of perspectives – some of which are new in the area. Many of the chapters are based upon papers that were presented at the London meeting. We have included some few invited contributions. Further, Professor Rowell's paper (chapter 1) was presented as the 1996 Osman Hill lecture of The Primate Society of Great Britain – a medal is given biannually to a distinguished primatologist. A talk by Dr Jane Goodall provided the opportunity to award her the first Conservation medal of the Primate Society of Great Britain.

Finally, it is a pleasure to thank our sponsors, the Zoological Society of London, The Mammal Society, and the Primate Society of Great Britain for their generous financial and logistical support. We also wish to record our special thanks to Unity Macdonnell of the Zoological Society of London for her sound and friendly advice during the early stages when the meeting was being planned. She also did a great deal towards the actual organisation of the meeting and helped substantially to make it such a pleasant occasion.

Of course, we thank all our participants – those that contributed to the meeting by presenting papers and chairing sessions, together with those who have contributed chapters to the volume. Their names and affiliations are listed separately. Further, apart from the names of referees that are given in particular chapters, a number of people kindly refereed specific papers, and we greatly acknowledge their help. They are the late Professor Peter Jewell, Dr Alick Jones, Dr Trevor Poole, Professor Leslie Rogers and Dr Andrew Smith.

HOB most warmly thanks Chris Martin and Victoria Mountford of the Department of Psychology at Reading University for much technical and secretarial help. They are both splendid colleagues. Victoria's highly able professional skills, combined with her friendly interest, well beyond the call of duty, have made an enormous difference in recent years. Special thanks also to Dorothy Townsend, my mother, for much help in many ways – especially in coping with a large garden – during a very busy period. KRG thanks the University of Texas for granting her sabbatical leave to work on this project and Wolfson College, Oxford, for providing her with working facilities and living accommodation. She is especially grateful to Dr Michael Argyle of

Wolfson College for arranging her appointment there, and to John, Andy and Tom for tolerating her long absences.

HOB and KRG

References

Box, H. O. (1994). Comparative perspectives in primate social learning: new lessons for old traditions. In *Current Primatology*, ed. J. J. Roeder, B. Thierry, J. R. Anderson and N. Herrenschmidt, pp. 321–8, Strasbourg: Universite Louis Pasteur.

Heyes, C. M. (1996). Introduction: identifying and defining imitation. In *Social Learning in Animals: The Roots of Culture*, ed. C. M. Heyes and B. G. Galef, Jr, pp. 211–20, San Diego: Academic Press.

Heyes, C. M. and Galef, B. G., Jr, eds. (1996). *Social Learning in Animals: The Roots of Culture*. San Diego: Academic Press.

Laland, K. N., Richerson, P. J. and Boyd, R. (1993). Animal social learning: toward a new theoretical approach. *Perspectives in Ethology*, **10**, 249–77.

Laland, K. N., Richerson, P. J. and Boyd, R. (1996). Developing a theory of animal social learning. In *Social Learning in Animals: The Roots of Culture*, ed. C. M. Heyes and B. G. Galef, Jr, pp. 129–154. San Diego: Academic Press.

Zentall, T. (1996). An analysis of imitative learning in animals. In *Social Learning in Animals: The Roots of Culture*, ed. C. M. Heyes and B. G. Galef, Jr, pp. 221–43. San Diego: Academic Press.

Zentall, T. and Galef, B. G., Jr, eds. (1988). *Social Learning: Psychological and Biological Perspectives*. Hillsdale, NJ: Lawrence Erlbaum Associates.

Part 1

New perspectives in studies of social learning

Editors' comments

HOB and KRG

These chapters add new perspectives for ways in which we understand and study mammalian social learning.

First, the primates is a prominent group in this domain, not least because they have had, and still do have, much appeal with primary reference to ourselves. This emphasis has certainly stimulated fruitful research in various ways, but it has biased our perceived comparative perspectives, and relatedly, the kinds of questions that we specifically ask. Thelma Rowell (chapter 1) considers these biases from a number of perspectives. For example, expectations based upon assessments of our own mental and social superiority are not supported by sound detailed comparative criteria. Moreover, given that primates are a hugely diverse group in size, shape, life histories and lifestyles, generalisations about behaviour among species on the basis that they are primates would, in some instances, appear to be less useful than comparisons among species that are niche equivalents. Importantly, Rowell suggests that criteria for measures of 'social sophistication' for instance, may be developed independently of the encumbrances of taxonomic status. Interestingly also, in the present context, she suggests that social learning might provide one set of the relative standards of comparison. Moreover, until recently the majority of studies of other mammals has not been undertaken with the same long term intensity of known individuals that has distinguished so many of the primate studies. Now that such information is increasingly available, new perspectives on mammalian behaviour are being appreciated. There does not appear, for instance, to be an obvious distinction in the quality of social behaviour between primates and other mammals. Further, Rowell raises important questions with regard to social learning and the relative behaviour of groups, in the sense of social entities, within populations of animals. Hence, although we persistently emphasise the success of individuals within social groups, it is important that social groups also vary in their successes; they have differential survival. Her point is, that if selection acts upon social groups, it acts upon the results of social learning within groups.

Second, a majority of studies in social learning among monkeys and apes has been primarily concerned with the mechanisms – the means whereby skill and information are acquired. This concern is clearly suited to studies that use well controlled experiments in captivity. With few exceptions, however (see Byrne, chapter 18) it is difficult to study mechanisms of social learning in natural situations. Barbara King (chapter 2) graphically illustrates this with her own field studies of foraging in yellow baboons (*Papio cynocephalus*). Her emphases here are upon additional – not alternative – perspectives. For example, her critical distinction between social information acquisition and social information donation circumvents common difficulties in field studies, and leads to robust distinctions between members of different generations of animals in the contribution that each makes to social information transfer. A variety of functional questions may be answered in this context. There are salient points both methodologically and comparatively. For instance, and importantly with reference to much apparently controversial evidence cited in the literature, tests of social information acquisition are appropriate in conditions in which food is hard to process 'or that requires special vigilance in some way'. Furthermore, great apes demonstrate much more evidence for the donation of information in foraging situations than has been found generally among monkey taxa. However, in other areas of behavioural development, such as in locomotor skills and the acquisition of social information, there is also evidence for the donation of information among a variety of monkey species, as well as among the great apes. However, and predictably, it is in the latter that there is also more evidence for the donation of information. Generally, there are clearly many comparative and contextual questions to address, the answers to which 'will depend on a complex of factors encompassing variation in phylogeny, ecology and cognition'. Importantly, King also considers interactive perspectives in studies of social learning, in which information is not conceived of as entities that are transferred intact and directly among individuals, but in which learning occurs when individuals act together. For instance, infants may alter and transform information donated to them as a result of their own behavioural repertoire, together with their perceptual, motor and memory skills. King emphasises the emergent nature of behaviour with the terms 'abstraction' and 'construction'. Moreover, she suggests ways in which these approaches may be examined empirically. This is clearly an important area for the future.

Third, little attention has been paid to biobehavioural propensities of individuals that influence their opportunities to acquire and use skill and information in social contexts. This is an important perspective, because it emphasises the social mediation of learning in the development of functionally competent behaviour rather than upon the acquisition of discreet units of

information. Once again, it is an interactive approach with particular emphases upon biological systems within individuals. For example, social attention
involves both attentional and emotional communication; it provides critical
means whereby individuals may take advantage of information that is available
from others, and learn to regulate their own behaviour. Hilary Box (chapter 3)
considers individual differences in the context of studies of temperament
among species of monkeys. This perspective has hitherto been unexplored but
it turns out to be highly productive in opening up potentially new areas of
understanding in social learning. Hence, although relatively few primate species have been studied in detail with regard to differences in temperament
within and between them, there is now a substantive database that is relevant
to issues in social learning. The point is that individuals within and between
species that are relatively inhibited behaviourally and easily disturbed physiologically respond differently to events in their physical and social environments, and acquire qualitatively different information about those environments, than individuals that are significantly and consistently less inhibited
behaviourally and less easily disturbed physiologically; they vary in the extents
to which they cope with, create and maintain social and other environmental
opportunities, including the acquisition of skill and information from other
animals. Importantly at this stage, there is a variety of hypotheses for empirical
research. For example, young animals acquire information about their environments within the security of their relationships with mothers and caretakers, and by their own initiatives in increasingly independent exploration.
Less fearful and less inhibited animals move away from their mothers earlier
and stay away for longer periods; they interact more positively with the
physical and social challenges of their environments; and they engage more in
activities such as play that facilitate the acquisition of information about other
individuals as such, as well as about appropriate patterns of social interaction.
These individuals are also more likely to maintain closer proximity to others
and to attend to what, where and how they are doing specific activities such as
foraging. There are implications for survival, health and fitness.

Major questions in the field of social learning relate to the evolutionary,
genetic and environmental contexts that render social transmission both
possible and advantageous. In the final chapter in this section (chapter 4), Sibly
addresses these issues by reviewing existing social learning models and extending them to the domain of life-history models. The result is a set of guiding
principles of our interpretations of the evolutionary contexts of animal social
learning.

Sibly begins by noting the general point that receiver animals may possess
the ability to monitor the behaviours of other animals and to learn from them

by simple classical conditioning, without either the transmitter or the receiver possessing specialised social learning mechanisms. Irrespective of the mechanisms involved, social learning has advantages over learning independently at intermediate levels of environmental variation or where environmental variation is regular and predictable (Boyd and Richerson 1988). Social learning is of no advantage in environments that are completely unpredictable. Social transmission of information and skills may be vertical from parent to offspring, oblique from adults to young other than their own offspring, or horizontal among members of a single generation. Vertical transmission is conservative in that it can support traditions within populations, such as with diet or nesting traditions, that extend over generations of animals. In contrast, horizontal transmission is rapid and ephemeral; hence, most appropriate for the transmission of information pertaining to rapidly changing aspects of the environment, such as transient food resources (Laland, Richerson and Boyd 1993). These two transmission pathways may, thus, achieve maximum advantages in different circumstances or with respect to the transmission of different types of information or skills.

The bulk of Sibly's contribution focuses on genetic and life-history considerations related to the evolution of the ability to transmit and receive information. Population genetic models such as those of Aoki and Feldman (1987) and their colleagues indicate that a genetic relationship between transmitters and receivers or other special circumstances must exist if animal populations are to evolve genes for the social transmission of information. In addition, for a communicative mechanism to spread, the selective advantage to receivers must be extremely large. Specifically, mathematical analyses indicate that receivers must have twice as many surviving offspring as non-receivers, unless mitigating circumstances exist such as biparental care and/or genes that provide not only for the social transmission of information and skills, but also for other abilities such as increased memory.

Life-history models extend genetic models by assuming that genes will spread in populations if they decrease mortality, permit breeding at an earlier age, or permit animals to have larger numbers of offspring. They also assume that the social acquisition and transmission of information and skills entails both costs and benefits. Only those genes with net fitness benefits will spread. Population genetic models indicate that transmitter genes that increase fitness directly in animals who possess them will spread to fixation in the population. In contrast, transmitter genes that affect fitness indirectly by benefiting others will only spread in the presence of a genetic relatedness between transmitters and receivers. Hence, if transmission is costly, genes affecting receivers will be more common than those affecting transmitters, and genes in recipients are always selected to receive more information than transmitters are selected to transmit. Advantageous receiver genes, however, do not necessarily spread.

References

Aoki, K. and Feldman, M. W. (1987). Toward a theory for the evolution of cultural communication: coevolution of signal transmission and reception. *Proc. Natl. Acad. Sci. USA*, **84**, 7164–8.

Boyd, R. and Richerson, P. J. (1988). An evolutionary model of social learning: the effects of spatial and temporal variation. In *Social Learning: Psychological and Biological Perspectives*, ed. T. R. Zentall and G. B. Galef, pp. 29–48. Hillsdale, New Jersey: Lawrence Erlbaum, Assoc.

Laland, K. N., Richerson, P. J. and Boyd, R. (1993). Animal social learning: toward a new theoretical approach. *In Behaviour and Evolution*, vol. 10, ed. P. P. G. Bateson and P. H. Klopter, pp. 249–77. New York: Plenum Press.

1

The myth of peculiar primates

Thelma Rowell

We are, today, a nested set of people interested in animals. The Zoological Society encompasses interest in 'all creatures great and small', the Mammal Society selects those with hair and milk, and the Primate Society confines itself to some 200 hairy, milky species, a single Order of a single Class of a single Phylum of animals. Why isn't the primate society a special interest section within the mammal society, of a size proportionate to the size of the order? Are Primates, and especially their social behaviour and learning ability, so different from other mammals? Is there really a valid entity called primatology, with its own methods, theories, and generalisations about its subject? Of course *we* are Primates, and there are an awful lot of *us* – but only a minuscule minority of people who study *us* would call themselves primatologists, so specific introspection is only a very minor part of the answer.

Logistical constraints on primate studies

The zoologists who began to study primate social behaviour in Europe were ethologists who had previously worked on the behaviour of birds or fish (not other mammals). Zoologists traditionally worked from type specimens: an animal represents its species, with some ordered variety due to sex, age and perhaps season. Beyond that, variation is a nuisance for a typologist. It can often be assumed to be the result of observational or experimental 'slop', and fortunately a nice large sample size allows statistics to tidy up such problems and reveal a 'true picture' – a mean or modal value. In a more modern approach, the variation in the sample becomes of interest in itself, and one might be interested in changes in variation as conditions change, or changes from generation to generation.

The new thing the early monkey watchers did, I think, was to recognise the individuality of their animals. Recognising individuality is not at all the same as recognising variation, which is a property of populations. I don't think we deliberately chose to recognise individuality – the step was forced on us because you never have enough primates to study in more traditional ways. They are extraordinarily expensive to keep in captivity, and in the wild they are mostly on endangered species lists, mostly live in dense vegetation high over your head, and in small groups at that. And they live long lives, very slowly. All this means that each precious specimen is studied intensively and for long periods, and the differences between them become inescapable: instead of a sample of specimens you have series of individuals, each with a history: a series of experiences which *cannot* be the same as those of any other animal, as well, of course, as a genome which is unlikely to be identical with any other.

For years we monkey-watchers have mostly seen this as a problem and tried to ignore it: we clandestinely enjoyed the unfolding soap operas in our monkey groups while striving to extract and present generalisations to our colleagues. Biology must surely be more than an endless series of individual biographies, and while individuals are not identical, they do have *similar* histories and *similar* genomes, so on the whole we played down the individuality and stressed the similarities. Nonetheless, studying individuals does provide different opportunities, as well as imposing some serious limitations, on research.

These are not concerns I detected as I started reading about the behaviour of sheep, which are also largish, longish-lived, and highly social mammals, and I began to wonder why. I can think of three possible reasons.

1 Perhaps because of the logistical reasons I have outlined, questions asked of sheep and monkeys have been different. More sheep are available and they turn over more rapidly so you can do more controlled experiments of short duration, and in the wild you can be concerned, practically, with population dynamics. The question of individuality hasn't arisen.
2 Perhaps the people are different: people who don't like soap operas don't do long-term studies of monkeys.
3 Perhaps the animals are different: sheep, and other nonprimates, might actually show less individuality than any primate.

Since the first two reasons are surely valid, we haven't adequate evidence to accept or reject the third yet, and I will come back to it later.

The people who study primates

Most zoologists find more productive subjects than primates – given the glorious diversity of animals, it is usually possible to find a more accessible and numerous subject group with which to answer almost any question in zoology. Most people who study primate behaviour have backgrounds in the social sciences, and in the US almost all of them are based in anthropology departments. This is mostly due, I think to the genius of one man, Sherwood

Washburn, an anthropologist at Berkeley, who in the early 1960s managed to parlay the general interest in human origins and human past into *actual money* for studies of primate social behaviour in the field and laboratory (Washburn and deVore 1961).

The interest he exploited was based in LeGros Clark's (1959) idea, that living primates could be classified into a series of grades, or steps. This is a closer look at the last few rungs of the old *échelle des êtres*: one after another, a series of primate taxa evolved an increasing distance towards achieving humanity, and then somehow ran out of steam and stopped. Thus they present a series of vignettes of the stages leading towards the emergence of humans. All we need to do is to walk back down the steps to go back in time and understand the origin of humanity. I am not imputing this simplistic view to Washburn himself, but I do think it is a fair caricature of the attitude he was able to exploit. It is an example of what I call the 'little furry people' approach to primates. People are too complicated, and often averse to being studied anyway. Other primates represent successively simpler versions of people, easier to analyse and available for experimentation. To an extent, they replace anthropology's lost hope of identifying primitive people to study. This interpretation is based on a completely outdated view of evolution and indeed of primate taxonomy: there is no evidence that any modern primates have stopped evolving, nor that they were striving to become people and can be seen to have failed in that endeavour to varying degrees, nor that differences between us and other primates can all be described as deficiencies in the latter (as in 'sub-human primates' – what a revealing phrase!).

Looking at primates because people are primates carries implications that need scrutiny. There is a clan pride which presumes that *our* relations are *better* – even quite distant ones. We might concede that elephants are brawnier, dolphins more agile, and both can hear things we cannot, but we consider ourselves socially superior and better at learning. So from the start we expected to find cognitive ability and social sophistication in our relatives, at least in nascent form – *and we looked for them.* So, it is students of primates that ask questions like: 'can we detect evidence of Machiavellian intelligence here, even in rudimentary form?', while a biologist trained in the reductionist paradigm would perhaps be asking of the same observations 'what are the minimal neural connections that would be necessary for this response pattern?'.

If you ask a stupid question, you get a stupid answer. If you phrase your question (that is, design your experiment or structure your observations) so as to be able to receive only a very simple answer, you will only get a very simple answer, even though it is only a part of the complete, complicated answer. There is also the danger of using entirely the wrong language when asking your question. If we ask our animals questions in an irrelevant framework, we may get an answer which can be interpreted in that framework but is still actually nonsense. Everyone here would agree that an anthropomorphic framework could be irrelevant. How many would be prepared to consider the possibility that an evolutionary framework might be irrelevant?

Fools step in where angels fear to tread. In the West, zoologists learned from social scientists to ask questions of primates that had not been asked of animals before, and perhaps we taught them how to ask the questions properly. Japanese scientists, with the advantage of their own indigenous monkeys on the doorstep to study, and traditional knowledge of monkeys from which to start, expanded primate studies rather earlier. A Buddhist/Shintoist philosophical background leads to a very different approach to animal behaviour from the Judaeo-Christian one (and I hope the reader is not under the impression that their studies of animal behaviour are uninfluenced by their own cultural background . . .). Without the sharp distinction which is drawn between humans and animals in Western culture, evolutionary theory was not so revelatory in Japan – it didn't challenge fundamental attitudes, so it was readily acceptable – pretty ho-hum, really (Asquith 1986, 1991).

Also, there is a strong Japanese interest in the functioning of small groups. Studies in Japan began with parallel studies of many groups of Japanese macaques, and they recognised differences in behaviour between groups within the same population of monkeys very early. This led them to consider the possibility of 'proto-culture' among monkeys (Itani 1958, Kawai 1965a,b) with similar properties to human cultures – and that put them far ahead of Western typologists in the study of social learning.

My understanding is that on the one hand there is intense Western interest in the individual (at least in its fitness) and how it is affected by its relations with others. On the other hand, the Japanese interest has been primarily in relationships, and how they are affected by the individual. It was inevitable that at first Japanese and Westerners had a lot of difficulty in understanding each other (apart from Western linguistic laziness) as each side appeared to be taking for granted what the other was excited about. Both have gained from the effort to develop a common language. From the Western side, I think the Japanese approach has lead Westerners to appreciate higher levels of social complexity in primates than we would have otherwise reached by now.

'The further out from England, the nearer is to France' as the whiting pointed out (Carroll 1865). Did we inevitably widen the gap between monkey-watchers and other students of animal behaviour as we reduced the gaps between zoologist and social scientist, between East and West?

To summarise, I am suggesting that, first, we were from the beginning forced to study primates in a different way because they are few and have slow life histories. We developed, and formalised, ways of handling the constrained data we could collect, methods which are equally applicable to other animals. Second, different and diverse expectations of and interests in primates have led to different questions being asked of them. These circumstances combined have been sufficient to give rise to a perception that primates are different from other mammals and so to extensive discussion of *why* they are different – and specifically, why they are cleverer and socially more sophisticated – rather like us.

Are primates different?

Much of the discussion about the peculiarity of primates has focussed on brain size and metabolic requirements of brains, and slow rates of development, and the possible adaptive significance of each and how they lead to each other (Pagel and Harvey 1993). It seems to me an argument which can be picked up at any point and carried on round in dizzying circularity.

From fieldworkers came suggestions of ecological factors: primates have a peculiarly varied and challenging diet to select, or a peculiarly complicated habitat to move through, or a special need for social sophistication, perhaps to enhance social learning. Any or all of these have provided a unique selection pressure on primates. Such discussions are generally written and read by primatologists, and have a lack of comparative depth. You are left asking, cleverer *in what way*, than what? a diet more diverse than whose?

There is a serious lack of comparative data, because people with nonprimate expertise have generally not asked the same questions. People who study elephants or dolphins simply do not accept the original premise of primate superiority; people who study rodents perhaps do accept it and simply draw the line which has always separated off people from other animals a little further out, taxonomically speaking, and ignored all primates – 'too difficult'.

Of course, primates are by no means the only mammals which are long-lived, difficult to observe, and available in small numbers. It is appropriate to treat all mammals to the pattern of intensive and sustained study of identified individuals as they pass through successive life-stages which has from the first characterised primate studies. Methods developed to deal with long-term primate observations are being applied to other animals, and we are at last getting comparable studies of group-living nonprimate mammals. We are hearing from some of the most exciting in this volume.

The descriptions coming in from all these studies have a basic ingredient in common with the average primate social behaviour study – they all include patterns of interaction which are explicable, not in the present assessment of individual qualities, but *only in terms of past social events*. This is what is necessary in order to recognise evidence of long-term relationships between individuals in practice (Rowell and Rowell 1993). This, I think, is the basic ingredient in what I am calling 'social sophistication'.

It surprised a lot of people, both primatologists and other mammalogists, who expected only primates to be socially sophisticated. It was in 1984, at the International Congress of Primatology in Nairobi, that Lawrence Frank presented an account of his study of spotted hyena's social organisation entitled 'Are hyenas primates?', and thus perhaps opened the question to debate (Frank 1986; his paper was not included in the proceedings volumes of that congress). The point was that hyenas lived in an ordered society with many of the characteristics, the rules of conduct, which have been described for macaques. That was early days, and the point could be made even more forcefully now with the continued studies of the same population of hyenas by Holekamp and Smale (1993).

From that and a growing number of other studies, I do not gain the impression that there is an obvious break in the quality of social behaviour between primates and other mammals. It is also clear that there are no agreed measures or even definitions of 'social sophistication'. If this is an interesting question, we need criteria for comparing and rating social systems which are not constrained by the features of one order or another. That will need dialogue between primatologists and everyone else. Marina Cords recently suggested to me (in litt) that methods of conflict resolution might provide key criteria. I think social learning might offer another set of standards.

I suggest we start by abandoning generalisation about primates. This is an extraordinarily diverse order for its size, and I simply do not expect useful generalisations about social behaviour that cover gorillas and tarsiers and exclude nonprimates, any more than there are equivalent generalisations about their diet, or locomotion. For example, the small carnivorous nocturnal primates, such as tarsiers (*Tarsius* spp.) or Demidoff's bushbaby (*Galago demidovii*), can surely be more usefully compared with their niche-equivalents than linked, in generalisations about primate behaviour, with large gregarious diurnal folivorous monkeys such as howlers (*Alouatta* spp.). We know a little about their social organisation only because, as primates, they were assumed from the start to *have* social organisation. In contrast, rodents of similar size, although far better studied, were for a long time taken to have only fluctuating population densities.

In fact, most popular generalisations about primate social behaviour are based on . . . most studies of primates, which have been of only a very few species. These are all species which spend much of the time on the ground and in fairly open spaces, and live in large cohesive groups, in which there is a lot of noisy squabbling and obvious makings-up. They give a high rate of return for observational effort, but they make up barely 5% of an order noted for its diversity of social behaviour, and of everything else.

We know by far the most about two to three species, each of just two genera – the macaques (*Macaca*) and the baboons (*Papio*). Chimpanzees (*Pan*) have attracted a lot of interest as our very closest relatives. The few other well-studied species are those most like the baboons and macaques within their own genera. Thus, among the African guenons we know most about the vervet (*Cercopithecus aethiops*) of open gallery forest and holiday resorts, rather than the 20-odd other *Cercopithecus* species which live quietly in dense forest. Similarly, the well-studied hanuman langur (*Presbytis entellus*) of Indian towns and temples is very different from the other 20-odd langur species which lead quietly boring lives high in the forests of southeast Asia. This handful of noisy extroverts, which is not taxonomically defined, consists of very peculiar animals indeed, not only among primates, but among all mammals. I marvel at their very existence, and I wonder why, and how.

I study guenons – blue monkeys (*Cercopithecus mitis*), in the Kakamega forest in Kenya, and occasionally baboon groups invade the area, looking for figs. My guenons and I hear them coming half a kilometre away. They are always squabbling among themselves, with screams and threat barks and grunts, and the moans of temporarily lost infants. My guenons make quiet alarm growls and freeze in creeper tangles, because baboons eat monkeys as well as figs, if they can get them. But surely baboons themselves are edible; certainly the local farmers are their sworn enemies, because they are effective crop thieves. How do they afford to devote so much time and concentration to interactions with other members of the group? I imagine my feral sheep allowing themselves to be so pre-occupied with intra-group squabbles, and the thought experiment produces instant mutton. Sheep would surely be wiped out by predators taking them by surprise while so distracted. Exploiting distraction in potential prey is a main tactic of predators: the sport of hawking with dogs relies on it, and in the suburbs cats make a killing at courting groups of sparrows in early spring.

The rate of overt social exchange within a forest guenon group is at least an order of magnitude less than in a baboon group. It is very quiet and peaceful, not to say a little boring, to follow a forest guenon group. Not much different, I find, from watching sheep, or deer. However, each group of related female guenons owns a territory, and nearly everything they eat grows within the territorial boundary. Nearly every day they defend their boundary against neighbouring groups of females (the males dont seem to take part, but that is another story). These inter-group encounters are noisy and energetic affairs, typically lasting a quarter of an hour or so, but during them the two groups make as much racket as a baboon group makes most of the time. Eagles, which are the main predator of forest monkeys, sometimes make an attack during an encounter: the eagles apparently recognise that such intense social interaction is a distraction of which advantage can be taken. So when these quietly inconspicuous monkeys begin to behave rather like their squabbling relatives, they do seem to increase their risk. This must count as a major cost of group territoriality.

The functions and functioning of groups

In concentrating on the entertaining baboons and other squabbling primates, we have also concentrated on the pattern of interaction within groups, and the supposed enhanced fitness of those who successfully compete for status within a group. We have unravelled the making of alliances and counter-alliances within groups. All this obvious, entertaining social manoeuvring and manipulation is so laughably similar to what we see ourselves do, it must be a higher level of sociality than that of other, more discrete primates and other mammals – mustn't it? It needs to be demonstrated that we are dealing with substance rather than style here. The key question might be whether the interspecific differences in style of within-group behaviour translate into substantial differences between behaviour of groups. The behaviour of the group is an emergent property of that of its members, with qualities which cannot be described at the lower level of individual relationships. It deserves study in its own right (Hinde 1983, Allen and Starr 1982). The explanation for the behaviour of the few squabbling primate species could lie in the resultant differences in the behaviour of their groups. It is difficult to assess this possibility because of the extremely long lifespan of groups, which far exceeds that of the individual.

I believe that the primary function of groups was, and still is, protection from predators, and for brevity I shall confine myself to that aspect of group life. There are several possible mechanisms for this, but they are all more effective the better the members of a group coordinate their behaviour.

Patas monkeys (*Cercopithecus* (*Erythrocebus*) *patas*) first drew our attention to this (Rowell and Olson 1983). They are long-legged guenons which live cryptically in very patchy open woodland. Though they have long legs for a monkey, they can't run as fast as dogs, so their only protection is first avoiding detection, and second, early warning and a sprint for the trees. Patas monkeys continuously scan their environment, moving their heads from side to side like little radar monitors, often standing on their hind legs to see over long grass. They watch for predators and they watch each other, and if they see anything suspicious they stop scanning and stare. The scanning neighbours instantly see the fixed stare and look the same way themselves. Thus, the presence of the possible danger is instantly communicated to the entire group, without anyone saying anything or making any flashy signal. Even during play or courtship the action is frequently broken off while both monkeys scan. The system only works if no-one is distracted. It is the perfectly cooperative mutual benefit system, not open to cheating. A monkey which did not play its part and keep scanning would only put itself at greater risk. It is a widespread system – once demonstrated so obviously by the patas, I could see it in other monkeys, and in sheep and in goats as well.

If guenons, and most other gregarious primates and nonprimates, keep distracting interactions to a minimum, baboons and the other noisy primates make a dramatic contrast which sets them apart. I ask again – how do they get away with it? They must either be unusually successful at evading predators, or for some reason be under very little threat from them. Or, they must gain some greater advantage from putting themselves at increased risk of predation. Are there comparably introverted nonprimates?

Once a group exists, it is going to function as an educational establishment. Even if you can imagine a gregarious animal with no interest at all in what fellow group members are doing, simply staying in the group as it moves around is going to mean acquiring the same home range and knowledge of local resources.

In the matter of predation, we see infant monkeys learning what not to be frightened of. When we arrive at our field site each year, a new crop of infants has been born, which react with alarm chirps when they first see us. Adults respond to their chirps by looking down at us, then return to their foraging or grooming. In effect they are saying 'Oh, that's all right dear, that's just Thelma' – and the infants never chirp at us again, indeed they come closer than the adults ever would.

Without pondering the exact mechanisms by which information is transferred between generations, then, a permanent social group accumulates local knowledge and, more interestingly, locally distinctive solutions to general problems, so developing the differences between groups which Japanese workers first described in macaques and called 'protocultural'. Are such differences confined to primates, or to some subset of primates? Has anyone asked the question of nonprimates?

Groups are only parts of populations, which also include individuals which live outside groups, often for much of their lives. Groups interact with each other both directly, as at territorial boundaries, and indirectly through the migrations of individuals between groups. If we can watch for long enough, we see that some groups do better than others, thriving and budding off new groups, while others dwindle, and sometimes their remnants get absorbed by neighbouring groups. There is a large element of luck, no doubt, in the relative success of groups. None the less, what we are actually looking at is the differential survival of groups. The phenomenon really ought to be called group selection, if the term had not been commandeered, and much derided, elsewhere.

The problem, of course, is in the definitions. The population geneticist uses the word group to mean a breeding unit, a deme. I am using it to mean a social entity. In the very beginning, long ago, we used to assume that the social entity was indeed a closed breeding unit, before we understood about male migrations and other systems seemingly designed to limit inbreeding. With closer observation, and paternity testing, it becomes increasingly clear that the social system is not necessarily, even usually, the mating system. Somehow we still accept the limitations inherent in the population geneticist's definition and meekly ignore differences in success between social groups, concentrating instead on looking for differences between individual's relative success within a group.

If selection can act on the behaviour of groups, it is acting on the result of social learning within groups. What is being selected is the ability of the individual members to engage in the transmission of information, and their ability to cooperate with others in mutually beneficial activities like predator detection.

My next question is whether the *style* of interaction within groups might enhance the environment for social learning – or otherwise. It might just be that the squabbling primates have severally invented a better system of within-group education.

But that is quite enough speculation.

References

Allen, T. F. H. and Starr, T. B. (1982). *Hierarchy*. Chicago: Chicago University Press.

Asquith, P. J. (1986). Anthropomorphism and the Japanese and western traditions in primatology. In *Primate Ontogeny, Cognition, and Social Behaviour*, ed. J. Else and P. C.

Lee, pp. 61–72. Cambridge: Cambridge University Press.

Asquith, P. J. (1991). Primate research groups in Japan: orientations and east–west differences. In *The Monkeys of Arashiyama: Thirty-five Years of Research in Japan and the West*, ed. L. Fedigan and P. Asquith, pp. 81–98. New York: SUNY.

Carroll, L. (1865). *Alice's Adventures in Wonderland.* London: Allan Wingate.

Frank, L. G. (1986). Social organisation of the spotted hyaena (*Crocuta crocuta*). II Dominance and reproduction. *Anim. Behav.* 35, 1510–27.

Hinde, R. A. (1983). A conceptual framework. In *Primate Social Relationships*, ed. R. A. Hinde, pp. 1–7. Oxford: Blackwell Scientific Publications.

Holekamp, K. E. and Smale, L. (1993). Ontogeny of dominance in free-living spotted hyaenas: juvenile rank relations with other immature individuals. *Anim. Behav.* 46, 451–66.

Itani, J. (1958). On the acquisition and propagation of a new food habit in the troop of Japanese monkeys at Takasakiyama. *Primates*, 1, 84–98. (In Japanese with English summary.)

Kawai, M. (1965a). Newly acquired pre-cultural behaviour of the natural troop of Japanese monkeys on Koshima island. *Primates*, 6, 1–30.

Kawai, M. (1965b). Japanese monkeys and the origin of culture. *Animals*, 5, 450–5.

Le Gros Clark, W. E. (1959). *The Antecedents of Man. An Introduction to the Evolution of the Primates.* Edinburgh: Edinburgh University Press.

Pagel, M. D. and Harvey, P. H. (1993). Evolution of the juvenile period in Mammals. In *Juvenile Primates*, ed. M. E. Pereira M.E. and L. A. Fairbanks, pp. 28–37. Oxford: Oxford University Press.

Rowell, T. E. and Olson, D. K. (1983). Alternative mechanisms of social organisation in monkeys. *Behaviour*, 86, 31–54.

Rowell, T. E. and Rowell, C. A. (1993). The social organisation of feral *Ovis aries* ram groups in the pre-rut period. *Ethology*, 95, 213–32.

Washburn, S. L. and DeVore, I. (1961). Social behavior of baboons and early man. In *Social Life of Early Man*, ed. S. L. Washburn and I. DeVore, pp. 91–105. London: Methuen.

2

New directions in the study of primate learning

Barbara J. King

The dominant approach to social learning

'The core of the field of social learning lies . . . in the analyses of the ways in which acquisition of behaviour by one animal can be influenced by social interaction with others of its species' (Galef 1996:8). Given this focus, work on nonhuman primates (hereafter 'primates'), especially the intensely social monkeys and apes, can be particularly fruitful for understanding how behaviour is acquired socially – and not only behaviour but also the information on which many behaviours presumably are based. Primates have extended periods of infancy and juvenility, during which they must acquire multiple skills in order to survive and reproduce. Young, inexperienced animals must learn about: species- or group-specific patterns of behaviour and communication; one's rank in the group relative to that of others; which animals are potential allies or potential competitors; which foods may be eaten and which should be avoided; how to find and process foods; and a host of other formidable 'problems' that animals encounter in life.

Yet we cannot simply assume that primates engage in social learning because they have problems to solve and appear intelligent to us in solving them. Primates in the wild may acquire species-typical social and foraging skills even when there are no adults present to act as social models (*Ateles geoffroyi*) (Milton 1993). In some species, young animals pay little attention in specific contexts such as foraging to the older, experienced animals that are present and could serve as models (*Saimiri oerstedi*) (Boinski and Fragaszy 1989).

These examples reinforce a point that Rowell (1993, chapter 1, this volume) has urged on us: It is best to be suspicious of relying too heavily on assumptions that fit closely with our (Western) cultural ideology. Western academics tend to emphasise and value teaching and learning very highly. Instead of assuming certain notions about social learning, we must turn assumptions into hypotheses for testing.

Hypothesis testing on this topic can be quite tricky, however. The dominant approach to understanding social learning in primatology right now is to focus on asking precisely *how* learning is accomplished, that is, by which mechanisms. Is a skill learned independently or socially, and if the latter, is it learned via observation, emulation, imitation, or some other process?

This focus on mechanism can provide a fascinating window into the problem-solving and learning abilities – and thus the cognitive powers – of the primates under study, if the research setting is appropriate to teasing apart the mechanisms in question (see e.g. Tomasello 1996, Whiten and Custance 1996). For research carried out under controlled conditions, the focus on mechanisms is quite appropriate. But in naturalistic or field settings, without controlled conditions, it is often problematic.

Some successes can be noted, of course. Byrne (1995, chapter 18, this volume) was able to show that imitation by wild mountain gorillas (*Gorilla gorilla beringei*) during foraging need not be of the slavish type in which details are exactly replicated across individuals. Rather, copying occurs at a broader, pattern-based level. This study describes in depth particular types of imitation rather than attempting a broad-based empirical comparison across a variety of mechanisms. Whitehead (1986) used precise criteria to differentiate between social and individual learning by mantled howling monkeys (*Alouatta palliata*), also during foraging. He demonstrated that infants learn to eat leaves, but not fruits, by social means. He was able to do this for only two mother–infant pairs, however, attesting to the intensive nature of this type of research in the field.

Just why should investigating mechanisms of social learning outside of controlled conditions be so difficult? One answer emerges from assessing the methods typically used in primatological field research, including my own study of social information transfer during foraging in yellow baboons (*Papio cynocephalus*). Two useful alternatives to the conventional approach to social learning are discussed below. The first alternative uses an intergenerational framework to assess whether immatures or adults bear the primary responsibility for social information transfer. The concepts involved are termed social information acquisition and social information donation. Emphasis in this section of the chapter is put on new studies that evaluate the acquisition–donation framework as it was first put forth several years ago (King 1991, 1994a). The second alternative questions the very nature of the term *social information transfer*. It urges primatologists to look beyond the individual learner to consider cases in which learning and knowledge may emerge from dyads or subgroups. My main aim in this chapter is thus to urge social learning theorists to tailor their goals and hypotheses to their research settings, and to realise that studying issues of learning and problem-solving in naturalistic and field settings is replete not so much with problems but with opportunities to explore new directions.

Problems with the dominant approach: a baboon example

A brief summary of important features of baboon social life will be helpful in understanding the points I wish to make. Baboons are cercopithecine monkeys from the Old World. The two groups I studied in 1985–86 were part of a population of savanna baboons in Amboseli National Park, Kenya (for details see King 1994a). Savanna baboons live in multi-male, multi-female groups with matrilineal social organisation. Males typically emigrate from their natal groups at puberty, whereas females remain in the group into which they were born and associate preferentially with members of their own matriline. Infant baboons, like most cercopithecine infants, have a strong bond with their mothers, and often grow up surrounded by female relatives from several generations (for details about baboon mothers and infants see J. A. Altmann 1980).

Savanna baboons have been characterised by S. A. Altmann (see Altmann and Altmann 1970) as eclectic omnivores. Although baboons have a very broad diet in terms of types of plants and animals eaten, they are selective about which parts they consume and which they ignore or discard. Further, they often manipulate, using their teeth and hands, the items that they do choose to ingest. Baboon foods are often encased in some way, whether underneath the ground or in some other hard-to-process matrix (Parker and Gibson 1977, King 1986). These foods must be manipulated, freed from their matrix, and made consumable. Taken together, these behaviours apparently ensure that baboons find adequate nutrients while avoiding most toxins.

These features of baboon infancy and foraging behaviour make this species an ideal candidate, in theory, for a study of social learning of foraging skills. Yet in practice, numerous challanges beset such research in the field. A hypothetical example based on my research can help explain why. In my study I used, among other sampling techniques, the widely employed method of focal-animal sampling (Altmann 1974) to collect 15-minute samples on each of the infants, in turn, in my study groups. Let's say that on 1 December, an infant, age two months, eats a yellow fruit from a green leafy bush just after her mother eats an identical fruit from an identical bush one metre away. On each of the five previous days, I had observed this group from 7 a.m. to 6 p.m. with a break for lunch. On November 28 I had noted the first seasonal appearance of the yellow fruit, and knew for certain that the two-month-old infant could not have encountered that food item before that date.

It's easy to see that my focal infant might have mouthed, manipulated or ingested the yellow fruit during focal-animal sampling of another infant, during my lunch break, before 7 a.m., or after 6 p.m. She might have done this via trial and error experimentation, either at some distance from her mother and other companions, or when oriented away from them, unable to observe them. The point is this: despite long hours of careful observation, there is no way for me to know for certain whether the observed social interaction between the infant and her mother on December 1 was either necessary for the infant to recognise the fruit as food and how to eat it, or was responsible for accelerating the infant's acquiring of those skills. In other words, there is no way for me to make a reliable claim about social learning (see Galef 1988, see also Russon 1996:172 for a related discussion of 'first attempts' in studies of imitation).

Of course, one could decide to observe only one infant all day long, to refuse a lunch break and to extend daily observation into the early morning and late evening hours. But most primatologists cannot operate this way, day after day, during the course of a long-term study. Further, primatologists need not work harder and harder to find a way to use a mechanism-based framework for understanding social learning when good alternatives are being developed.

Social information acquisition and social information donation

A growing number of researchers interested in how behaviour is acquired and transmitted advocate functional approaches of one sort or another (e.g. Box chapter 3, this volume, Caro and Hauser 1992, King 1994a, Parker 1996, Fragaszy and Visalberghi 1996). As Parker (1996:349) notes, a hallmark of using a broad functional approach is a 'focus on the results of the actions rather than the means by which the results are achieved'. Focusing on results of actions is how, after my fieldwork, I went about trying to make sense of how Amboseli baboons acquire foraging skills.

Assessing social information acquisition and social information donation, rather than social learning, helps to tease apart reliably the relative contributions of adults and immatures in any context in which the immatures might extract or receive information from adults. (Animals other than immatures may need to acquire new information and skills, and may acquire them from non-adults, but for the sake of simplicity I will continue to refer only to immatures and adults.) Social information acquisition includes any behaviour that helps immatures gain, or potentially gain, information from adults. When information transfer does occur, it is the result of the immature's own behaviour in the absence of any action directed overtly to it by adults. Although an immature's observation of adult behaviour may be a type of social information acquisition, quite often the immature's action goes beyond observation.

At Amboseli I observed 19 infant baboons, ages 2–32 weeks, when they were still dependent nutritionally on their mothers. Even at these very young ages, baboons behaved during foraging in ways that increased their opportunities for finding out what specific food parts adults ate, what condition (e.g. of ripeness and colour) the items were in, and how the adults processed them. Infants co-fed with their mothers, for example, at varying rates according to how easy or difficult to process the food items in question were. (I defined co-feeding as feeding on the same basic type of food at the same time as another baboon.) Infants co-fed at a much higher rate on below-ground corms than on easy-to-pluck foods such as leaves (see King 1994a for details). Amboseli infants also sniffed the muzzles of adults during foraging, apparently to receive sensory cues about the foods being eaten, and scrounged scraps from adults during foraging. Although the infants were tolerated at close proximity by the adults, adults did not offer guidance or intervention during foraging, even when an infant chose a food not eaten by adults in the group.

Social information acquisition contrasts cleanly with social information donation, in which adults direct some action or behaviour at immatures, enabling them to receive, or potentially to receive, more information than they otherwise would. Amboseli baboons did not donate information to immatures during foraging. Judging from a survey of the literature on the topic of primate foraging behaviour, the dominant pattern in primate foraging is social information acquisition, not social information donation. Primate infants seem to have been selected to be information extractors (King 1994b); they direct energy and effort at setting up and maintaining social interactions that might yield information relevant to acquiring foraging skills.

This last statement is best viewed as a hypothesis for testing. My Amboseli observations were completed before I devised my functional framework: when I was in the field I did not look specifically for social information acquisition or social information donation. Work by others using these concepts is just beginning. Two particularly important questions for future research are: Do the concepts of social information acquisition and social information donation adequately describe what primates actually do? If so, does the pattern suggested for foraging – more social information acquisition than donation – hold up more generally across contexts and species?

Recent experimental work is yielding some intriguing hints that bear on these questions. To date these studies have been carried out only under controlled laboratory conditions, but the results enable us to make predictions that might guide research in other settings as well. In a series of experiments using the tufted capuchin monkey (*Cebus apella*), Fragaszy and her colleagues have investigated whether inexperienced animals really do set out to acquire information socially during foraging. In a summary of their work, Fragaszy and (Visalberghi 1996:97) write that their research results are not consistent with the idea of infants as motivated to seek information from others about foraging. My observations of what Amboseli baboon infants do, they note, are 'strikingly similar' to what they themselves saw infant tufted capuchins do in terms of initiating interactions in the absence of adult guidance, but:

> *It is not clear whether or to what extent infants are 'seeking' information through their activities, even though they are clearly the initiators of the inter-actions and the transfers of food that occur. Infant baboons may simply be try-ing to obtain some of what adults are eating, especially when it is a favoured food that they cannot obtain themselves, as corms are. We are more inclined to view the social interactions occurring between infants and others as im-mediately motivated by the infants' interest in the food, and desire to eat it themselves. This parsimonious interpretation does not rule out the possibility that acquisition of information may be a product of the process, but it deem-phasises the acquisition of information as a proximate motivator of the in-fant's behaviour.*

The distinction drawn here, between infants *seeking information* in order to solve problems and *obtaining information as a byproduct of* a less focused social process whose main end is food acquisition, can be valuable in thinking about how to test hypotheses based on the concepts in the acquisition/donation framework. My framework is meant, however, as only a functional one. It suggests that primate infants have been selected, over many generations, to set up social interactions that result in more information transfer than otherwise would occur. This claim is unconcerned with issues of information *seeking*, a term that implies intent to gather information on the part of the infant.

Working within Fragaszy and Visalberghi's own experimental framework, another concern arises. The specific parameters of the capuchin research may not have elicited the fullest expression of information acquisition: a close examination of two of the experiments reported by Fragaszy and her colleagues indicates that this possibility is worth investigating, because the novel foods presented to the infants required no special skills of assessment or processing. In one study (Fragaszy *et al.* 1997b), novel foods were presented to 11 capuchins aged 4.5 to 12 months that were housed in a larger social group. These foods are elsewhere (Fragaszy *et al.* 1997a:198) described as requiring no specific actions on the part of the holder before being eaten. Infants expressed interest in what others were eating more often when food was novel than familiar, but this interest was not selectively expressed before eating, as would be expected in a model based on social information acquisition. In 24% of test sessions interest did occur before eating, but in another 25% interest occurred after eating. In 51% of sessions the infants showed no interest at all in what others were eating. Taken alone, these numbers support the view that infants display interest in food rather than 'seek' information about foraging.

In another study (Fragaszy *et al.* 1997a), however, 'difficult to process' unshelled pecans were presented (with chow as a control) to capuchins aged 2 months to adulthood. Here, results were more mixed. Infants significantly more often fed near others, ate from others' hands and manually took food from others' hands or mouths when the pecans were available. Infants approached adults more often than peers when nuts were present, perhaps because adult monkeys were more likely to be eating or acting on a nut. Attempts to take foods from peers, and interest in peers' food, were directed toward competent openers of nuts in two-thirds of cases, even though non-openers made up half of the peer group. 'Thus it seems that young monkeys may be actively seeking food, but they may not be selectively approaching experienced individuals. We cannot determine from these findings whether infants are also seeking "information".' (Fragaszy *et al.* 1997a:198). But because infants contacted older experienced foragers or competent peers most often, it would seem that these results *are* consistent with a claim for seeking of information – a claim that is not part of my own framework, but that others may wish to make. Efforts to test hypotheses related to the idea of social information acquisition may be best carried out using foods that are hard to process or that require special vigilance in some other way. Genz and Snowdon (personal communication) presented five family groups of captive cotton-top tamarins (*Saguinus oedipus*) with a range of foods including tuna laced with white pepper, meant to mimic spoiled food in the wild, and a novel food, blackberries. Three of the family groups contained infants ranging in age from 8 to 12 weeks at the start of the study. Infants learned about the foods through both social information acquisition and social information donation.

From previous research by Snowdon and colleagues, it was known that adult tamarins give a special infant-directed call to infants of this age in the presence of familiar, nonaversive foods. This stage of infancy is characterised by an intense period of food-sharing between adults and infants, and infants have a much higher success rate of getting food from adults when adults use these special calls (Roush 1996). In the new study, the range of foods presented to the tamarins was broadened. Upon eating the tuna, adult tamarins gave head shakes that were apparently involuntary and that did not occur during eating of other foods. Juveniles and subadults attended to these head shakes, and most avoided the tuna on the first and subsequent presentation, for up to three weeks.

Infant tamarins received alarm calls directed specifically at them by adults in the presence of the tuna. These directed calls were more intense than calls given in the families without the infants. The calls were scored as 'directed' because the adult caller's head and body were oriented toward the infant, and because after hearing such a call, an infant who had picked up food immediately dropped it and did not approach it on subsequent presentation for at least two weeks. This prolonged food avoidance was not seen when infants only observed head shakes but did not receive directed alarm calls. As Genz and Snowdon say, 'The lack of any type of directed calls among the two families without infants, and the presence of infant-directed alarm calls in all three families with infants suggests that donated information about a potentially toxic food is communicated only to infants.' (One tamarin family, however, did upon presentation of tuna give alarm calls that were not infant-directed.)

Infants also learned about a novel food, blackberries, by paying attention to adults' alarmed state in the presence of that food. In two of the families, adults gave many alarm calls upon presentation of the blackberries, but did not direct the calls to infants specifically. In one case, however, an adult physically blocked an infant's access to this food. In sum, this study yielded strong evidence in captive tamarins for both social information acquisition (juveniles and subadults paid attention to adult head shakes and non-directed alarm calls) and social information donation (adults gave infant-directed calls and, in one instance, physically prevented an infant's access to food).

Other examples of adult guidance during foraging come also from the food-sharing tamarins and marmosets (the callitrichids). In these species, voluntary food donation from adults to infants can be seen to include transfer of information about diet along with transfer of the food item itself. More examples are known of social information acquisition than donation when looking at monkeys in the foraging context, but for a full understanding of these patterns we should look also at apes and at non-foraging contexts. Given the prediction that information transfer about food might correlate with foods requiring special vigilance, it is no surprise that higher levels of information donation during foraging seem to exist for apes than for monkeys. Great apes are superb extractive foragers both with and without tools (Parker and Gibson 1977). They extract edible food items from a variety of encasements or matrices, including hard shells of fruits or nuts, termite mounds and the ground. Gorillas and bonobos extract foods manually and with their teeth; orangutans and chimpanzees use these methods but also use tools (van Schaik *et al.* 1996, Goodall 1973, McGrew 1992). Nut-cracking as found among chimpanzees at Tai, Ivory Coast, for instance, is mastered only after a long period of apprenticeship during which immatures receive both maternal food donations (via sharing of nuts) and maternal guidance about which tools to use and how to use them (Boesch 1991).

Not only social information acquisition but also social information donation apparently occurs at higher rates when food is hard to process. In some populations, infants solicit more from mothers during feeding when the food is difficult to process (Hiraiwa-Hasegawa 1990). In others, mothers share more food under these conditions (McGrew 1975). Even when food is not hard to process, field reports show that ape adults intervene at higher rates than monkeys do, as when ape mothers remove food items from their offsprings' hands or mouth (see King 1994a, Byrne chapter 18, this volume for examples). In general, apes forage using more 'technical complexity' (Byrne chapter 18, this volume) than do monkeys. Apes need to master more foraging-related skills than even such complex foragers as Amboseli baboons, and thus in the past probably experienced greater selection pressures for social information donation than did most monkeys.

Social information donation is not limited to the foraging context. Primate mothers encourage their infants to crawl, climb, walk or ride in a certain position. Such sequences are richly described for both monkeys (e.g. Altmann 1980) and apes (e.g. Yerkes and Tomilin 1935, van de Ritj-Plooij and Plooij 1987:25) (see Maestripieri 1995 for a review). In some cases, the signals as well as the physical behaviours are taught (Maestripieri 1995), as is clear in a passage describing behaviour in a Gombe chimpanzee mother (FF) – infant (FD) pair:

> FD was walking on a rock (about 50 cm high) while FF was sitting out of contact but within arm's reach. Then FF stood up slowly, turned her back towards him and approached him while flexing her knees slightly and looking back at him with her lowered back closest to him. FD did not cling immediately and FF waited motionless while looking at him. Finally, he clung and FF walked a few paces to turn around and come back to the same rock to return him onto the rock. Over and over she started to travel in this way and when FD did not walk onto the rock himself, she placed him there. Finally, FD clung immediately whenever FF 'signalled' by flexing her knees and looking back.
>
> Next, FF started the whole procedure again with FD on the ground. She lowered her back by crouching onto the ground in front of him while looking back at him over her shoulder. Whenever FD seemed not to be looking at FF she would wait and, ultimately, gain his attention by touching him. When touched in this way he always responded by clinging immediately (van de Ritj-Plooij and Plooij 1987:25).

This event occurred when FD was in his fifth month of life. Great variability exists in the extent to which individual primate mothers encourage their infants in these behaviours; differential maternal experience is one likely cause of such variation (Maestripieri 1995). A captive bonobo mother with significant maternal experience encouraged her infant to crawl when the infant was less than two months of age (personal observation, Language Research Center). The infant was born on 28 June 1997; on 18 August 1997, the following incident occurred:

> *The mother, MA, broke contact with her infant, EL. Holding EL's arm, she moved away a step and sat down within arm's reach, then let go of EL's arm. EL now lay alone on a blanket, on her stomach, holding her head up (although her head wobbled). After a brief interaction with her older sister, EL pushed up off the blanket slightly with her arms, and turned her head a bit more toward MA. Immediately MA touched EL's upper back. EL then oriented even more toward MA, pushed up and tried to move forward toward MA. MA moved the blanket a bit (it was wrinkled up underneath and in front of EL, possibly obstructing her). She touched EL again. Another bonobo approached, and MA gathered EL to her own body.*

My interpretation of this event, in context with many other events of the same time period (still being analysed), is that MA structured a situation to encourage EL to crawl; MA also routinely used her hands and feet to hold EL, by EL's hands, out from her own body. In these instances, EL assumed a wobbly standing position, with legs extended. Savage described a similar maternal behaviour in captive chimpanzees, noting that in the resulting position 'the chimpanzee infant exercises its legs by repeatedly straightening the knees with a pushing motion, just as do human infants. Additionally, since the chimpanzee infant is held by its hands instead of under its armpits as is the human infant, the infant strengthens its arms by flexing the elbows and pulling upward.' (Savage 1975:136, see Yerkes and Tomilin 1935).

When mothers are in body contact with infants, they may transmit information to those infants through their posture and the tension in their muscles (Russell *et al.* 1997:191, Savage-Rumbaugh pers. commun.). A clinging infant probably notices different maternal states via sense impressions from the mother's body. When the mother participates in a tense social encounter, for example, her body probably 'feels different' to the infant than when the mother is relaxed and grooms an associate. When certain maternal postures, positions or levels of tension occur over and over again in certain contexts, the infants may attend to them and use them as a source of information when observing or participating in social situations. This would be an example of social information acquisition, not donation, given that the maternal response is presumably automatic, not voluntary. This entire research area is unexplored within primatology, however, and the possibility of directed guidance through the body should not be ruled out.

Other contexts for social information donation may include predator avoidance and social behaviour. Primatologists have collected much data on maternal rescuing of infants in dangerous situations, and punishment of infants for inappropriate behaviour. Do these behaviours ever include directed action, i.e. social information donation? Further, when vervet monkey adults use their own vocal responses to encourage infants to make proper alarm calls to predators (Caro and Hauser 1992), is this a form of social information donation? When mother rhesus macaques grasp their own infants to their ventrum together with infants of higher-ranking females, using the so-called double-hold (de Waal 1990), is this social information donation, perhaps about how to choose future allies? We don't yet know.

At present, a reasonable hypothesis is that apes engage in higher levels of social information donation than monkeys, across contexts. It is reasonable because it fits not only with data on patterns of social information donation itself, but also with the growing consensus that apes are cognitively more advanced than monkeys in a variety of realms (e.g. Byrne chapter 18, this volume, chapters in Russon *et al.* 1996). It fits particularly well with Parker's (1996) suggestion of an adaptive complex within apes (but not monkeys) that includes intelligent tool use, true imitation, teaching by demonstration and self-awareness. In Parker's model, this complex of behaviours emerged in the common ancestor of great apes as an adaptation for apprenticeship in specific forms of extractive foraging, particularly extractive foraging by tools. Whether one accepts the foraging basis of the model or prefers an explicit integration of social and technological factors, a link with the social information donation model is clear. Both suggest that apes must learn more complex skills than monkeys in many areas of life, and that selection has occurred to supplement the primate skill of social information acquisition with a behavioural package that includes a form of direct guidance.

Much research needs to be done before this research framework yields enough data to allow firm conclusions about the distribution of social information acquisition and donation across species and contexts. Yet some predictions have emerged from even the preliminary results, for example that both social information acquisition and donation may be selected for in situations where inexperienced animals must master some skill, ranging from processing of foods to a new form of locomotion. Whether acquisition or donation occurs, or both, will depend on a complex of factors encompassing variation in phylogeny, ecology and cognition.

Emergent behaviour and social learning: some preliminary thoughts

The social information acquisition versus donation framework just described avoids many of the problems associated with the study of social learning using a mechanism-focused approach, yet it is not free of concerns and problems itself. Let's shift perspective and look at primate social learning and problem-solving through another lens, in order to assess these concerns and how to address them.

This shift questions the very foundation of the concepts social information transfer, social acquisition of information and social information donation. These terms imply that information is a thing, a neatly packaged entity that can be handed over from one individual to another, just as an item of food can be handed over. The acquisition versus donation framework implies that information can be acquired, received, or donated intact and unchanged, and that information flows directly from one individual to another individual.

Such a picture is a very linear one. Its suggests that an act of social behaviour, including any instance of communication, occurs when one actor initiates and another actor responds (or fails to respond). It claims that discrete start and stop points, and possibly even stimuli and responses, can be identified for the behaviour in question. Yet is this an accurate description of complex behaviour? Behaviour, again including communication, might be better thought of as a product of interactional synchrony (e.g. Byers 1976, Slavinoff unpublished data) in which information flows in a nonlinear way around the participants who interact in a coordinated (but unconscious) rhythm. Behaviour is thus not composed of stimulus and response, initiator and recipient. Instead, any act can be regarded as either stimulus or response or both – according to how the total behavioural sequence is punctuated.

This nonlinear view is closely wedded to a systems approach, which although not new (see Bateson 1972) is newly strengthened by links with recent developments in complexity theory (see Waldrop 1992). The heart of this integrated approach is that a system – whether of brain cells, honeybees or humans – is not reducible to its component parts (see Fogel 1995). Learning thus may occur by systems themselves, not just by their members. In humans, for instance, learning can occur not only socially but collectively, with some skill or knowledge emerging as a product of a group working as a group. Hutchins (1995) found in a study of workers on a US Navy ship, for instance, that solutions to some problems of navigation were discovered by a group of workers, as a group, before they were clearly understood by any of the individual participants.

We don't know yet whether true collective learning of this nature occurs in nonhuman primates, although the evidence to date suggests it may not (Tomasello *et al.* 1993). The model of emergent behaviour nevertheless has relevance for what happens during social interaction in monkeys and apes. On this view, a mother doesn't so much donate information to her infant, and an infant doesn't so much acquire information from her mother, as they co-construct behaviour. Learning and problem-solving may take place not primarily in an individual's head or as the product of a linear transaction centered around information transfer, but rather as an emergent process from two or more individuals acting as a unit (see Vygotskian perspectives such as those in Rogoff 1990, Lave and Wenger 1991).

An infant chimpanzee trying to master skills of nut-cracking behaviour, for example, does pay attention to its mother as she cracks, and may incorporate some aspects of the nut-cracking sequence that she performs or demonstrates. Yet the infant contributes something too. Idiosyncratic perceptual, motor or memory skills, and previous experience in playing with or manipulating nuts or tools, may affect which elements of maternal knowledge the infant can perceive and use, and in turn, what the final behavioural product looks like. The 'donated information' from the mother may be rather unlike the 'information received' by the infant, because the infant will alter and transform it. The process I am describing may be termed, borrowing from Russon's (1996:156) discussion of social imitation, 'abstraction and reconstruction', or alternatively, in order to emphasise the emergent nature of the behaviour, abstraction and co-construction.

At this point, the reader may be forgiven for giving in to exasperation: how could a study of social learning based on these principles of emergence and co-construction be carried out? If my stated goal is to improve reliability of social learning studies, how can I advocate such a seemingly nebulous approach? Admittedly, this whole enterprise is in its infancy, at least within primatology. Yet it has something important to contribute, for it insists on a nonlinear view, and attempts to understand more fully the richness of what happens during socially based acquisition of skills and behaviour. Precisely how to design studies to test hypotheses about emergent behaviour is still being worked out, but progress is being made (e.g. Strum *et al.* 1997, Slavinoff unpublished data). Studies tailored to hypothesis-testing about abstraction, co-construction and emergent learning are now needed.

One approach would be to look for different levels at which these processes are expressed by nonhuman primates. At one level, testable hypotheses could be fashioned to include a focus on the process of abstraction and co-construction in contexts where (to revert to earlier terms) immatures are said to acquire information socially or to receive donations of information socially. Researchers could richly describe dyadic interactions in the context of problem-solving, for example mother–infant interaction when infants are first using new skills. They could untangle which aspects of maternal performance are incorporated into the infant's performance and which are not.

At another level, researchers could look for truly emergent behaviour that is produced by the dyad acting as a dyad, or the subgroup acting as a subgroup. Here, the chief hurdle is accurately teasing apart the difference between cases where 1. behaviour truly emerges from a level above the individual, and 2. behaviour results because different individuals make sequential but related contributions. This problem is akin to distinguishing the difference between two types of collective behaviour seen in chimpanzees hunting in groups. In some cases, truly cooperative behaviour seems to occur, as when chimpanzees succeed in bringing down prey because collectively they produce a certain strategy that is more than the sum of the individuals each working alone. In other cases, hunting success seems to stem from the fact that individuals take independent but simultaneous actions that happen to work. In other words, the challenge in understanding emergent behaviour is not of a different order of magnitude than other challenges more routinely faced by primatologists.

A concluding note

In discussing the traditional mechanism-focused approach to social learning by primates, and offering some alternatives to it, I want to emphasise that none of the approaches is inherently better or worse. Indeed, my point has been to suggest that each approach has value in the appropriate research context, but that it is time to move beyond an insistence on a mechanism-dominated framework.

Acknowledgements

Special thanks to Kathleen Gibson and Hilary Box for urging me to rethink the structure of an earlier version of this chapter, and to Talbot Taylor for helpful comments on yet another draft version. I am grateful to Sue Savage-Rumbaugh and Dan Rice of the Language Research Center, Georgia State University, and to Erin Selner of the College of William and Mary, for help in understanding bonobo behaviour and in collecting data and videotape samples, respectively.

Special thanks to Grey Gundaker for sharing a fierce commitment for non-linearity.

References

Altmann, J. A. (1974). Observational study of behavior: sampling methods. *Behaviour*, **49**, 213–66.

Altmann, J. A. (1980). *Baboon Mothers and Infants*. Cambridge: Harvard University Press.

Altmann, S. A. and Altmann, J. A. (1970). *Baboon Ecology: African Field Research*. Chicago: University of Chicago Press.

Bateson, G. (1972). *Steps to an Ecology of Mind*. NY: Ballantine Books.

Boesch, C. (1991). Teaching among chimpanzees. *Anim. Behav.*, **41**, 530–2.

Boinski, S. and Fragaszy, D. M. (1989). The ontogeny of foraging in squirrel monkeys, *Saimiri oerstedi. Anim. Behav.*, **37**, 415–28.

Byers, P. (1976). Biological rhythms as information channels in interpersonal communication behavior. In *Perspectives in Ethology*, Vol. 2, ed. P. P. G. Bateson and P. H. Klopfer, pp. 135–64. New York: Plenum Press.

Byrne, R. (1995). *The Thinking Ape: Evolutionary Origins of Intelligence*. Oxford: Oxford University Press.

Caro, T. and Hauser, M. (1992). Is there teaching in nonhuman animals? *Q. Rev. Biol.*, **67**, 151–74.

Fogel, A. (1995). Development and relationships: a dynamic model of communication. *Adv. Stud. Behav.* **12**, 259–90.

Fragaszy, D. and Visalberghi, E. (1996). Social learning in monkeys: primate 'primacy' reconsidered. In *Social Learning in Animals: The Roots of Culture*, ed. C. M. Heyes and B. G. Galef, pp. 65–84. San Diego: Academic Press.

Fragaszy, D. M., Feuerstein, J. M. and Mitra, D. (1997a). Transfers of food from adults to infants in tufted capuchins (*Cebus apella*). *J. Comp. Psychol.*, **111**, 194–200.

Fragaszy, D. M., Visalberghi, E. and Galloway, A. (1997b). Infant tufted capuchin monkeys' behaviour with novel foods: opportunism, not selectivity. *Anim. Behav.*, **53**, 1337–43.

Galef, B. G. (1988). Imitation in animals: history, definition, and interpretation of data from the psychological laboratory. In *Social Learning: Psychological and Biological Perspectives*, ed. T. Zentall and B. G. Galef, pp. 3–28. Hillsdale, NJ: Erlbaum.

Galef, B. G. (1996). Social learning and imitation. In *Social Learning in Animals: The Roots of Culture*, ed. C. M. Heyes and B. G. Galef, pp. 3–15. San Diego: Academic Press.

Goodall, J. (1973). Cultural elements in the chimpanzee community. In *Precultural Primate Behavior*, ed. E. W. Menzel, pp. 144–84. Basel: S. Karger.

Hiraiwa-Hasegawa, M. (1990). Role of food sharing between mother and infant in the ontogeny of feeding behavior. In *The Chimpanzees of the Mahale Mountains: Sexual and Life History Strategies*, ed. T. Nishida, pp. 267–75. Tokyo: University of Tokyo Press.

Hutchins, E. (1995). *Cognition in the Wild*. Cambridge, MA: MIT Press.

King, B. J. (1986). Extractive foraging and the evolution of primate intelligence. *Hum. Evol.*, **1**, 361–72.

King, B. J. (1991). Social information transfer in monkeys, apes, and hominids. *Yearb. Phys. Anthropol.*, **34**, 97–115.

King, B. J. (1994a). *The Information Continuum: Social Information Transfer in Monkeys, Apes, and Hominids*. Santa Fe: School of American Research Press.

King, B. J. (1994b). Primate infants as skilled information gatherers. *Pre- and Perinatal Psychol. J.*, **8**, 287–307.

Lave, J. and E. Wenger. (1991). *Situated Learning: Legitimate Peripheral Participation*. Cambridge: Cambridge University Press.

McGrew, W. C. (1975). Patterns of plant food sharing by wild chimpanzees. In *Contemporary Primatology: Proceedings of the Fifth Congress of the International Primatological Society*, ed. M. Kawai, S. Kondo and A. Ehara, pp. 304–9. Basel: S. Karger.

McGrew, W. C. (1992). *Chimpanzee Material Culture*. Cambridge: Cambridge University Press.

Maestripieri, D. (1995). Maternal encouragement in nonhuman primates and the question of animal teaching. *Hum. Nature*, **6**(4), 361–78.

Milton, K. (1993). Diet and social organization of a free-ranging spider monkey population: the development of species-typical behavior in the absence of adults. In *Juvenile Primates*, ed. M. E. Pereira and L. Fairbanks, pp. 173–181. New York: Oxford University Press.

Parker, S. T. (1996). Apprenticeship in tool-mediated extractive foraging: the origins of imitation, teaching, and self-awareness in great apes. In *Reaching into Thought: The Minds of the Great Apes*, ed. A. E. Russon, K. A. Bard and S. T. Parker, pp. 348–70. Cambridge: Cambridge University Press.

Parker, S. T. and Gibson, K. R. (1977). Object manipulation, tool use and sensorimotor intelligence as feeding adaptations in cebus monkeys and great apes. *J. Hum. Evol.*, **6**, 623–41.

Ritj-Plooij, H. H. C. and van de Plooij, F. X. (1987). Growing independence, conflict, and learning in mother–infant relations in free-ranging chimpan-

zees. *Behaviour*, **101**(1–3), 1–86.

Rogoff, B. (1990). *Apprenticeship in Thinking*. New York: Oxford University Press.

Roush, R. (1996). Food-associated calling behavior in cotton-top tamarins (*Saguinus oedipus*): environmental and developmental factors. Unpublished Ph.D. dissertation, University of Wisconsin, Madison.

Rowell, T. (1993). Foreword. In *Juvenile Primates*, ed. M. E. Pereira and L. Fairbanks, pp. vii–ix. New York: Oxford University Press.

Russell, C. L., Adamson, L. B. and Bard, K. A. (1997). Social referencing by young chimpanzees (*Pan troglodytes*). *J. Comp. Psychol.*, **111**(2), 185–93.

Russon, A. E. (1996). Imitation in everyday use: matching and rehearsal in the spontaneous imitation of rehabilitant orangutans (*Pongo pygmaeus*). In *Reaching into Thought: The Minds of the Great Apes*, ed. A. E. Russon, K. A. Bard and S. T. Parker, pp. 152–76. Cambridge: Cambridge University Press.

Russon, A. E., Bard, K. and Parker, S. T. (eds.) (1996). *Reaching into Thought: The Minds of the Great Apes*. Cambridge: Cambridge University Press.

Savage, E. S. (1975). Mother–infant behavior among captive group-living chimpanzees (*Pan troglodytes*). Unpublished Ph.D. dissertation, University of Oklahoma, Norman.

Strum, S. C., Forster, D. and Hutchens, E. (1997). Why Machiavellian intelligence may

not be Machiavellian. In *Machiavellian Intelligence* II, ed. A. Whiten and D. Byrne, pp. 50–85. Cambridge: Cambridge University Press.

Tomasello, M. (1996). Do apes ape? In *Social Learning in Animals: The Roots of Culture*, ed. C. M. Heyes and B. G. Galef, pp. 319–46. San Diego: Academic Press.

Tomasello, M., Kruger, A. C. and Ratner, H. H. (1993). Cultural learning. *Behav. Brain Sci.*, **16**, 495–552.

van Schaik, C. P., Fox, E. A. and Sitompul, A. F. (1996). Manufacture and use of tools in wild Sumatran orangutans: implications for human evolution. *Naturwissenschaften*, **83**, 186–8.

de Waal, F. B. M. (1990). Do rhesus mothers suggest friends to their offspring? *Primates*, **31**, 597–600.

Waldrop, M. M. (1992). *Complexity*. London: Penguin.

Whitehead, J. M. (1986). Development of feeding selectivity in mantled howling monkeys, *Alouatta palliata*. In *Primate Ontogeny, Cognition, and Social Behaviour*, ed. J. Else and P. C. Lee, pp. 105–17. Cambridge: Cambridge University Press.

Whiten, A. and Custance, D. (1996). Studies of imitation in chimpanzees and children. In *Social Learning Animals: The Roots of Culture*, ed. C. M. Heyes and B. G. Galef, pp. 291–318. San Diego: Academic Press.

Yerkes, R. M. and Tomilin, M. I. (1935). Mother–infant relations in chimpanzee. *J. Comp. Psychol.*, **20**(3), 321–59.

3

Temperament and socially mediated learning among primates

Hilary O. Box

Introduction

The functions of socially mediated learning concern the acquisition of skills and information that enable individuals to adjust competently to the demands of their environments.

Young primates of the two hundred and fifty or so species of living primates are born into a wide diversity of habitats in predominantly tropical and subtropical regions. As a group, they eat many different foods, and although there are species that primarily select insects, fruits and flowers or foliage, and are thus relatively specialised in their diets, many species are rather loosely described as 'generalist omnivores'; they forage opportunistically on leaves, fruits, flowers, seeds, grass corms, bulbs, gums, insects, eggs and a variety of vertebrates.

Primates are also born into a wide diversity of social organisational systems; they may live predominantly with their mothers, in family groups or in units that are made up of many individuals. Primate social systems include some of the most complex among mammals. Further, compared with most other mammals of similar body size, simian primates have large brains, long life-spans, long interbirth intervals, complex modes of parental care, and long prereproductive phases of development, as in periods of adolescent sterility (Eisenberg 1981). The comparative relevance of these characteristics raises a variety of issues. For instance, the fact that a notably high proportion of the lifespan involves development after weaning and before reproduction in the vast majority of species, leads to interesting questions as to why individuals should delay reproduction for so long (Pagel and Harvey 1993). Although there are other hypotheses, many people (e.g. Gavan 1982, Pereira and Altmann 1985) emphasise this as a time in which individuals learn about the complexities of their habitats and their social organisational contingencies before they invest in the demands of reproduction. However, behavioural development also continues over the lifespan, and involves the continual adjustment of individuals to the diversity of social and ecological features of their environments. Competence in dealing successfully with these demands involves acquiring knowledge that includes that of other group members. There is much to learn about the social complexities of the species, sex and age, about cooperative and competitive behaviour, mating and rearing offspring, about avoiding potential hazards such as predators and inclement weather, and about obtaining food and other maintenance behaviours. Mistakes can be extremely costly. Moreover, social and physical competence is matched against dynamic social and ecological conditions.

So far, and although relatively few species of the wide diversity of primate taxa have been studied in sufficient detail, there have been excellent wide-ranging discussions of the opportunities for, and functions of, socially mediated learning in various species of simian primates in nature (e.g. Pereira and Altmann 1985). Naturally, major topics of discussion concern ecological, social, developmental and cognitive influences. However, there is also marked behavioural flexibility, and individual differences in behavioural strategies are central to functional questions (e.g. Altmann 1980). Moreover, the principle concern of this present chapter is that the organisation of behaviour among individuals influences their opportunities for behaviour in the contexts of socially mediated learning. For example, there is a close interrelationship between emotional responsiveness and development as expressed in social and technical skills (Box and Fragaszy 1986). Critically, animals of significantly different biobehavioural dispositions react with their environments, and learn different things about their environments in different ways that, in turn, have significant consequences for their life strategies. Socially mediated learning involves the extents to which individuals cope with, maintain and create social and other environmental opportunities. Hence, it is relevant to consider behavioural and other biological correlates of responsiveness to challenging environmental events; to consider dispositions of individuals interactively – as with emotion, attention and activity; and to emphasise self-regulatory behaviour, as with selective attention towards or away from environmental conditions. These are propensities that facilitate positive and negative responses that are associated with the uptake and use of skill and information from other individuals. In these regards, the study of temperament has fertile but as yet unexplored potential. The aims of the chapter are to open up some relevant questions and stimulate further research. Moreover, one of many good reasons for taking examples of primates (in this way) is that there has been so much emphasis upon the potential influences of mental abilities on social learning (e.g. Box 1994), but relatively little upon interactions among emotional, motivational and social influences. Individual differences in these domains open up potentially new areas of understanding in social learning, and generate hypotheses for empirical research. Further, the literature supports a sound framework within which consistent differences, in behavioural and physiological responsiveness to environmental events, may be described between individuals and species (Clarke and Boinski 1995). There is a variety of techniques by which such individual differences may be assessed. There is a significant interest in relationships between these dimensions behaviourally and physiologically, together with their functional outcomes.

Concepts and dimensions of temperament

Temperament may be described generally as the characteristic style of emotional and behavioural response of an individual in a variety of different situations (Prior 1992) that is often, but not invariably, demonstrated very early in life. It is the stance that an individual takes towards its environment across time and situations. It refers to styles of responsiveness and not to specific acts. It involves both behavioural and other biological correlates of responsiveness to unfamiliar and challenging environmental events in all aspects of the environment both animate and inanimate. Hence, it influences the ways in which individuals cope with everyday environmental events and the competence of their natural development.

The study of temperament in human behaviour has a very long and varied history; it has included interest in both behavioural and other biological correlates (Rothbart 1989). There has also been an upsurge of studies in recent years. For example, in a series of very well-known studies with children, Kagan *et al.* (1992) have described distinct and opposing temperamental styles that are demonstrated by clusters of measures based around responses to avoid or approach, when individuals are challenged by events such as unfamiliar surroundings and objects, and/or unfamiliar people (Kagan *et al.* 1988, Kagan and Snidman 1991). Around 15% of their American Caucasian samples can be divided into extremes of responsiveness based upon both behavioural and physiological measures. Hence, some children are classified as shy; their behaviour is restrained and inhibited. They are reactive physiologically. For example, they have high and minimally variable heart rates; they have greater pupillary dilation, together with high cortisol and norepinephrine concentrations with challenge (Kagan *et al.* 1988, 1990). The opposing temperamental disposition is that of bold, outgoing individuals who are uninhibited and unreactive. These extremes remain stable and are described as being based upon constitutional variations in physiological organisation that bias such temperamental responsiveness from birth; its expression however, may require particular environmental conditions. Children with these extreme dispositions have highly predictable responsiveness to environmental events. Moreover, it is with these extreme and distinct dispositions that such studies are primarily concerned. More usually, in animal studies for instance, the range of individual differences is considered. Much of the literature on human temperament is related to applied and clinical questions as in those associated with social adjustment and health. With reference to children, a good deal of the impetus for studies of temperament came, and comes from, researchers in developmental psychology and child psychiatry; it is much concerned with clusters of characteristics that relate to identifying a 'difficult temperament'. Very inhibited, shy responsiveness has also been associated with immune-mediated illnesses (Kagan 1994), victimisation (Olweus 1993) and an increase in the occurrence of psychiatric illness (Hirshfield *et al.* 1992). Interestingly, there is an interest in studies of temperament in the management and health of domestic animals such as those that are farmed (e.g. Lyons *et al.* 1988, Mendl *et al.* 1992).

Temperament is a very broad concept, and it is unsurprising that there is no generally agreed scientific definition. In practice, researchers vary considerably in the extents to which they select and study behavioural, genetic and physiological dimensions of temperament both independently and interactively (Rothbart 1989). In fact, the value of the area lies not in a potential generality of the concept, but in specific domains of study as they relate to particular areas of interest (Bates 1989). The term 'temperament' is also sometimes used interchangeably with that of 'personality'. Indeed, there is considerable overlap, but there are also important distinctions. Hence, personality refers to specific social dispositions of individuals, such as their relative confidence, aggressiveness or playfulness. These are commonly determined by detailed and repeated observations in different social situations. Although temperamental dispositions have commonly included multiple dimensions – with responsiveness to environmental challenges as one of these, most recent *research* has focused on responsiveness to challenge (Suomi pers. commun.) Hence, in some ways temperament is more useful to consider individual differences with regard to propensities for socially mediated learning. For example, temperamental dispositions can be considered apart from their immediate social context.

Studies of temperament and personality cover a very wide range of animal groups. They include fish (e.g. Francis 1990) and octopus (Mather and Anderson 1993) for instance. Among mammals, observations of individual differences in temperament within wolf litters have been associated with predatory success and social status, for example (Fox 1972). Breeds of dogs have been extensively studied for temperamental differences among them (Scott and

Fuller 1965, Goddard and Beilharz 1985, Willis 1995). There has also been selective breeding, as for emotional responsiveness, among species of rodents (e.g. Gray 1971).

Studies of individual differences among species of monkeys and apes have included innovative studies by Stevenson-Hinde and her co-workers (e.g. Stevenson-Hinde and Zunz 1978). For example, Stevenson-Hinde *et al.* (1980) described individual ratings of dimensions of personality among the everyday interactions of group-living captive rhesus monkeys (*Macaca mulatta*) that were correlated over years. Behaviour included ratings of excitability, confidence and sociability. Hence, it was of interest to establish that behavioural traits among individuals could be identified reliably and remain stable over considerable periods. In addition, Stevenson-Hinde and Zunz (1978) found that their most social individuals were also the least fearful. Subsequent studies with rhesus monkeys (cf. Higley and Suomi 1989) have shown an increasing number of dimensions of responsiveness to environmental events that are associated with temperamental dispositions. There are stable differences in physiological responsiveness as in the hypothalamic–pituitary axis and the sympathetic nervous system. Physiological reactivity is frequently measured by heart rate variability, cortisol and central amines such as norepinephrine and serotonin. Reactive individuals are also much more likely to have significantly higher baseline levels in the turnover of catecholamines – especially norepinephrine, and to have higher, less variable heart rates than unreactive individuals. Behaviourally, individual rhesus monkeys that are reactive are less likely to approach the unfamiliar; they have longer latencies to respond when they do approach; and they are more inhibited socially and less likely to initiate social interactions with unfamiliar animals. In brief, they are more fearful and less likely to attempt all forms of challenges in their environments (Higley and Suomi 1989). There are substantial differences among species, including closely related species of monkeys. These are of interest in their own right and will be referred to more specifically later on. For now, it is pertinent to take some examples from a species on which a good deal of work has been done. The examples are used to consider the issues involved rather than to infer generalities. Hence, rhesus monkeys have been used extensively in this area, and with a variety of environmental challenges; they have been studied intensively in captivity with a variety of environmental challenges as individuals, and in social groups. They are also increasingly studied in nature (see Figure 3.1).

Significant differences in temperamental responsiveness may be identified at the time when young monkeys begin to leave their mothers and explore their environments; differences are stable over time and have evidence of heritability (Suomi 1987, 1991). Critically, it is the presence of environmental challenges

Figure 3.1. Rhesus macaque female and offspring. Photo by David H. Hill.

that exacerbate differences among individuals (Suomi 1985). Furthermore, individual differences among animals that are fearful and inhibited behaviourally, compared with those that are not, are further exaggerated by more challenging situations. Suomi (cf. 1997) discusses two subgroups of individuals that have aberrant developmental prognoses for competence in natural conditions. Hence, the majority of young monkeys within populations of rhesus explore and take interest in a whole range of unfamiliar conditions, to which they also respond with minimal physiological arousal. They adjust quickly and show less intensity of behavioural responses. However, some 20% of the monkeys with the same background and environmental challenges show consistent and unusual behavioural disruption, together with elevated physiological responsiveness. They have significantly elevated and frequently prolonged activation of the hypothalamic–pituitary–adrenal axis of sympathetic nervous system arousal and increased noradrenergic turnover (Suomi 1986). They are behaviourally withdrawn in initial interactions with their age mates; they are more reluctant to leave their mothers, and explore their environments less. They show withdrawal and behavioural depression to events such as maternal separation. By contrast, individuals that are not reactive and behaviourally inhibited begin to move away from their mothers earlier. They stay away for longer, and at greater distances; they take advantage of new inanimate situations. Further, individuals that move away from their mothers earlier are usually also those that are first to initiate interactions among their peers when they first encounter them. Importantly, field studies have confirmed and extended results obtained in captivity. Significant differences among individuals tend to be stable over development, and show evidence for heritability (e.g. Rasmussen and Suomi 1991) in both behavioural and physiological dimensions of responsiveness. Field studies have also shown, for example, that high reactive females appear to show less adequate care of their offspring, particularly of their first offspring, than do unreactive primiparous females. Among adolescent males, highly reactive individuals usually migrate from their natal groups when they are significantly older than the non-reactive males in their birth cohort (Rasmussen and Suomi 1989). Furthermore, the reactive males typically use more 'conservative strategies' to become part of new social groups than do their less reactive peers (Suomi *et al.* 1992).

The second subgroup that Suomi (cf. 1997) describes accounts for some 5–10% of a population and is generally described as 'impulsive', particularly in the context of aggressive interactions. Both behavioural and physiological dimensions of responsiveness show up early in life and remain stable into adulthood. There are interesting parallels with human studies, and there is evidence for significant heritability.

Critically, impulsive individuals have low concentrations of central nervous system serotonin activity as measured by concentrations of its major metabolite in primates, namely 5-hydroxyindoleacetic acid (5-H1AA), in cerebral spinal fluid. Various risks are associated with low concentrations of serotonic turnover in the central nervous system. For example, whereas the normal functions of aggressive behaviours concern competitiveness and assertiveness as in social status, and very rarely escalate into extremely violent behaviour, impulsive males are predictably more likely to escalate inappropriate aggressive behaviour and suffer wounding. Interestingly, these individuals are no less coordinated or skilled in their fighting behaviour, for instance, than their peers; neither are they substantially victimised. The point is that they initiate aggressive behaviour, including that towards individuals that are much larger and of high social status. In these circumstances injurious outcomes are more probable. Hence, impulsive individuals, and especially males, seem to be less able to regulate their responses. This may also be seen in rough and tumble play where again, escalated interactions may lead to injury. These are animals that also take risks in their locomotor behaviour, as by leaping dangerously among trees (Mehlman *et al.* 1994). Unsurprisingly then, they are also more likely to die prematurely. Hence, males with low CSF 5-HIAA concentrations are much more likely to die than their peers during the period of high mortality when prepubertal males leave their natal groups (Higley *et al.* 1996a).

Impulsive individuals migrate when they are younger than normal; they are less social and lack social skills, and they are not as capable of defending themselves physically. Social incompetence has much to do with their being expelled from their natal groups well before puberty (Mehlman *et al.* 1995). Most of these males live solitarily and die within a year or so (Higley *et al.* 1996a). Impulsive females are not generally expelled from their natal groups. However, studies in captivity have shown that they are not competent mothers, and are consistently very low in social status (Higley *et al.* 1996b). By contrast, there are positive correlations with serotonin turnover and affiliative behaviour in free living rhesus monkeys. For example, Mehlman *et al.* (1995) have shown that males with higher CSF 5-HIAA concentrations migrate later. There are also positive correlations with the time spent in grooming others, in being close to them in proximity and in the average number of monkeys observed within a 5 m radius. Individuals with significantly higher CSF 5-H1AA concentrations are not only much less likely to be impulsive in risky behaviour but they are additionally protected by receiving more social support within their group. Individuals with low concentrations are both impulsive risk takers, and lack social support. Interestingly, individuals with very high concentrations of CSF 5-HIAA are prone to rigidity and inhibition of many behaviours (Soubrie 1986, cf. Mehlman *et al.* 1994).

The examples given so far emphasise robust differences in temperamental responsiveness among individuals with regard to environmental challenges. There is evidence for early and consistent developmental identification, together with evidence for heritability. It is also relevant to examine evidence for the social and ecological conditions under which characteristic patterns of responsiveness among individuals may be modified. These are of interest in the present context because, apart from an interest in measures of responsiveness as creating opportunities for socially mediated learning, it is also important to appreciate different kinds of events that may influence individual propensities for responsiveness to the environment including that of socially mediated learning. There are good examples from both prenatal and postnatal influences. For example, although the mechanisms remain open to question, various kinds of environmental stressors such as brief removal from the home group, together with unfamiliar bursts of noise experienced by females during gestation, significantly influenced long-term behavioural and physiological responsiveness in their offspring (Schneider 1992). These individuals were more reactive behaviourally and physiologically to mild environmental challenge throughout their development. Hence, compared with infants whose mothers had not experienced prenatal stressors, when these young monkeys were given access to an unfamiliar playroom that contained toys and unfamiliar peers, they were less likely to manipulate the toys, explore the room and initiate interactions with the other monkeys. They also showed greater and longer-lasting elevations of plasma cortisol and adrenocorticotropic hormone (ACTH) during the sessions in the unfamiliar area. The critical fact is that such individuals have increased reactivity and behavioural inhibition to subsequent environmental challenges.

Experimental evidence with macaques on postnatal experiences of the young includes that in which females treat their offspring aversively under conditions of change, as with relatively restricted access to, and availability of, food (Andrews and Rosenblum 1994). Again, changes in the social status of mothers within their social units, particularly where a drop in status is substantial, but may also include general social instability, also produces adverse treatment of the young. Furthermore, although there are robust experiments in captivity, there is also much potential for postnatal disturbances in natural conditions. These include occasions when females consort during the breeding season (Berman *et al.* 1994) and their offspring are temporarily separated from them. All young monkeys show agitation initially but most do settle down. Importantly, however, there are significant individual differences, and for some individuals the consequences are long lasting; they respond to environmental challenges with exaggerated responsiveness both physiologically and behaviourally; once again, they are generally more fearful and avoid unfamiliar situations. Further, Suomi and Levine (1998) consider the potential consequences to pregnant rhesus monkey females and their offspring of being exposed to severe environmental disruptions such as violent thunderstorms. The disruptions may include the destruction of vegetation and the death of group members including relatives. 'One immediate consequence would be acute and dramatic elevation of the female's cortisol levels during the storms, producing effects on the fetus that would be reflected postnatally in increased reactivity to subsequent stressful situations'. Infants of stressed mothers overreact to changes in their environments. Moreover, there may be consequences of the environmental disruptions for the mother long after her infant is born. In particular, results of the environmental disruption may include a less predictable and more variable supply of food; the social group may be more vulnerable to predation and her social status may decrease by loss of the support of her kin. All these conditions, as Suomi and Levine point out, would influence maternal care and further effect the responsiveness of her infant to environmental events. Interestingly, by these means there may be substantial differences in reactivity among offspring born to the same mother. Food shortages and social disruption are good examples of environmental constraints that potentially influence individual differences in responsiveness that have implications for fitness and the acquisition of behavioural strategies.

Further, although they do not concern natural conditions, the results of cross fostering experiments are relevant to our understanding of the potential for behavioural flexibility, and of the potential impact of social opportunities upon responsiveness. For example, Suomi (cf. 1997) describes a series of well controlled experiments in which highly reactive neonates, and those within the normal range of reactivity, were fostered during their first six months, with mothers that were either extremely nurturing, or normally so. No differences in behavioural responsiveness was found for normally reactive individuals, independent of the mothering style of their foster mothers. However, the behaviour of the reactive monkeys was significantly different. Hence, reactive monkeys reared with normal mothers predictably responded to even minor environmental challenges with exaggerated biobehavioural responsiveness, whereas those reared with nurturing females were actually more precocious. Levels of ventral contact declined earlier, they were more active and explored the environment more, and did not show as much disturbance during weaning as any other group. Further, when they were taken to live in a larger social group with many young monkeys at 6 months old, this particular group of individuals became especially 'adept at recruiting and retaining other group members as allies in response to agonistic encounters and, perhaps as a consequence, they subsequently rose to and maintained top positions in the group's dominance hierarchy' (p. 180). Reactive individuals that had been reared by normal females fell in rank and did not change. No such differences were observed among any of the control animals.

It is particularly interesting in the context of socially mediated learning to note that when cross-fostered monkeys subsequently gave birth, they adopted the maternal style of their foster mothers, regardless both of their own temperamental disposition, and the mothering style of their biological mothers. We may emphasise two important points here. One is that at least in principle, reactivity is not always an adverse disposition in either the short or long term. The second is that the social opportunities of mothering styles, even when provided by unrelated females, may be transmitted to the next generation and include animals of very different temperamental dispositions.

Implications for socially mediated learning

The information given in the previous section draws attention to aspects of individual differences in which biobehavioural dispositions influence responses to environmental events. We may now consider some implications of

these differences for socially mediated learning, and in particular with reference to examples of mediating processes. For example, social attention is
central to socially mediated learning; it provides critical means whereby individuals may take advantage of the information that is available from other
individuals, and learn to regulate their behaviour with reference to others. It
involves attentional and emotional communication by individuals of varying
dispositions to respond to features of their environments – in different social
contexts. For example, there is much emphasis upon the importance of the
mother as a secure base from which infants gradually begin to interact independently with, and learn about, aspects of their physical and social environments (Bowlby 1973). This is functionally very important because, as with all
mammals, this is a period of very high mortality. Interestingly, however,
Boccia and Campos (1987) suggest that in order to act as a secure base, the
mother also serves as a reliable source of information about how to respond to
the environment. They consider that from an infant's perspective, it is advantageous to seek information about unfamiliar environmental conditions that
create uncertainty. Despite slight differences in terminology (Campos and
Stenberg 1981, Novak 1973) we may refer to this process generally as 'social
referencing'. It is also important to point out, and without undermining the
general value of the argument, that the use of Barbara King's distinction
between social information acquisition and social information donation has a
number of methodological and conceptual advantages when considering such
material (chapter 2, this volume). Hence, by acquiring information, an individual, in this case an infant, gains both knowledge about the environment,
and the emotional security to act upon it. Boccia and Campos (1987) also
emphasise the potential advantage to the mother of attracting the attention of
her infant and communicating information about environmental events – as in
hazardous situations. Boccia and Campos (1987) report experiments that are
directly relevant to this issue. Hence, experiments with human infants and
their mothers showed that infants acquired information about the emotional
responsiveness of their mothers in ambiguous circumstances – a visual cliff
(Sorce et al. 1984). Importantly, infants were not found to act so closely, for
example, upon the affective negative information given by their mothers when
the environmental conditions did not pose obvious hazards. Moreover, experiments by Boccia and Campos (unpublished data) also manipulated positive
and negative emotional responses of mothers to strangers. These were found to
influence significantly the responses to their offspring in the same affective
direction. The results demonstrated that infants monitored the affective signals
of their mothers and appeared to regulate their responses to a stranger both in
expressive and attentional behaviour. This is important because monitoring

and/or acquiring information about the social environment influences socialisation. In particular, emotional communication, mediated by social attentional processes, can influence styles of interactions. It is also the case, however, as Boccia and Campos emphasise, that emotional communication is part of day to day interactions among individuals of all ages in the context of social attention.

With some few notable and influencial exceptions (Chance 1967, Chance and Jolly 1970), there have been few studies of social referencing among monkeys and apes. Early studies by Novak (1973) with monkeys were suggestive of social referencing processes, but were not related primarily to natural situations. They are interesting in their own right as in the influences of different conditions of rearing for instance. Subsequently, however, Evans and Tomasello (1986) found that young captive common chimpanzees discriminated the relationships of their mothers with other adult females of the social group, and adjusted their own behaviour to them accordingly. The social and mental ability to make such discriminations and regulate behaviour as concluded by Evans and Tomasello (1986), is of interest in discussions of mental abilities and socially mediated learning among simian primates generally. In fact, there is good evidence that great apes are qualitatively different from monkeys in this regard (Parker 1996, Byrne chapter 18, this volume). Moreover, there are many possibilities for experiments in social referencing. There are comparative perspectives, as well as in attempts to distinguish the relative significance of emotional and attentional influences upon behaviour (Itakura 1995). Most importantly, experiments in this domain provide means whereby we can study the elusive acquisition of social information among primate species. There is so much emphasis upon the acquisition of skill such as in foraging strategies, but relatively little as yet upon the acquisition of social information. For species that live in long-lasting social groups with relatively complex patterns of social interaction that are central to their life strategies studies of social learning of social information are critical, but relatively neglected. Moreover, studies of individual differences in temperamental disposition add further perspectives. As we have seen, reactive and behaviourally inhibited infants interact with their mothers in ways that are significantly different from those that are relatively unreactive and behaviourally uninhibited. They 'use' their mothers differently in her potential 'role' of providing a secure base in emotional security and relevant information. Unreactive and uninhibited individuals move away from their mothers earlier; they stay away longer; they take advantages of new situations and initiate social interactions when they first come across unfamiliar individuals. Much may depend upon the disposition of the mother of course, but it is an advantage of being an outgoing unfearful individual to develop greater behavioural flexibility in coping with environmental events and to have a wider potential nexus of social information in dealing with social and physical events. (This may be potentially hazardous in the most dangerous environments, however, Suomi pers. commun., Altmann 1980.) By some contrast, reactive, inhibited individuals may well be animals that are at considerable risk from members of their own social group for example. Hence, there are clear advantages in their reluctance to leave their mothers. They may depend upon the information that she provides as well as upon her relative immediate protection.

For a physically independent but naive animal, learning how to deal with the social and physical environment is facilitated by proximity to experienced individuals. In many situations, social attention depends upon physical proximity among individuals. For instance, Coussi-Korbel and Fragaszy (1995) emphasise that tolerance of spatial proximity supports visual observation of both the affect and activities of other individuals, together with physical access to the places and objects that are used by the other animals. This increases the likelihood that individuals can augment socially acquired information with their own behaviour, as with the same objects, in the same place. When primate infants are very young they are carried by, and subsequently follow, their mothers into proximity with sources of food, resting places and social companions. Gradually, they independently determine the time that they spend in proximity and interactions with their social and physical environment (Pereira and Altmann 1985). Critically then, spatial proximity involves seeking and tolerating inter-individual distances within which information may be communicated (Altmann 1980). Hence, once again, individual differences in temperamental disposition are important in creating opportunities for socially mediated learning. Individuals that can be bolder and less reactive will have advantages from both seeking and tolerating physical proximity with others more frequently, in more positive emotional responsiveness, and at an earlier time. For example, Pereira and Altmann (1985) suggest that, apart from gaining physical protection *per se*, time spent by young juveniles in proximity to adults provides a context in which they may readily react to, and learn about, the alarm and cautious behaviour of the adults in dangerous situations. Additional advantages of protection may accrue when adult males form affiliative relationships with particular individuals. The adults may subsequently intervene on the juveniles' behalf in aggressive encounters, for instance. Competition may also be avoided with adult females and older immatures during feeding, when younger juveniles stay in close proximity to adult males (Pereira and Altmann 1985). Further, because younger juveniles are permitted close proximity to older experienced animals, they are able to feed on high quality food patches and learn 'to identify and process different food items' (p. 240).

There has been much discussion about activities of social play in the development of physical and social competence. In socially mediated learning play provides opportunities for individuals to acquire information about other individuals as individuals, as well as about patterns of social interaction. There are significant quantitative differences among species, and among individuals of different ages and sex in different ecological and social conditions. Sex differences in play behaviour, for example, may be predicted where males and females have significantly different life histories as adults (e.g. Harcourt *et al.* 1976, Brown 1988). Young rhesus monkeys begin to associate predominantly with their own sex in play activities from around four to five months old if not before. This continues through puberty, if not beyond (Suomi pers. commun.). Males engage in much rough and tumble play with all monkeys that will participate. However, females stop participating in rough and tumble play from about six months old, and they rarely initiative such activities. They spend more time in grooming and chasing with other females. Further, the developmental complexity of social play is well illustrated by work with rhesus monkeys. Hence, when the monkeys are around four months old their bouts of play are very short; they occur with one other animal, and with very simple behavioural sequences. Later on, between two and three years of age, social play occupies a similar amount of time but occurs in much longer bouts, and frequently engages two or more animals. Importantly, social play is very much more complex; coalitions are formed and reversed among individuals. Critically, social play includes sequences of behaviour that 'are what appear to be precursors, or elementary forms, of adaptive behaviours, that, in adulthood, are used to coordinate reproductive and social activities' (Suomi 1991, p. 34). In well-known species such as rhesus monkeys, play behaviour not only provides information for socially mediated learning in a variety of natural functional contexts, but has been found to be a predictive dimension of temperamental responsiveness among individuals in their development. Hence, levels of spontaneous day-to-day play in home cage interactions in later infancy have been found to predict individual differences into childhood and adolescence (Higley 1985). In brief, more reactive and behaviourally inhibited patterns of responsiveness inhibit play behaviour (Higley 1985) and influence opportunities for socially mediated learning. Individual juveniles that are insecure in their attachments to their mothers for instance, both play less frequently and demonstrate less sophisticated patterns of play interactions than infants that have secure attachments. Moreover, environmental challenges are shown to be advantageous to the development of normal behaviour among relatively unreactive and uninhibited individuals. For example, in field situations, and as was noted earlier, reactive individuals may be disadvantaged by their responsiveness to brief separations from their mothers when the mothers resume mating activities. Relatively unreactive and uninhibited individuals may be advantaged by such conditions. Hence, Berman *et al.* (1994) found that when separated from their mothers, free ranging male rhesus monkeys on Cayo Santiago played more. Females played less and engaged in more social grooming. Berman *et al.* (1994) suggest that early experiences of separation may play a role in the development of normal gender differences in behaviour.

The development of social skills involves a variety of dimensions of social competence that reflect the demands of complex and changing social interactions. These include competitive and cooperative strategies; they involve cooperative alliances, reconciliation and social status. A good deal of work is needed here with reference to individual differences in temperamental dispositions. However, there are some interesting findings with regard to social status. This is a critical area of social competence that involves priority of access to resources that are necessary for survival and fitness. We may note, for example, from various studies in captivity, and in the field, that although we need fine grain analyses, there is an association between lack of high social status and fearful reactive responsiveness in the work with rhesus monkeys. Scanlan (1987) has shown that, in captivity, at least in peer reared monkeys, more reactive individuals, namely those that were more disturbed by separation from their regular social units, were subsequently less likely to show leadership skills that are characteristic of monkeys that have high status. He also found (Scanlan 1987) that reactive animals were more likely to give way, to acquiesce, to other monkeys in social interactions, or in situations where there was competition for resources. Alternatively, Stevenson-Hinde and Zunz (1978) had found that outgoing, less fearful rhesus monkeys were those of high status. Further, Rasmussen and Suomi (1989) found that low reactive free ranging juvenile rhesus males had profiles of behaviour that would facilitate success in competitive interactions. Moreover, an interesting set of studies by Sapolski and his group (e.g. Sapolski and Ray 1989, Ray and Sapolski 1992) draws attention to additional perspectives, albeit in a different species. For example, his long-term field studies on savannah baboons (*Papio cynocephalus*) that examined associations between social status and stress responsiveness showed that testosterone responses of high status and subordinate males differed significantly when they were exposed to the environmental stressor of capture. Testosterone titres of subordinates dropped markedly, whereas those of the high status individuals elevated rapidly and remained high for an hour or so.

Further, the work shows that dominant males encompass both individuals that have a high reactive style, and those that have a low reactive style.

> *Those dominant males who consistently exhibited a low-key behavioural re-*
> *sponse to psychological stressors and were also able to calibrate their response*
> *to different intensities of stress (such as distinguishing between the mere pres-*
> *ence of another adult male vs. being threatened by that male) also had low*
> *basal concentrations of cortisol. In contrast, dominant males who were highly*
> *reactive to identical stressors had basal cortisol levels comparable to those of*
> *subordinate adult males. As Rasmussen and Suomi (1989) suggested for low-*
> *reactive male rhesus, Sapolsky (1990) argued that a non-reactive style may fa-*
> *cilitate attainment and/or maintenance of high rank in baboons. (Clarke and*
> *Boinski 1995, p. 110).*

This kind of interactive study draws attention to clusters of responsiveness among individuals that affect different functional implications between individuals that may superficially have closely similar social 'characteristics', such as that of high social status. Social status is clearly a complex domain with regard to individual differences in temperamental reactivity. Correlations between particular dimensions of responsiveness are useful to generate hypotheses, but the concept of status involves many aspects of interactive responsiveness, as Higley and Suomi (1989), for example, have emphasised. Importantly, the work by Sapolski associates levels of endocrine activity and styles of social information processing. He also argues (Sapolski 1990) that profiles of hormonal activity in both high status males that have a reactive style of responsiveness as well as in the subordinate males have consequences for their fitness and survival. In the latter context immune responsiveness is lowered in these individuals, with attendant health risks such as cardiovascular disease. Hence, the very important lessons to be learned from this work interrelate individual differences in temperament and socially mediated learning. A functional example is demonstrated in that some individuals are able to predict and control the outcome of competitive social interactions. They also have a temperamental disposition that allows for social tension to be dissipated.

Differences in temperament among species

Studies of individuals in temperamental responsiveness within species demonstrate significant differences that concern propensities for socially mediated learning. It is also relevant to consider individual differences in association with differences among species in their ecological and social systems – in order to widen our functional perspectives. There is a great deal to do in this regard (see for example, Wilson *et al.* 1994). As we have noted already, the diversity of lifestyles among species of the simian primates is considerable. However, although relatively few species have been investigated so far, there is clear evidence for significant differences in both physiological and behavioural responsiveness among species (cf. Clarke and Boinski 1995). Specifically and in parallel with the findings for individuals within species, the shy–bold/reactive–unreactive dimensions of comparison are apparently the most appropriate (Clarke and Boinski 1995). Species have been studied with a variety of unfamiliar social and nonsocial challenges. Further, there are robust data to support hypotheses that associate interrelationships among species-typical temperamental responsiveness with their social and ecological systems (Clarke and Boinski 1995). For example, long-term exemplary studies on South American titi monkeys (*Callicebus malach*) and squirrel monkeys (*Saimiri sciureus*) show that the latter are more active, opportunistic, impulsive and bold (Mendoza and Mason 1984, 1986). These differences are consistent with differences between them in their physiological responsiveness to environmental stressors as well as to significant ecological and social characteristics. There is a variety of less systematic studies that have demonstrated significant differences in responsiveness among widely different primate taxa. Interestingly, similarities across species show that temperamental dispositions are apparently more closely related to differences in lifestyles than to close taxonomic status (Clarke and Boinski 1995). Tentative hypotheses for species differences in temperament include defence against predation, feeding and habitat specialisations. Clarke and Boinski (1995) note, for example, the hypothesis that species that are more curious, bold, unfearful and instrumental in their approach to unfamiliar situations are also those that utilise relatively high-energy foraging strategies and/or complex foraging, and tend to have 'omnivorous' diets. There is a long-standing interest in feeding strategies and responsiveness to environmental challenges (Jolly 1964, Glickman and Sroges 1966).

Further, there are suggestions that differences among species in their group composition, patterns of dispersal and adult sex ratios may be associated with differences among them in the expression of social tolerance, affiliation and aggressive behaviour (cf. Clarke and Boinski 1995).

In view of the considerable amount of work on temperamental responsiveness in rhesus monkeys, however, it is interesting to consider comparisons among species of the same genus (*Macaca*). The macaques are potentially a good model for such comparisons. Hence, the 20 or so species live in a wide diversity of habitats across Asia; there is a diversity of dietary adaptations

(Melnick and Pearl 1987); there are differences in ranging behaviour, habitat use and foraging behaviour, and there are differences in sociodemographics, sexual and social behaviour (cf. Clarke and Boinski 1995). So far there have been few studies that have made comparisons directly, and the majority of the species are still to be studied in this perspective. However, there is now good evidence on the basis of a variety of social and inanimate challenges that a number of species are behaviourally and physiologically distinct (e.g. Clarke and Mason 1988, Clarke *et al.* 1988, Clarke and Lindburg 1993, de Waal 1989, Sackett *et al.* 1976, Thierry 1985). Responsiveness varies greatly in different challenging situations. There are significant interspecific variations, for example, in aggressiveness, reconciliatory behaviour, curiosity and exploratory behaviour. Hence, among a number of species of macaques, rhesus monkeys are bold and unreactive in their approach to unfamiliar environmental events. They are relatively curious and exploratory; they are most aggressive and the least conciliatory; they are less passive socially. Interestingly, this is a species that is very widely distributed, and thrives in a diversity of ecological conditions including, of course, commensal relations with humans (Richard *et al.* 1989). Individuals that are relatively reactive, and inhibited, may be comparatively disadvantaged in such a species (except maybe for certain environments (Suomi, pers. commun.)). Moreover, comparative studies among macaques highlight the influence of 'social support' in behavioural development, namely the amount of affiliative behaviour directed towards an individual (Boccia *et al.* 1997). It is well known that bonnet macaques (*M. radiata*), for example, have a different social developmental network than pigtail macaques (*M. nemistrina*), with which they are often studied. The bonnet macaques have a relatively open social system; mothers are less restrictive and infants interact frequently, as in allomaternal interactions and play behaviour for instance, with a number of group members, including adult females. Social attachments are formed with more individuals than among young pigtail macaques, whose mothers are more restrictive. There are immediate implications for social learning. Differences in social networks are important for routes of transmission of behaviour, as in social attention to a wider social nexus, that also provides information about the environment in affiliative contexts. Moreover, individuals of these two species respond differently behaviourally and physiologically to environmental challenges. An important point is that the natural availability of alternative social attachments can modulate against the behavioural and physiological consequences of aversive events. Importantly, infant bonnet macaques without such social support are not buffered against adverse environmental events. However, the different social organisational style of bonnet macaques provides the opportunity to ameliorate expressions of negative responsiveness, and provide increased opportunities for socially mediated learning of positive responsiveness to environmental events. Social support both modulates physiological responsiveness and facilitates social attachment, which in turn may facilitate security, exploration and exposure to information that is available from other group members. We need comparative research into these issues. We might also note that social opportunities of social support influence the ability of individuals not only to cope with their environments, but also to remain healthy and reproductively fit. Hence, there are also issues that are relevant to health, welfare and conservation in both natural and captive conditions. For example, developing research in the field of psychoneuroimmunology is of interest in this context (Laudenslager *et al.* 1993, Laudenslager and Boccia 1996).

Concluding remarks

Individuals, as among simian primates in this case, vary in the extents to which they interact with their environments in ways that inhibit and facilitate the acquisition of ecologically relevant information, including that which is socially mediated. Moreover, studies of differences in their temperamental dispositions present provocative implications for socially mediated learning. Well-substantiated information for some few species of simian primates provides good cases for discussion. The crux of the matter is that the organisation of behaviour among individuals influences opportunities for, and potential use of, information about the environment that is acquired in the context of interactions with other individuals. Hence, it is generally relevant that consistently and relatively reactive animals, as in fearful/inhibited animals such as rhesus monkeys, may acquire qualitatively different information about their environments, and respond to events within them, in qualitatively different ways, than individuals that are significantly and consistently less reactive and inhibited. They may explore and manipulate less. Unreactive and uninhibited animals are more active; they move around and explore more. They also become less disturbed physiologically. In other words, they cope better, especially with unfamiliar circumstances. Importantly, these are responses that increase the probability that less fearful individuals will confront, attend to and become familiar with previously unfamiliar features of their environments. When relatively fearful individuals do confront challenging situations, they are also more likely to avoid them and become physiologically disturbed. Moreover, less fearful and active animals interact more, and in emotionally more positive ways with other animals. They are more likely to maintain closer physical proximity to others, to attend more to what others are doing and where. Hence, they may have greater chances of facilitating advantageous responses to environmental challenges – as in feeding strategies. Again, in the acquisition of social skills, less fearful animals engage in play activities more than relatively fearful animals. Such interactions facilitate the development of information about other individuals, and about the quality of a range of social behaviours that is developed. This includes the development of competitive and cooperative behaviour as in cooperative alliances and reconciliation behaviour. ('It can also be argued that reactive individuals *generally* pay more attention to "subtle" environmental events, so that they have a *different* set of opportunities to "learn" about their environments' (Suomi, pers. commun.).)

Further, individual variation in temperamental disposition among such animal populations represents the range of individual differences that has been selected for. Hence, it is interesting to consider, for example, that in large socially complex social units such as are found among many species of the simian primates, a wide range of temperamental dispositions may allow for the exploitation of a wider range of social success, whilst at the same time reducing direct competition among individuals (Clark 1991). Fox (e.g. 1972) provides convincing evidence, and a relevant discussion in this general context, from his studies of temperamental reactivity within litters of wolf cubs.

Given that the study of mammalian temperament in mediating influences upon social learning is an unexplored area, the evidence and arguments will need a good deal of specific examination and development. There is also a number of caveats. It is critical that temperamental dispositions involve clusters of responsiveness to environmental demands. Moreover, studies of temperamental dispositions include some aspects of behavioural organisation; others include neurological organisation (e.g. Bateson 1991). There are no simple correlations. This is a complex domain that involves interactions among biological systems as with emotion, attention, activity and mental ability, and social contexts, as some studies of the acquisition of social status demonstrate, for example (see pp. 47–8). Overall, however, an interactive approach has substantive and realistic appeal. There is so much emphasis in social learning studies upon the acquisition of 'acts' as 'packages' and the mechanisms that are involved. A biobehavioural interactive approach draws attention to the fact that learning how to deal competently with the demands of the environment is socially *mediated*.

Acknowledgements

It is my pleasure to thank Barbara King, Steve Suomi and James Steele for their very helpful constructive comments on this chapter. My thanks also to Victoria Mountford for all her excellent secretarial assistance.

References

Altmann, J. (1980). *Baboon Mothers and Infants.* Cambridge, MA: Harvard University Press.

Andrews, M. W. and Rosenblum, L. A. (1994). The development of affiliative and agonistic social patterns in differentially reared monkeys. *Child Devel.,* **65**, 1398–404.

Bates, J. E. (1989). Concepts and measures of temperament. In *Temperament in Childhood,* ed. G. A. Kohnstamm, J. E. Bates and M. K. Rothbart, pp. 3–26. Chichester, UK: John Wiley and Sons.

Bateson, P. P. G. (1991). *The Development and Integration of Behaviour: Essays in Honour of Robert Hinde.* New York: Cambridge University Press.

Berman, C. M., Rasmussen, K. L. R. and Suomi, S. J. (1994). Responses of free-ranging rhesus monkeys to a natural form of social separation. Parallels with mother-infant separation in captivity. *Child Devel.,* **65**, 1028–41.

Boccia, M. L. and Campos, J. J. (1987). Social attentional processes in human and non-human primate infants. *Primate Report,* **18**, 3–10.

Boccia, M. L., Scanlan, J. M., Laudenslager, M. L., Broussard, C. L. and Reite, M. L. (1997). Presence of 'friends' mitigates behavioral and immunological responses to maternal separation in bonnet macaques. *Physiol. Behav.,* **61**, 191–8.

Bowlby, J. (1973). *Attachment and Loss, Vol. 2, Separation.* New York: Basic Books.

Box, H. O. (1994). Comparative perspectives in primate social learning: new lessons for old traditions. In *Current Primatology,* ed. J. J. Roeder, B. Thierry, J. R. Anderson and N. Herrenschmidt, pp. 321–27. Strasbourg: Universite Louis Pasteur.

Box, H. O. and Fragaszy, D. M. (1986). The development of social behaviour and cognitive abilities. In *Primate Ontogeny, Cognition and Social Behaviour,* ed. J. G. Else and P. C. Lee, pp. 119–28. Cambridge: Cambridge University Press.

Brown, S. G. (1988). Play behaviour in lowland gorillas: age differences, sex differences, and possible functions. *Primates,* **29**, 219–28.

Campos, J. J. and Stenberg, C. (1981). Perception, appraisal, and emotion: the onset of social referencing. In *Infant Social Cognition: Empirical and Theoretical Considerations,* ed. M. Lamb and L. Sherrod, pp. 273–314. New Jersey: Erlbaum Associates.

Chance, M. R. A. (1967). Attention structure as the basis of primate rank orders. *Man,* **2**, 503–18.

Chance, M. R. A. and Jolly, C. J. (1970). *Social Groups of Monkeys, Apes and Man.* London: Cape.

Clark, A. B. (1991). Individual variation in response to environmental change. In *Primate Responses to Environmental Change,* ed. H. O. Box, pp. 91–110. London: Chapman and Hall.

Clarke, A. S. and Boinski, S. (1995). Temperament in nonhuman primates. *Am. J. Primatol.,* **37**, 103–25.

Clarke, A. S. and Lindburg, D. G. (1993). Behavioural contrasts between male cynomologous and lion-tailed macaques. *Am. J. Primatol.,* **29**, 49–59.

Clarke, A. S. and Mason, W. A. (1988). Differences among three macaque species in responsiveness to an observer. *Int. J. Primatol.,* **9**, 347–64.

Clarke, A. S. and Schneider, M. L. (in press). Prenatal stress alters social and adaptive behaviour in adolescent rhesus monkeys. *Ann. New York Acad. Sci.*

Clarke, A. S., Mason, W. A. and Moberg, G. P. (1988). Differential behavioural and adrenocortical responses to stress among three Macaque species. *Am. J. Primatol.,* **14**, 37–52.

Colmenares, F. and Rivero, H. (1982). Social attention, information-acquisition behaviour patterns and maturing processes. *Bio. Behav.,* **7**, 271–6.

Coussi-Korbel, S. and Fragaszy, D. M. (1995). On the relation between social dynamics and social learning. *Anim. Behav.,* **50**, 1441–53.

de Waal, F. B. M. (1989). *Peacemaking Among Primates.* Cambridge, MA: Harvard University Press.

Eisenberg, J. F. (1981). *The Mammalian Radiations.* Chicago: University of Chicago Press.

Evans, A. and Tomasello, M. (1986). Evidence for social referencing in young chimpanzees (*Pan troglodytes*). *Folia Primatol.*, **47**, 49–54.

Fox, M. W. (1972). Socio-ecological implications of individual differences in wolf litters: a developmental and evolutionary perspective. *Behaviour*, **41**, 298–313.

Francis, R. C. (1990). Temperament in a fish: a longitudinal study of the development of individual differences in aggression and social rank in the Midas Cichlid. *Ethology*, **86**, 265–352.

Gavan, J. A. (1982). Adolescent growth in nonhuman primates: an introduction. *Human Biology*, **54**, 1–5.

Glickman, S. E. and Sroges, R. W. (1966). Curiosity in zoo animals. *Behaviour*, **26**, 151–88.

Goddard, M. E. and Beilharz, R. G. (1985). A multivariate analysis of the genetics of fearfulness in potential guide dogs. *Behav. Genet.*, **15**, 69–89.

Gray, J. A. (1971). *The Psychology of Fear and Stress.* Toronto, NY: McGaw Hill.

Harcourt, A. H., Kelley, K. J. and Fossey, D. (1976). Male emigration and female transfer in wild mountain gorillas. *Nature*, **263**, 226–7.

Higley, J. D. (1985). Continuity of social separation behaviours in rhesus monkeys from infancy to adolescence. Unpublished doctoral dissertation, University of Wisconsin, Madison, WI.

Higley, J. D. and Suomi, S. J. (1989). Temperamental reac-tivity in non-human primates. In *Temperament in Childhood*, ed. G. A. Kohnstamm, J. E. Bates and M. K. Rothbart, pp. 153–67. Chichester, UK: John Wiley and Sons Ltd.

Higley, J. D., Mehlman, P. T., Fernald, B., Vickers, J. H., Lindell, S. G., Taub, D. M., Suomi, S. J. and Linnoila, M. (1996a). Excessive mortality in young free-ranging nonhuman primates correlates with low CSF 5-HIAA concentrations. *Arch. Gen. Psychiatry*, **53**, 537–43.

Higley, J. D., King, S. T., Hasert, M. F., Champoux, M., Suomi, S. J. and Linnoila, M. (1996b). Stability of interindividual differences in serotonin function and its relation to severe aggression and competent social behaviour in rhesus monkey females. *Neuropsychopharmacology*, **14**, 67–76.

Hirshfield, D. R., Rosenbaum, J. F. Biederman, Bolduc, E. A., Faraone, S. V., Snidman, N., Reznick, J. S. and Kagan, J. (1992). Stable behavioral inhibition and its association with anxiety disorder. *J. Am. Acad. Child Adolescent Psychiatry*, **31**, 103–11.

Itakura, S. (1995). An exploratory study of social referencing in chimpanzees. *Folia Primatol.*, **64**, 44–8.

Jolly, A. (1964). Prosimians manipulation of simple object problems. *Anim. Behav.*, **12**, 560–70.

Kagan, J. (1994). *Galen's Prophecy.* New York: Basic Books.

Kagan, J. and Snidman, N. (1991). Infant predictors of inhibited and uninhibited pro-files. *Psychol. Sci.*, **2**, 40–4.

Kagan, J., Resnick, J. S. and Snidman, N. (1988). Biological basis of childhood shyness. *Science*, **240**, 167–71.

Kagan, J., Resnick, J. S. and Snidman, N. (1990). The temperamental qualities of inhibition and lack of inhibition. In *Handb. Devel. Psychopathol.*, ed. M. Lewis and M. Miller, pp. 219–26. New York: Plenum Press.

Kagan, J., Snidman, N. and Arcus, D. M. (1992). Initial reactions to unfamiliarity. *Curr. Dir. Psychol. Sci.*, **1**, 171–4.

Laudenslager, M. L. and Boccia, M. L. (1996). Some observations of psychosocial stressors, immunity, and individual differences in nonhuman primates. *Am. J. Primatol.*, **39**, 205–21.

Laudenslager, M. L., Boccia, M. L. and Reite, M. L. (1993). Biobehavioural consequences of loss in nonhuman primates: individual differences. In *Handbook of Bereavement: Theory, Intervention and Research*, ed. M. S. Stroebe, W. Stroebe and R. Hansson, pp. 129–42. Cambridge: Cambridge University Press.

Lyons, D. M., Price, E. O. and Moberg, G. P. (1988). Individual differences in temperament of domestic dairy goats: constancy and change. *Anim. Behav.*, **36**, 1323–33.

Mather, J. A. and Anderson, R. C. (1993). Personalities of octopuses (*Octopus rebescens*). *J. Comp. Psychol.*, **107**, 336–40.

Mehlman, P. T., Higley, J. D., Faucher, B. A., Lilly, A. A.,

Taub, D. M., Vickers, J. H., Suomi, S. J. and Linnoila, M. (1994). Low CSF 5-HIAA Concentrations and severe aggression and impaired impulse control in nonhuman primates. *Am. J. Psychiatry*, **151**, 1485–91.

Mehlman, P. T., Higley, J. D., Faucher, B. A., Lilly, A. A., Taub, D. M., Vickers, J. H., Suomi, S. J. and Linnoila, M. (1995). Correlation of CSF 5-HIAA concentrations with sociality and timing of emigration in free ranging primates. *Am. J. Psychiatry*, **152**, 907–13.

Melnick, D. J. and Pearl, M. C. (1987). Cercopithecines in multimale-groups: genetic diversity and population structure. In *Primate Societies*, ed. B. B. Smuts, D. L. Cheney, R. M. Seyfarth, R. W. Wrangham and T. T. Struhsaker, pp. 121–34. Chicago: University of Chicago Press.

Mendl, M., Zanella, A. J. and Broom, D. M. (1992). Physiological and reproductive correlates of behavioural strategies in female domestic pigs. *Anim. Behav.*, **44**, 1107–21.

Mendoza, S. P. and Mason, W. A. (1984). Rambunctious *Saimiri* and reluctant *Callicebus*: systemic contrasts in stress physiology. *Am. J. Primatol.*, **6**, 415.

Mendoza, S. P. and Mason, W. A. (1986). Contrasting responses to intruders and to involuntary separation by monogamous and polygynous new world monkeys. *Physiol. Behav.*, **38**, 795–801.

Mendoza, S. P. and Mason, W. A. (1989). Primate relationships:

social dispositions and physiological responses. In *Perspectives in Primate Biology*, Vol. 2, ed. P. K. Seth and S. Seth, pp. 129–43. New Delhi: To-day and To-morrows Printers and Publishers.

Novak, M. A. (1973). Fear attachment relationships in infant and juvenile rhesus monkeys. Unpublished doctoral dissertation. University of Wisconsin, Madison, WI.

Olweus, D. (1993). Victimization of peers: antecedents and long-term outcomes. In *Social Withdrawal, Inhibition, and Shyness in Children*, ed. K. Robin and J. B. Asendorpf, pp. 315–41. Hillsdale, NJ: Erlbaum.

Pagel, M. D. and Harvey, P. H. (1993). Evolution of the juvenile period in mammals. In *Juvenile Primates. Life History, Development and Behaviour*, ed. M. E. Pereira and L. A. Fairbanks, pp. 28–37. Oxford: Oxford University Press.

Parker, S. T. (1996). Apprenticeship in extractive foraging: the origins of imitation, teaching, and self-awareness in great apes. In *Reaching into Thought: The Minds of Great Apes*, ed. A. E. Russon, K. A. Bard and S. T. Parker, pp. 348–70, Cambridge: Cambridge University Press.

Pereira, M. E. and Altmann, J. (1985). Development of social behaviour in free-living nonhuman primates. In *Nonhuman Primate Models for Human Growth and Development*, ed. E. S. Watts, pp. 217–309. New York: Liss.

Prior, M. (1992). Childhood tem-

perament. *J. Child Psychiatry*, **33**, 249–79.

Rasmussen, K. L. R. and Suomi, S. J. (1989). Heart rate and endocrine responses to stress in adolescent male rhesus monkeys on Cayo Santiago. *Puerto Rico Health Sci. J.*, **8**, 65–71.

Rasmussen, K. L. R. and Suomi, S. J. (1991). Mothers and sons: somatic and physiological characteristics of adult female rhesus and their adolescent offspring. *Am. J. Primatol.*, **24**, 129.

Ray, J. C. and Sapolski, R. M. (1992). Styles of male social behaviour and their endocrine correlates among high ranking baboons. *Am. J. Primatol.*, **28**, 231–50.

Richard, A. F., Goldstein, S. J. and Dewar, R. E. (1989). Weed macaques: the evolutionary implications of macaque feeding ecology. *Int. J. Primatol.*, **10**, 569–94.

Rothbart, M. K. (1989). Biological processes in temperament. In *Temperament in Childhood*, ed. G. A. Kohnstamm, J. E. Bates and M. K. Rothbart, pp. 77–110. Chichester, UK: John Wiley and Sons Ltd.

Sackett, G. P., Holm, R. A. and Ruppenthal, G. C. (1976). Social isolation rearing: species differences in behaviour of macaque monkeys. *Dev. Psychol.*, **12**, 283–8.

Sapolski, R. M. and Ray, J. C. (1989). Styles of dominance and their endocrine correlates among wild olive baboons (*Papio anubis*). *Am. J. Primatol.*, **18**, 1–13.

Sapolski, R. M. (1990). Adrenocortical function, social rank, and personality among wild baboons. *Biol. Psychiatry*, **28**, 862–78.

Scanlan, J. M. (1987). Social dominance as a predictor of behavioural and pituitary-adrenal response to social separation in rhesus monkey infants. Paper presented at the meeting of the Society for Research in Child Development. Baltimore, MD, April 1987.

Schneider, M. L. (1992). Prenatal stress exposure alters postnatal behavioural expression under conditions of novelty challenge in rhesus monkey infants. *Devel. Psychobiol.*, **25**, 529–40.

Scott, J. P. and Fuller, J. L. (1965). *Dog Behaviour: The Genetic Basis*. Chicago and London: University of Chicago Press.

Sorce, J. F., Emde, R., Campos, J. J. and Klinnert, M. D. (1984). Maternal emotional signalling: its effect on the visual cliff behaviour of one-year-olds. *Devel. Psychol.*, **21**(1), 195–200.

Soubrie, P. (1986). Reconciling the role of central serotonin neurons in human and animal behaviour. *Behav. Brain Sci*, **9**, 319–64.

Stevenson-Hinde, J. and Zunz, M. (1978). Subjective assessment of individual rhesus monkeys. *Primates*, **19**, 473–82.

Stevenson-Hinde, J., Stillwell-Barnes, R. and Zunz, M. (1980). Subjective assessment of rhesus monkeys over four successive years. *Primates*, **21**, 66–82.

Suomi, S. J. (1985). Biological response styles: experiential effects. In *Biologic Response Styles: Clinical Implications*, ed. H. Klas and L. J. Siever, pp. 2–17. Washington, DC: American Psychiatric Press.

Suomi, S. J. (1986). Anxiety-like disorders in young primates. In *Anxiety Disorders of Childhood*, ed. R. Gittelmann, pp. 1–23. New York: Guilford.

Suomi, S. J. (1987). Genetic and maternal contributions to individual differences in rhesus monkeys biobehavioural development. In *Perinatal Development: A Psychobiological Perspective*, ed. N. A. Krasnegar, E. M. Blacs, M. A. Hofer and W. P. Smothersson, pp. 397–420. New York: Academic Press.

Suomi, S. J. (1991). Uptight and laidback monkeys. In *Plasticity of Development*, ed. S. E. Brauth, W. S. Hall and R. J. Dooling, pp. 27–56. Cambridge, MA: MIT Press.

Suomi, S. J. and Levine, S. (1998). Psychobiology of intergenerational effects of trauma: evidence from animal studies. In *International Handbook of Multigenerational Legacies of Trauma*, ed. Y. Daniel, pp. 623–37. New York: Plenum Press.

Suomi, S. J. (1997). Early determinants of behaviour: evidence from primate studies. *Br. Med. Bull.*, **53**, 170–84.

Suomi, S. J., Rasmussen, K. L. R. and Higley, J. D. (1992). Primate models of behavioural and physiological change in adolescents. In *Textbook of Adolescent Medicine*, ed. E. R. McAnarney, R. E. Kreipe, D. P. Orr and G. D. Comerci, pp. 135–9. Philadelphia: Saunders.

Thierry, B. (1985). Patterns of agonistic interactions in three species of macaques (*Macaca mulatta, M. fascicularis, M. tonkeana*). *Aggressive Behav.*, **11**, 223–33.

Willis, M. B. (1995). Genetic aspects of dog behaviour with particular reference to working ability. In *The Domestic Dog: Its Evolution, Behaviour and Interactions with People*, ed. J. Serpell, pp. 54–64. Cambridge: Cambridge University Press.

Wilson, D. S., Clark, A. B., Coleman, K. and Dearstyre, T. (1994). Shyness and boldness in humans and other animals. *Trends Ecol. Evo.*, **9**, 442–6.

4

Evolutionary biology of skill and information transfer

Richard M. Sibly

Introduction

The term social learning refers to learning that is influenced by observation of or interaction with another animal or its products (Box 1984, Galef 1988). At least two animals are necessary for skill or information transfer to occur, one animal that learns or receives the information, here called the *receiver*, and another providing the skill or information, here called the *transmitter*. I shall consider that skill or information transfer occurs whenever a transmitter influences a receiver. This chapter deals with the evolutionary biology of such skill or information transfer. Although the scope of the chapter is therefore wider than social learning, its conclusions apply to social learning in particular.

Genes or behaviours specifically evolved for transmission or reception are not necessary for skill or information transfer to occur. For instance, receivers may learn to associate the presence of transmitters with the nearby presence of food, and so use the location of conspecifics as a cue when foraging. Mechanistically, this could be a product of classical conditioning. Once transmitters and receivers occur in the vicinity of each other, for whatever reason, the stage is then set for the coevolution of transmitting and receiving should genes affecting transmitting and receiving arise through mutation or otherwise. To motivate and guide discussion of the evolutionary biology, four examples will now be introduced. The examples have been chosen so that each can include genes specifically evolved for transmission.

When butterflies open their wings, some species reveal large sham predatory eyes. These butterflies transmit the false information that they are predators, causing small birds, that otherwise might kill them, to flee. The birds' flight responses are innate, and it seems clear that the genes that produce the birds' flight responses evolved before eyespots evolved in butterflies. In this case the genes affecting the receivers (the birds) existed first, and the genes affecting the transmitters (the butterflies) evolved second.

Male–male fights for access to breeding sites or females are often preceded by periods of mutual assessment in which antagonists size each other up. If such assessments reliably predict the outcome of fights, then it will pay antagonists to act on the results of assessments, thereby avoiding fights and the chances of serious injuries or death. It is easy to see, therefore, how assessment strategies could originally have evolved, since genes that caused assessment and appropriate subsequent action would have been selected. When assessment strategies had become common in the population there would then be selection on antagonists to advertise, perhaps exagerate, their fighting prowess, and this could lead on to further coevolution introducing more sophisticated assessment strategies.

Alarm calls may have evolved from escape responses in group-living animals. One can imagine that originally there was an advantage in fleeing when another group member fled. If this strategy – 'flee when others flee' – became common in the population, there could then be selection to evoke the escape response in others by giving a suitable call. In these situations, however, individuals giving alarm calls might thereby attract the attention of predators. Such transmissions could be costly to the transmitter. The transmitter genes would only spread if the selective disadvantage of attracting the predator's attention was outweighed by advantages accruing to copies of the transmitter genes in recipients. In this case the direction of selection is affected by the number of recipients and the proportion containing copies of the transmitter gene (Hamilton's rule).

Transmission of feeding skills may have begun with receivers following or imitating the movements of transmitters. There are many examples in this volume. Transmitters may, however, pay a cost if as a result of transmission they feed more slowly or encounter more feeding competition. As with alarm calls, the cost of transmission may be offset by advantages obtained by copies of transmitter genes present in recipients.

Some general principles are suggested by these four examples. First, in many instances receivers evolve abilities to monitor feeding, migratory and other behaviours of other animals without those other animals evolving any specialised transmitter behaviours. In these cases, only the receivers have specialised genes or behaviours. Second, if transmitters do have specialised genes or behaviours, then since genes affecting receiving are hardly likely to be the same as those affecting transmitting, coevolution between at least two loci is generally involved. Third, since it is unlikely that both genes come into existence by mutation simultaneously, one evolves before the other. In the above examples the receiver gene evolves before the transmitter gene. It may be that this is a general rule. It appears also that modification of receiver behaviour may generally benefit receivers directly, e.g. by improving foraging success, whereas modification of transmitter behaviour generally benefits transmitters only indirectly, via receivers who are related genetically to them. In this case the indirect benefits to transmitters occur through copies of the transmitter genes that are carried by receivers.

Once genes for receiving are present in the population, the situation is changed from how it was at the start of the evolutionary process. Now the transmitters are in a new environment insofar as their actions may affect the behaviour of other individuals. There may then be an advantage to genes that modify transmitters' behaviours, either to benefit receivers or, conceivably, to reduce receivers' success if this somehow results in a compensating advantage for transmitters. This identifies another key issue – under what conditions are genes affecting transmitting selected for?

Once the coevolutionary ratchet has turned and there exist genes that affect transmitting as well as genes that affect receiving, the environment of receivers has changed, and there may be further evolution of receiver behaviour. This in turn may produce further evolution of transmitter behaviour, and this may lead on, perhaps, to continuing coevolution.

In the discussion of different modelling approaches that follows, some definitions and distinctions are necessary. Note first, we are dealing with any modifications of receiver behaviour that result from the presence, behaviour, or products of the transmitter. These include social learning but also include other social behaviours. In several of the models the genetic relationship of transmitter and receiver turns out to be important, and it has proved useful to distinguish between vertical (i.e. parent–offspring) transmission, oblique (i.e. other between-generation) transmission, and horizontal (i.e. within-generation) transmission (Feldman and Cavalli-Sforza 1986, Laland *et al.* 1993). Also, note that the scope of the discussion potentially includes cultural evolution and the evolution of language, provided only that one takes, as is natural for a biologist, an atomist approach to the study of culture (Maynard Smith and Warren 1982, Dunbar 1996, Hauser 1996, Kroodsma and Miller 1996).

The above preamble provides a framework within which to characterise the existing approaches to modelling the evolution of skill and information transfer. The key issues are:

1   How are genes thought to affect receiving and/or transmitting?
2   What types of receiving and transmitting behaviours are favoured by natural selection?
3   Are transmissions that benefit receivers costly to transmitters?
4   How is selection of transmitting and receiving affected by the genetic relatedness, if any, between transmitter and receiver? Genetic relatedness

is important if the behaviour of one animal has fitness costs or benefits for others (Hamilton 1964).

5   How is selection of transmitting and receiving affected by environmental variation (Boyd and Richerson 1988)?

Learning in general, and social learning in particular, is only advantageous at intermediate levels of environmental variation. This is because where there is no environmental variation, or where environmental variation is regular and predictable by the animal (from e.g. time-of-day, season, or phase of the moon) then learning is unlikely to be the way to achieve an optimal response. The reason is that compared with genetic determination or hormonal switching, learning is complicated, expensive to engineer genetically, and prone to error. Thus, learning has no advantage in the absence of environmental variation, but at the other extreme learning is also of no use in a completely unpredictable environment, since there past and present states provide no useful information about the future, so there is nothing that can usefully be learnt. We would therefore only expect to see learning processes where environments are variable but to some extent predictable (Stephens 1991, Bergman and Feldman 1995). Given there exists environmental variation and that learning is selected, it remains to consider in what circumstances *social* learning is advantageous.

Using these key issues as reference points the three main approaches to modelling the evolution of skill and information transfer will now be briefly characterised. These can be described as the 'selfish gene' approach, an 'evolutionarily stable strategy' (ESS) approach, and a 'formal population genetics' approach. At the end of the chapter I describe a life-history approach that in principle allows field testing of explanatory hypotheses.

The main modelling approaches

The style of analysis used in this chapter so far is a product of the selfish gene approach (Hamilton 1964, Trivers 1971, Dawkins 1976, Wilson 1976). The selfish gene approach grew out of ethology, and so has its origins in the early writings of Tinbergen and Lorentz (reviews will be found in Dawkins 1995, a special issue of *Phil. Trans. Roy. Soc.* (**340**:161–255, 1993), and Halliday and Slater 1983). Early analyses had construed communication as benefiting both transmitter and receiver, but in a mould-breaking paper Dawkins and Krebs (1978) pointed out the potential importance of manipulation and deception. Thus animals may transmit 'false' as well as or instead of 'true' information (see also Zahavi 1977, Grafen 1990, Maynard Smith 1991).

The selfish gene approach has been extended in the 'evolutionarily stable strategy' (ESS) analyses of Boyd and Richersen (reviewed in Laland *et al.* 1993). Whereas the genetics are not modelled explicitly, learning mechanisms are. A major focus of interest is to establish the conditions under which environmental variation selects for social learning mechanisms.

The following account is based mostly on Boyd and Richerson (1988). Only selection acting on receiver behaviour is considered, and it is supposed that genes affect the split between receivers' reliance on social and individual learning. For simplicity the environment is considered as consisting of two habitats in equal proportions, and a fixed fitness increment accrues from behaviour appropriate to the habitat in which an animal finds itself. Some animals learn appropriate or inappropriate behaviour individually, and the rest learn by imitating others.

In a constant uniform environment Boyd and Richerson showed that within the terms of their model social learning is the optimal strategy. The reason is that under the terms of the model a small proportion of individual learners end up producing inappropriate behaviour, but at evolutionary equilibrium this is avoided by social learners. The model shows how behaviours spread in the population, and how this depends on two parameters, representing the proportions of animals that end up with appropriate and inappropriate behaviours as a result of their reliance on individual learning.

Boyd and Richerson then examined how their results are modified in varying environments. In the case of spatial variation, migration rates between habitats are critical. The results of the analysis suggested that there is a wide range of migration rates, and a wide range of qualities of individual information, where it is optimal to employ a mixture of social and individual learning. More social learning should be used if individual learning is unreliable, because at evolutionary equilibrium social learners avoid the mistakes made by individual learners, at least in a constant environment (see above). However, as the environment becomes less predictable, more individual learning should be used. (Within the terms of the model, a decrease in predictability is equivalent to an increase in migration rates between dissimilar habitat.) Similar conclusions apply to temporal variation (i.e. fluctuating environments) but migration rate is then replaced by the rate of temporal fluctuation. This result is supported by a recent population-genetic analysis (Feldman *et al.* 1996). Thus, whether the environmental variation is temporal or spatial, a decrease in predictability selects for increased individual learning. Individual learning is needed to 'calibrate' the system, but too great a dependence on individual learning can be costly because individual learners can make mistakes. Equivalently, however, too great a dependence on social learning can be costly, since it adds hysteresis to the system. In extreme cases whole populations could, after environmental fluctuations, persist in inappropriate behaviours, if too many animals simply imitated others.

Boyd and Richerson's model suggests that the adaptiveness of social learning relative to individual learning depends on

- the accuracy of individual learning;
- the chance that transmitters experienced the same environment as receivers.

If individual learning is inaccurate, it often pays to use social learning instead; conversely, if transmitters are unreliable, individual learning is more attractive. To these factors Laland *et al.* (1993) suggested adding

- the importance of vertical as opposed to highly horizontal social transmission.

Vertical transmission is 'traditional' and conservative; highly horizontal transmission is rapid and ephemeral. Laland *et al.* (1993) speculated that rapid horizontal transmission will be found to be advantageous in highly unpredictable environments in cases including small population size, multiple transmitters, and biassed transmission. By biassed transmission they mean that receivers are more likely to imitate some transmitters than others. Boyd and Richerson (1989) found that receivers should imitate the more common behaviour, and that social learning should become less important as discrepancies between transmitters increase.

Whereas the above approaches can be used to obtain a wider view, a strong school of classical population genetic modelling initiated by Feldman and Cavalli-Sforza (1976) has obtained more rigorous answers, albeit to less general questions. These studies of the coevolution of genes and culture are mostly classical two-locus population genetic models, but with the added feature that receiver individuals may or may not receive advantageous information. For simplicity the advantageous information is considered to be discrete – either you get it or you don't. It is therefore necessary to distinguish not only genotypes – AB, Ab, aB and ab in a haploid two-locus model – but also 'phenogenotypes', which distinguish between genotypes that have and those that do not have the advantageous information. Using a bar notation to distinguish advantaged individuals there are eight phenogenotypes: $\overline{AB}$, $\overline{Ab}$, $\overline{aB}$, $\overline{ab}$, AB, Ab, aB and ab. Darwinian fitness is assumed to depend on the phenotype, and this depends on whether or not the advantageous trait has been acquired. Genotypes influence whether or not this happens, and, therefore, fitness depends only indirectly on genotypes. The models analysed are generally haploid, but diploid models would probably give similar results. Topics considered include the transmission of altruistic traits (Feldman *et al.* 1985), and the effects of vertical transmission as opposed to oblique or horizontal transmission (Feldman and Cavalli-Sforza 1986). Recently, a multiple locus model has been analysed (Feldman and Zhivotovsky 1992). A general result is that some population structure, such as genetic relatedness between transmitters and receivers, is needed before the state of genetic determination can be invaded (e.g. Cavalli-Sforza and Feldman 1983a,b).

A representative paper from this school is that by Aoki and Feldman (1987), entitled 'Toward a theory for the evolution of cultural communication: co-evolution of signal transmission and reception'. This paper modelled explicitly the coevolution of genes affecting transmission and genes affecting reception, in a two-locus model in which one locus controls the transmission and the other the reception of information that may have adaptive value. It was assumed, as is reasonable, that there are fitness costs to transmission and reception. Thus, there are likely to be, for instance, physiological costs of supporting the biological machinery needed for communication, and noisy transmissions may attract predators or competitors, and so on. For simplicity only vertical (i.e. parent–offspring) transmission was analysed. Various other assumptions and parameters are needed to obtain a full population genetics model; thus a parameter was included to specify the frequency of recombination between the two communication genes, and it was assumed that the population is sexual and haploid, although similar results are likely to obtain in diploid populations.

Aoki and Feldman supposed that for a communication system to spread in a population three things are required: genes for transmission, genes for reception and something worth communicating. They supposed that if all three are present a selective advantage accrues to the receiver, and they showed that the selective advantage has to be very large if the communication system is to spread. To be precise, the selective advantage has to be so large that advantaged individuals have more than twice as many surviving offspring as non-advantaged individuals. This is a surprising result since few examples are known of information this valuable. It is therefore instructive to understand the reason why this two-fold advantage is required.

When communication genes are rare in the population, individuals with both the transmitter gene and the receiver gene and the advantageous information ($\overline{AB}$ individuals) will be very rare. Aoki and Feldman calculate the conditions under which these individuals would increase in the population, as follows. Since they are rare in the population, $\overline{AB}$ individuals will, almost certainly, mate with individuals possessing neither the communication genes nor the advantageous information (i.e. *ab* individuals). Thus, the mating can be written $\overline{AB} \times ab$. If the communication genes are linked (the most advantageous case) then the offspring of such a mating receiving one communication gene will also get the other. Thus, ignoring recombination for a moment, half the offspring get both communication genes, and the other half get the alternate alleles. Next, Aoki and Feldman assumed that offspring learn from just one parent, and, further, that *ab* parents cannot possess the advantageous information, since they had no means of acquiring it. Thus, half the offspring with the communication genes learn from the 'wrong' (i.e. *ab*) parent and as a result do not obtain the advantageous information. This results in a loss of half the potential number of $\overline{AB}$ individuals in the next generation. It is to outweigh this loss that the advantageous information has to more than double the reproductive output of the $\overline{AB}$ individuals.

Aoki and Feldman also took account of a number of further obstacles to the spread of the communication system, including the assumed fitness costs of communication, attrition of the transmitted information, loss of the conjunction of communication genes through recombination, and so forth, and these were built into the model by the inclusion of further parameters.

In a later paper, Aoki and Feldman (1989) showed that the two-fold condition can be ameliorated somewhat if the communication genes have separate independently advantageous effects. Thus, greater memory capacity could be a prerequisite for speech but have other advantages independent of its association with speech. The two-fold condition can also be reduced under certain conditions of biparental care (Takahasi and Aoki 1995).

Another way to remove the two-fold cost is by allowing learners to evaluate the communicated information, and only act on the most advantageous. Such evaluation skills might have very general advantages to learners and thus might evolve independently or prior to social learning and communication. Population genetic frameworks within which to analyse the evolution of learning have been devised by Harley and Maynard Smith (1983), Feldman and Cavalli-Sforza (1986), Houston and Sumida (1987), Stephens (1991) and Bergman and Feldman (1995).

Whereas these modelling approaches give insight into how skill and information transfer may have evolved, they do not allow ready testing in the field. In the next section it is shown that fitness costs and benefits may, at least in principle, be calculated in field conditions, using a life-history approach.

A life-history approach allows field testing

Selection pressures may act at various points in the lives of animals; these age effects are explicitly taken into account in studies of life-history evolution, which analyse the ways in which life histories are moulded by natural selection. The factors which cause genes affecting the life history to increase in frequency are well understood; those genes spread which reduce the mortality rate of their carriers, or cause them to breed earlier or to produce more offspring, other things being equal (Sibly and Antonovics 1992). Other things being equal means that the gene only affects one of the life-history characters, that is, it only affects one of mortality rate, age at breeding, or fecundity. If one character is affected beneficially but another adversely there is said to be a *trade-off* between those two characters. Genes which cause animals to breed earlier or to produce more offspring often achieve their effects by increasing production, i.e. the net rate of obtaining resources. There is therefore very generally selection to maximise the net rate of obtaining resources, and so to maximise foraging success, if foraging behaviours do not differ in their mortality implications.

In summary, those genes are selected which cause their carriers to produce more offspring, to incur lower mortality, or to grow faster if this results in earlier breeding. The rates at which such genes will spread can be measured, and their rates of increase can be used as measures of *fitness*. The existence of a quantitative fitness measure makes it possible to calculate *fitness gains*. These quantify, in units of fitness, the advantages conferred by genes with specified effects. To say there is a fitness gain is equivalent to saying that the gene is favoured by natural selection. When there is a trade-off between two characters, one being beneficially affected and the other adversely affected by a gene, the character which is adversely affected has a negative fitness gain, and this too can be quantified in units of fitness. Only if the net fitness gain is positive will the gene spread in the population. For communication genes to be maintained in the population, therefore, they must confer a net fitness gain. In the next sections we consider how this net fitness gain is calculated in practice. A distinction is made between the effects of a gene on its carrier, here called *direct effects*, and its effects on other individuals, here called *indirect effects*.

Calculation of direct fitness effects

Direct fitness effects are relatively straightforward to calculate. It is necessary to have measurements of any fecundity increase, growth increase (if this will result in earlier breeding) or mortality decrease that results to a particular carrier from a given transfer of skill or information. In the perhaps unlikely event of a trade-off – e.g. rate of feeding increased but at a cost of increased mortality rate – then net fitness gain depends on the partial selective pressures (PSPs) on fecundity, growth and mortality (formulae for PSPs (λ-sensitivities) are available in e.g. Caswell 1989). Net fitness gain is then calculated as:

Net fitness gain = (PSP fecundity) × (fecundity increase) + (PSP growth) × (growth increase)
+ (PSP mortality) × (mortality increase)   (1)

Theory shows that genes whose net fitness gain is positive spread to fixation irrespective of their levels of dominance (provided they are not overdominant) if their life-history effects do not change during selection (Sibly and Curnow 1993). In practice, however, fitness gains may reduce as the number of carriers increases (e.g. if individuals compete for limited resources), and this may reduce the rate of increase of the gene.

Calculation of indirect fitness effects

For simplicity we only consider genes whose direct effects are on transmitters. Indirect effects are more complicated to calculate than direct effects because the advantages are obtained indirectly, here via those copies of the gene that are carried by recipients. The terms benefit and cost will now be introduced. *Benefit* is defined as the net fitness gain experienced by a receiver in consequence of a transmission caused by the transmitter gene. This benefit is calculated as in eqn (1). *Cost* is defined as minus the net fitness gain experienced by the transmitter, calculated as in eqn (1). The overall net fitness gain of the transmitter gene is now calculated from Hamilton's rule:

Net fitness gain = r benefit – cost   (2)

where r is the coefficient of relatedness between transmitter and recipient. Only if the net fitness gain is positive will the transmitter gene increase in the population (see e.g. Sibly 1994).

Transmitter–receiver conflict

For simplicity, we have not so far considered the case that genes affecting receivers have indirect effects on transmitters. However, if genes affecting receivers do have indirect, possibly adverse, effects on transmitters, a more complicated analysis is needed. Using benefits and costs as in the previous section to indicate net fitness consequences for receivers and transmitters respectively, Hamilton's rule shows that the net fitness gain of the *receiver* gene is:

Net fitness gain = benefit − r cost   (3)

Because net fitness gain measures the selection pressure on a gene, the selection pressures on transmitter and receiver genes can now be compared. From eqn (2) the selection pressure on transmitter genes is r benefit − cost; from eqn (3) the selection pressure on receiver genes is benefit − r cost. It follows that selection on receiver genes is always stronger than selection on transmitter genes (assuming benefit and cost are positive, and $0 < r < 1$). The general conclusion is that if transmission is costly, genes affecting receivers are likely to evolve more widely and so to be more common than genes affecting transmitters.

Further analysis shows that in evolutionary terms there always exists a conflict of interest between transmitters and receivers. This is because from eqn (2) a gene directly affecting transmitters spreads provided benefit/cost > $1/r$, but according to eqn (3) a gene directly affecting receivers spreads if benefit/cost > r. To see the practical implications of this consider the case that $r = 1/2$, as with transmission of parental skills to offspring. In this case genes are selected that cause parents to transmit the skill if benefit/cost > 2, but genes are selected that cause offspring to try to obtain it provided only that benefit/cost > 1/2. Thus, genes in recipients are selected to obtain more help than genes in transmitters are selected to supply. By analogy with parent–offspring conflict, this is transmitter–receiver conflict. This is analysed in more detail in Figure 4.1.

In the extreme case that transmitter and receiver are unrelated, so that $r = 0$, receivers are always selected to try to obtain help, whatever the cost to transmitters. By contrast transmitters are selected never to transmit if transmitting is costly, whatever the benefit might be to receivers. In such circumstances there is strong selection on each to deceive the other.

Finally, it is interesting to note a curious asymmetry between transmitters and receivers. Whereas advantageous transmitter genes always spread to fixation, whatever their levels of dominance, the analogous result does not hold for receivers. Initially advantageous receiver genes do not necessarily spread to fixation (this follows from the results of Sibly 1994). Thus, while transmitter behaviours are expected to be uniform throughout the population, it could be that there exists genetic polymorphism in receiver behaviours.

Conclusion

This chapter has considered the conditions under which mechanisms promoting the transfer of skill and information are favoured by natural selection. At

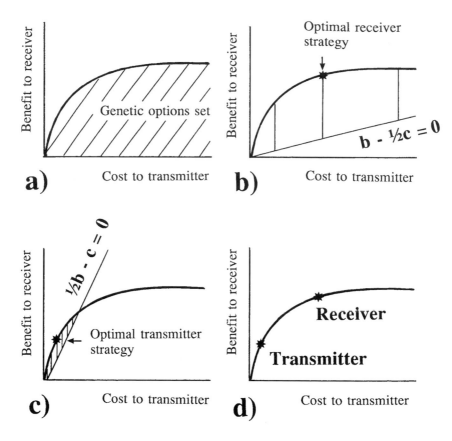

Figure 4.1. Schematic representation of transmitter–receiver conflict. (*a*) The 'genetic options set' shows what can be achieved genetically in a particular environment. The points making up the genetics options set indicate genetic possibilities, as follows. If a given transmission results in a cost *c* to the transmitter and a benefit *b* to the receiver, then this outcome is plotted as the point with coordinates (*c*, *b*). It is assumed there is a genetic basis to the transmission, so the point represents something that is genetically possible. Other genetically based possibilities may result in different outcomes, and the set of all such outcomes is the 'genetic options set'. There are, however, likely to be limits to the benefit that can be obtained however much effort is put into transmission, so benefits, relative to costs, may follow a diminishing law of returns, as indicated by the curve bounding the options set. (*b*) shows how the optimal receiver strategy is identified. Consider three possible genes that might act on receivers, allowing them to obtain benefits, but at some cost to the transmitter. The costs and benefits of the three genes are indicated by the positions of the tops of the three vertical lines. The lengths of the vertical lines are equal to the net fitness gains obtained by the genes, calculated from eqn (3) for the case $r = 1/2$. The optimal receiver strategy is the one giving the largest fitness gain (longest vertical line). (*c*) The optimal transmitter strategy is identified similarly, though the mathematics are slightly more complicated. (*d*) It follows from (*b*) and (*c*) that optimal strategies for transmitters and receivers are different, and this gives rise to transmitter–receiver conflict.

least two animals are necessary for skill or information transfer to occur, one animal that learns or receives the information, here called the *receiver*, and another providing the skill or information, here called the *transmitter*. Skill or information transfer are considered to occur whenever a transmitter influences a receiver. The evolution of skill or information transfer involves the coevolution of genes affecting transmitter and genes affecting receiver behaviours. The main issues are: 1. How are genes thought to affect receiving and/or transmitting? 2. What types of receiving and transmitting behaviours are favoured by natural selection? 3. Are transmissions that benefit receivers costly to transmitters? 4. How is selection of transmitting and receiving affected by the genetic relatedness, if any, between transmitter and receiver? 5. How is selection of transmitting and receiving affected by environmental variation?

There are three main approaches to modelling and understanding these coevolutionary processes. The approaches overlap in some respects but can be roughly distinguished as formal population genetics, selfish gene, and ESS analysis. Using life-history techniques fitness costs and benefits of transmitting and receiving can be measured in the field, taking account as necessary of the genetic relatedness of the individuals involved. Calculation formulae are available (eqn (1)). Using this approach it is shown that, if transmission is costly, selection on receiver genes is always stronger than selection on transmitter genes. It follows that genes affecting receivers are likely to evolve more widely and so to be more common than genes affecting transmitters. If transmission is costly it will only be selected when directed towards close relatives. Even so, evolutionary transmitter–receiver conflict is generally predicted. However, whereas advantageous transmitter genes always spread to fixation, whatever their levels of dominance, the analogous result does not hold for receivers. Thus, while transmitter behaviour is expected to be uniform throughout the population, it could be that there exists genetic polymorphism in receiver behaviour.

Acknowledgements

I am very grateful to H. O. Box and K. N. Laland for help in the literature search, and to K. Aoki, R. Boyd, K. Gibson and P. Richerson for comments on an earlier version of the manuscript.

References

Aoki, K. and Feldman, M. W. (1987). Toward a theory for the evolution of cultural communication: coevolution of signal transmission and reception. *Proc. Natl. Acad. Sci. USA*, **84**, 7164–8.

Aoki, K. and Feldman, M. W. (1989). Pleiotropy and preadaptation in the evolution of human language capacity. *Theor. Popul. Biol.*, **35**, 181–94.

Bergman, A and Feldman, M. W. (1995). On the evolution of learning: representation of a stochastic environment. *Theor. Popul. Biol.*, **48**, 251–76.

Boyd, R. and Richerson, P. J. (1985). *Culture and the Evolutionary Process*. Chicago: University of Chicago Press.

Boyd, R. and Richerson, P. J. (1988). An evolutionary model of social learning: the effects of spatial and temporal variation. In *Social Learning: Psychological and Biological Perspectives*, ed. T. R. Zentall and G. B. Galef, pp. 29–48. Hillsdale, NJ: Erlbaum.

Boyd, R. and Richerson, P. J. (1989). Social learning as an adaptation. *Lectures on Mathematics in the Life Sciences, Am. Math. Soc.*, **20**, 1–26.

Box, H. O. (1984). *Primate Behaviour and Social Ecology*. London: Chapman and Hall.

Caswell, H. (1989). *Matrix Population Models*. Sunderland, MA: Sinauer.

Cavalli-Sforza, L. L. and Feldman, M. W. (1983a). Paradox of the evolution of communication and of social interactivity. *Proc. Natl. Acad. Sci. USA*, **80**, 2017–21.

Cavalli-Sforza, L. L. and Feldman, M. W. (1983b). Cultural versus genetic adaptation. *Proc. Natl. Acad. Sci. USA*, **80**, 4993–6.

Dawkins, M. S. (1995). *Unravelling Animal Behaviour*, 2nd edn. Harlow, Essex: Longman.

Dawkins, R. (1976). *The Selfish Gene*. Oxford: Oxford University Press.

Dawkins, R. and Krebs, J. R. (1978). Animal signals: information or manipulation? In *Behavioural Ecology: An Evolutionary Approach*, ed. N. B. Davies and J. R. Krebs, pp. 282–309. Oxford: Blackwell Scientific Publications.

Dunbar, R. (1996). *Grooming, Gossip and the Evolution of Language*. London: Faber & Faber.

Feldman, M. W. and Cavalli-Sforza, L. L. (1976). Cultural and biological evolutionary processes; selection for a trait under complex transmission. *Theor. Popul. Biol.* **9**, 238–59.

Feldman, M. W. and Cavalli-Sforza, L. L. (1986). Towards a theory for the evolution of learning. In *Evolutionary Processes and Theory*, ed. S. Karlin and E. Nero, pp. 725–41. New York: Academic Press.

Feldman, M. W. and Zhivotovsky, L. A. (1992). Gene–culture coevolution toward a general theory of vertical transmission. *Proc. Natl. Acad. Sci. USA*, **89**, 11935–8.

Feldman, M. W., Cavalli-Sforza, L. L. and Peck, J. R. (1985). Gene–culture coevolution models for the evolution of altruism with cultural transmission. *Proc. Natl. Acad. Sci. USA*, **82**, 5814–18.

Feldman, M. W., Aoki, K. and Kumm, J. (1996). Individual versus social learning: evolutionary analysis in a fluctuating environment. *Athropol. Sci.*, **104**, 209–32.

Galef, B. G. (1988). Imitation in animals: history, definition and interpretation of data from the Psychological laboratory. In *Social Learning: Psychological and Biological Perspectives*, ed. T. R. Zentall and G. B. Galef, pp, 3–28. Hillsdale, NJ: Erhlbaum.

Grafen, A. (1990). Biological signals as handicaps. *J. Theor. Biol.*, **144**, 517–46.

Halliday, T. R. and Slater, P. J. B. (1983). *Communication*. Oxford: Blackwell Scientific Publications.

Hamilton, W. D. (1964). The genetical theory of social behaviour. I, II. *J. Theor. Biol.*, **7**, 1–52.

Harley, C. B. and Maynard Smith, J. (1983). Learning – an evolutionary approach. *Trends Neurosci.*, **6**, 204–8.

Hauser, M. D. (1996). *The Evolution of Communication*. Cambridge, MA: MIT Press.

Houston, A. I. and Sumida, B. H. (1987). Learning rules, matching and frequency dependence. *J. Theor. Biol.*, **126**, 289–308.

Kroodsma, D. E. and Miller, E. H.

(eds.) (1996). *Ecology and Evolution of Acoustic Communication in Birds*. Ithaca, NY: Cornell University Press.

Laland, K. N., Richerson, P. J. and Boyd, R. (1993). Animal social learning: toward a new theoretical approach. In *Perspectives in Ethology: Behaviour and Evolution*, Vol. 10, ed. P. P. G. Bateson and P. H. Klopfer, pp. 249–77. New York: Plenum Press.

Maynard Smith, J. (1991). Theories of sexual selection. *Trends Ecol. Evol.*, **6**, 146–51.

Maynard Smith, J. and Warren, N. (1982). Models of cultural and genetic change. *Evolution*, **36**, 620–7.

Rauno, V. A. and Mappes, J. (1996). Tracking the evolution of warning signals. *Nature*, **382**, 708–9.

Sibly, R. M. (1994). An allelocentric analysis of Hamilton's rule for overlapping generations. *J. Theor. Biol.*, **167**, 301–5.

Sibly, R. M. and Antonovics, J. (1992). Life-history evolution. In *Genes in Ecology*, ed. R. J. Berry, T. J. Crawford and G. M. Hewitt, pp. 87–122. Oxford: Blackwell Scientific Publications.

Sibly, R. M. and Curnow, R. N. (1993). An allelocentric view of life-history evolution. *J. Theor. Biol.*, **160**, 533–46.

Stephens, D. W. (1991). Change, regularity, and value in the evolution of animal learning. *Behav. Ecol.*, **2**, 77–89.

Takahasi, K. and Aoki, K. (1995). Two-locus haploid and diploid models for the coevolution of cultural transmssion and paternal care. *Am. Nat.*, **146**, 651–84.

Trivers, R. L. (1971). The evolution of reciprocal altruism. *Q. Rev. Biol.*, **46**, 35–57.

Wilson, E. O. (1976). *Sociobiology: The New Synthesis*. Cambridge, MA: Harvard University Press.

Zahavi, A. (1977). Reliability in communication systems and the evolution of altruism. In *Evolutionary Ecology*, ed. B. Stonehouse and C. M. Perrins, pp. 253–9. London: Macmillan.

Zentall, T. R. and Galef, G. B. (eds.) (1988). *Social Learning: Psychological and Biological Perspectives*. Hillsdale, NJ: Erlbaum.

Part 2

Social learning among species of terrestrial herbivores

Editors' comments

HOB and KRG

We have a wide range of interests, presenting a diverse collection of chapters under this general heading. We did not want to contribute a book on mammalian social learning without including a discussion of marsupials. Given that one of our central aims is to raise new perspectives, it would be inappropriate to leave out such a major group. It will be important subsequently to make comparisons with placental mammals. Moreover, although as it happens, marsupials have been a neglected group in social learning contexts, we can already see that this clearly reflects a lack of study rather than of biological substance. There are certainly difficulties in studying marsupials. For instance, as with many of the groups referred to in this volume, most species are small, cryptic and nocturnal. Karen Higginbottom and David Croft make a substantial contribution, however, by opening up relevant issues. Another attractive feature of their chapter is that it indicates practical applications of social learning studies among marsupials. Overall, inclusion of this chapter will place marsupials firmly within the social learning domain of interests. The authors consider both information that is currently available and, very importantly, a wide range of hypotheses for future research. For example, although mothers exclusively give parental care in the majority of species, significantly different patterns of maternal care provide different developmental contexts within which the young may acquire information. Hence, compared with species in which the young are left in a nest very early in life, species with large pouches and single offspring that gradually spend more time moving independently and following the mother have greater opportunities for social learning. Furthermore, the low sociality of most marsupial taxa provides very little opportunity for socially mediated learning among adults. Again, in species that live relatively long lives and in which the offspring stay in the home range of their mothers, the offspring have more time to learn from the mother, especially when they continue their association – rather than disperse. For example, frequent association with their mothers may enable young females to gain information about such things as the selection and distribution of food, the location of shelters and maternal skills. Interestingly, Higginbottom and Croft predict that body size will indirectly influence opportunities for social learning among marsupials. It has a negative effect on litter size, and a positive effect on the duration of variables that include the pouch life, the time until weaning and until sexual maturity, and longevity, as well as an effect on diurnal activity, opportunities for long-term social bonds and the intensity of maternal care of individual offspring. Hence, body size should be correlated positively with the availability of opportunities for socially mediated learning. Moreover, and although there are potential opportunities for social learning among a diversity of marsupials, the vast majority of information to date that is detailed enough to indicate clearly opportunities for social learning relates to kangaroos and wallabies of the genus *Macropus*. Hence, a variety of examples is given from work on these species. It is also the reason why this chapter is included in this section of the book.

The behaviours for which there is most circumstantial evidence involve associations of young with their mother in the development of appropriate responses to the environment, together with learning about aspects of the social and physical environment, and social play.

For example, good circumstantial evidence is given in different functional contexts that include the selection of food, the recognition and avoidance of predators, and the idiosyncratic use of their habitats. For instance, some species have highly individual and consistent patterns of home range use, including preferred areas for feeding and routes between feeding and resting areas. There is a number of intriguing issues. It is of interest, for example, that mothers, as in red-necked wallabies, have been observed to spend more time with their sons and interact with them more frequently than with their daughters; males are also more active in reuniting with their mothers. Hence, sons and daughters may have quantitatively different opportunities for social learning that may assume functional significance in the context of significantly different sex differences in life histories and adult social experiences.

Questions about sex differences and behavioural development that include opportunities for social learning are addressed specifically by Lee and Moss in their chapter on savannah African elephants (chapter 6). In this subspecies, family units are larger than among African forest elephants, and provide a quantitatively different social context for learning and development. There are temporary associations among families as well as more consistent associations. There are also large aggregations of elephants at times. Hence, elephant calves grow up in a rich and dynamic social environment. They are dependent nutritionally from two to four years, and are socially dependent for about ten to sixteen years. Moreover, although they share the same families, males and females are shown to be substantively different as they develop. The methodological perspective that Lee and Moss provide is important for studies of socially mediated learning. They consider detailed information from a large cohort of individuals over a long period; critically, they are concerned with a number of physical and social developmental parameters interactively – as a developmental complex. Hence, sex differences in the processes of physical growth and reproductive maturation have consequences for the context of learning, for opportunities to learn environmental and social skills, and for the ways in which males and females achieve their different needs. Moreover, the material clearly emphasises that sex differences in behavioural strategies develop because males and females differentially influence 'apparently similar social environments during early development' by their interactive behaviour with their mothers, and other individuals of each sex. For example, mothers treat their male offspring differently than they do their females. Male calves are more persistent and demanding in feeding from their mothers and they grow faster. Behaviour of both the mother and her calf are influenced by experience. There is an interaction between sex specific rates of growth, maternal experience, and the mother's physiological condition, that have consequences for growth and survival.

The data also show, for example, that as males grow up they concentrate their exploration and play interactions in the context of novel experiences – outside their own family groups. This has further implications for the subsequent migration of males, although the complexity of influences that determine when, and how long it takes individuals to migrate remains undetermined as yet. The maturation of females is both physically and socially more advanced than that of males. Their social behaviour, for example, is much concerned with interactions among individuals within their families. For instance, they play with young calves – both male and female – and as juveniles they make a substantial contribution to the survival of young calves. Further, the presence and behaviour of mothers apparently facilitates the behaviour of their inexperienced daughters in a number of reproductive activities that includes assistance with the first birth and the demands of the new calf.

In addition to behaviour that shows early sexual differentiation, the development of foraging skills poses some interesting questions that directly implicate the acquisition of information from other animals. Hence, given that their early dietary intake is limited by their size, strength and competence, experience with additional foods is clearly facilitated by sampling food from, and observing, other animals. Moreover, Lee and Moss note that the skills that are necessary for foraging among elephants are probably much greater than for most grazing ungulates. The possession of trunks and tusks allows them to exploit a wide variety of food material and demonstrate selectivity.

The majority of studies in social learning concentrate upon single species. Although this is clearly reasonable on many grounds, it is also valuable to consider together a number of species with different physical characteristics and lifestyles, living in the same environment, and to compare opportunities for, and potential functions of, social learning. David Klein's chapter on the caribou, muskox and Arctic hare (chapter 7) provides a good example of this perspective. It is also of general interest because there are so few attempts to consider functions of social learning among Arctic mammals. With reference to feeding behaviour, for example, we may consider that the social acquisition of information about food in young caribou is functionally more salient than among young muskoxen. The point is that relatively large bodied, well insulated muskoxen are the most energy conservative of the ungulates. They are able to forage rapidly on large quantities of low quality plant material. By contrast, the slender, less well insulated caribou lead energetically expensive lives that involve more time feeding over greater distances. Dietary selection of high quality food is important. It is relevant that mother caribou and their calves form close social bonds and maintain close proximity in their daily activities. Careful consideration of the evidence, including that from captive animals, however, indicates that social learning is most likely to be related to the recognition of good microhabitats from which to feed, rather than about specific plant material as such. Further, in order to make good use of the seasonal components of their range lands, caribou migrate, an activity that requires coordination in time, group cohesion and migrational abilities. By contrast, muskoxen do not usually migrate and live in clearly defined home ranges. There has been a long standing hypothesis that caribou learn specific routes of migration by interactions with experienced individuals. There is certainly strong evidence that young females learn their calving grounds by association with older females. The same grounds are used in successive years. Moreover, these grounds are the one 'fixed area' in the annual routes of migration.

Anti-predator behaviour raises additional comparisons between these species with regard to opportunities for social learning. In this case, muskox adults demonstrate more influence upon the development of avoidance behaviour in their calves than do caribou. Caribou generally avoid predators – mainly wolves – by being able to run fast. Muskoxen are less dispersed. They live in cohesive social units of females and young, and some adult males. The vulnerable young are guided to the centre of the static defensive grouping of muskoxen when attacked: both adult males and females nudge and sometimes hook the young with their persistent, sharp and up-turned horns to force them into the centre of the group.

Arctic hare differ substantially from both muskoxen and caribou in their physical characteristics and in their sociality. Critically, for example, leverets are born in secluded areas and left for long periods between suckling bouts. Moreover, the young hare obtain full body size by the beginning of the winter and do not have the close association with adults that is the case in the other species. Young hares associate at first with other juveniles, and do not associate with adults until later in winter. Hence, compared with muskoxen and caribou there are very different opportunities to learn from experienced adults; opportunities concern very brief interactions with their mothers and interactions with their litter mates. Very little is known about either type of interaction in this species, and we may be tempted to think that such a lifestyle offers very limited opportunities to socially acquire functionally significant information. However, these animals are selective foragers for example, and questions of social learning are potentially important. Moreover, the chapter by Robyn Hudson and her colleagues on the transmission of olfactory information from mother to young in the European rabbit (chapter 8) emphasises a variety of perspectives in this context. Hence, anatomical and physiological evidence shows that new born pups are able to perceive and process olfactory information. Moreover, recent studies demonstrate a wide range of olfactory information that mothers provide in the extreme situations of limited maternal care and no direct assistance in the transition to an abrupt weaning. Appropriate responses to olfactory information are both critical for the survival of newborns and facilitate their transition to independence. Experimental evidence is discussed in a number of contexts. For example, rabbit pups learn odours associated with the diet of their mothers. For instance, when different foods that occur naturally in the wild were given as supplements to laboratory chow during pregnancy and lactation, subsequent choices at weaning showed that the preferences of the pups were in accordance with their olfactory experience, and that these responses were strong and persistent. Again, and consistent with other mammalian groups such as sheep and rats, cross fostering experiments of infants from birth also demonstrate olfactory learning when exposure to information about diet occurs only *in utero*.

In fact, olfactory learning about diet from the mother may be obtained *in utero*, during nursing and via faecal pellets that are left in the nest by the doe. By such means pups can acquire a preference for various foods that the mother eats at different times during the two month period of gestation and nursing.

Importantly then, the potential for transmission of information in these animals with very young nervous systems is substantial. The experimental evidence in various domains is impressive. However, Hudson and her colleagues are concerned to emphasise a difficulty of knowing precisely what is learned. They raise a very important point in studies of socially mediated learning, namely, that without preference tests in various 'alternative' contexts we cannot say that animals have acquired a preference specifically for a given food. Hence, it may be that the young regard stimuli associated with their mother, and with their nests, as positive or at least familiar, and when they find such stimuli in a feeding situation, they respond preferentially.

These are issues that may be empirically addressed, however. What is of general importance in the domain of socially mediated learning is that, in keeping with similar results for an increasing number of mammals, the transmission of olfactory information from their mothers is part of the natural repertoire of young rabbits, and that enhancement of receptor sensitivity that is induced by experience may function to 'ensure sensory continuity across developmental stages and to detect biologically relevant odours'. Moreover, the authors emphasise the importance of bridging a gap between knowledge from experimental studies, and work in the field in order to demonstrate that the phenomena discovered have consequences for the survival of individuals or their offspring. Interestingly, the authors have evidence that food preferences are acquired by similar mechanisms in nature. Further, functional information from their field site in Hungary shows that rabbits have at least forty different plant species available. However, many of these are, to some extent, toxic or are of low nutritive value. Hence, information from a mother whose success in choosing the most suitable foods is already proven is advantageous for individuals living in a complex environment during a hazardous post weaning period.

The influence of the mother on food preferences of young mammals is also mentioned by Don Broom in his chapter on social transfer of information in domestic animals (chapter 9). Interestingly, for example, even brief exposure by lambs during the suckling period, to mothers who had consumed food that is not normally eaten, is found to influence significantly the subsequent independent acceptance of that food by the lambs. There is a growing interest in studies of socially mediated learning among species of domestic animals, not least because there are many practical implications for husbandry and health. There is certainly reference to the practical importance of social learning in this chapter, as in studies of the social facilitation of feeding by calves, but Broom is also concerned to consider the propensities of different species for socially mediated learning. Hence, in addition to detailed evidence for the specificity of individual recognition, for example, he emphasises studies that control for situations in which the presence of another animal may reduce fear responses in an untrained animal, and thereby facilitate learning, compared with the positive influence of the performance of a trained animal upon that of an untrained one. This, again, is a very important point in social learning studies, and in experimental circumstances these social controls require careful consideration. This chapter is much concerned with sheep, pigs and cattle, and although reference is made to dogs, cats and chickens, its inclusion in this section of the book is the one that is the most appropriate.

5

Social learning in marsupials

Karen Higginbottom and David B. Croft

Introduction

No studies of marsupials to date have demonstrated or explicitly investigated social learning. This largely reflects the paucity of information on social interactions in most taxa. However, it also reflects a persisting view that marsupials are behaviourally 'inferior' to eutherians, a lack of input by comparative psychologists into the study of marsupial behaviour, and a lack of awareness by marsupial biologists of the field of social learning research. This chapter aims to stimulate research by showing the potential importance of social learning as an adaptive process in at least some marsupial taxa.

Marsupials (sub-class Metatheria) are not primitive precursors of placental mammals (sub-class Eutheria); rather, both groups diverged from the therian line about 100 million years ago and adapted equally well to a broad range of environments (Dawson 1983, Kirsch 1984, Lee and Cockburn 1985). Where marsupial and placental taxa occupy similar habitats, they display considerable convergence in ecology and behaviour (e.g. Charles-Dominique 1983, Jarman 1991, Winter 1996). The extant marsupial fauna includes at least 266 species; these are now restricted to the Americas and Australasia, and reach their greatest diversity in the latter area (Table 5.1). There was considerable loss of marsupial diversity through the Pleistocene and Recent periods, especially of wolf- and leopard-like carnivores and large bovid-like herbivores.

Extant marsupial carnivores (100+ species) are predominantly small shrew-like terrestrial insectivores (Dasyuridae and Caenolestidae) occupying habitats from rainforest to desert, or arboreal opossums in forested habitats. The largest carnivores are the small fox-like quolls and the small hyaenid-like Tasmanian devil. The predominant omnivores are bandicoots and bilbies (17 species), weighing no more than 5 kg, and occupying terrestrial habitats including desert, forest and heath. Some potoroos and arboreal opossums are also omnivorous, while the majority of marsupials is predominantly herbivorous. Arboreal species number 50 or more and range from the diminutive nectar-exudate feeding pygmy possums, through folivorous possums and gliders, to the sloth-like koala and the larger tree-kangaroos. Habitats range from dry forest through to rainforest. A number of the Petauridae and

Phalangeridae are lemur-like in their habits. The terrestrial herbivores range from rabbit-like hare wallabies and fossorial wombats through to the marsupial equivalents of small to medium-sized deer, antelope, sheep and goats amongst the wallabies and kangaroos. These are found from the rainforest floor to the desert grasslands. No extant marsupial is larger than 100 kg and no species is fully aquatic. Most species are nocturnal; a few are crepuscular and one is diurnal.

Marsupials have a rich repertoire of communicatory acts involving all sensory modalities, and elicited in a wide range of contexts commensurate with placentals. Olfactory communication is probably of greatest importance in most species, as expected in animals that are predominantly nocturnal and solitary (Russell 1985, Salamon 1996). This includes marking of individuals and objects in the environment, often using specialised glandular secretions, which function in social behaviour (Biggins 1984, Salamon 1996). Nasal touching is one of the most common forms of interaction in a range of marsupial taxa. Odours facilitate individual recognition, denote group membership and allow detection of female sexual condition by males (Salamon 1996, Schultze-Westrum 1965). The use of visual signals is, not surprisingly, most marked in the more diurnal species such as kangaroos and wallabies, which use various stereotyped displays, particularly in agonistic encounters (Gansloßber 1989). Auditory signals are more widespread and include use of the vocal apparatus and making sounds on the substrate. The arboreal gliders have particularly rich vocal repertoires (Biggins 1984). Contact and distress calls between mother and young marsupials are common, probably more so than reported, since they are difficult to hear unless observed at close quarters (Russell 1984, 1989, Higginbottom, pers. obs.). Tactile communication, such as allogrooming, is most conspicuous between mothers and their young and during courtship behaviour, but is relatively rare between adult conspecifics.

Most marsupials forage solitarily and do not display territoriality (Jarman and Kruuk 1996, Russell 1984). A few species typically aggregate in open groups, and only the gliding possums (Petauridae) form closed membership groups with defended ranges (Jarman and Kruuk 1996).Various taxa, including the Macropodidae and Didelphidae, have less diverse social structures than their placental counterparts (Charles-Dominique 1983, Croft 1989). However, sociality is to some extent related to ecology. For example, among the kangaroos, wallabies and rat kangaroos (macropods), a tendency to form groups is associated with large body size, an open habitat, diurnal activity and feeding on grass, whereas the more solitary species tend to be smaller, cryptic, nocturnal and selective feeders on browse, fruit and fungus (Croft 1989, Jarman 1991). This relationship is similar to that found in ungulates of a similar size range, although macropods are generally less social than their ungulate counterparts (Jarman 1991).

The most important distinction between marsupials and placentals in terms of ecological and behavioural consequences is their different modes of reproduction. Compared with placentals, marsupials give birth to an extremely immature neonatus after a very short gestation, and have a much longer period of lactation and postpartum dependency (Eisenberg 1981, Lee and Cockburn 1985). They are probably physiologically equivalent to neonate placentals at about the time they achieve homeostasis, which is approximately the time of first exit from the pouch in species with well-developed pouches (Russell 1982). Growth rates of young are slower than in placentals, especially among the smaller species (Lee and Cockburn 1985). The period from first exposure to the environment to weaning, during which intensive learning can potentially occur, is typically longer than that of equivalent placentals. In most species the neonate young is enclosed in a pouch, which protects it and maintains its close contact with the mother. In species without or with reduced pouches the young are left in a nest at a relatively early stage of development.

Mothers provide all parental care in most marsupial taxa. Russell (1982, 1984) distinguished three patterns of such care (patterns A–C) and these are relevant to the potential for social learning. In A and B the young are left in a nest either as soon as they begin to release the teat (A), or soon after their eyes open and they become able to actively interact with the environment (B). The mother returns to the nest from time to time, suckling her young and keeping them warm. As they develop further, the young begin to leave the nest with their mother, on foot or riding on her back (the latter particularly in the arboreal species). They eventually leave the nest alone, returning to the mother occasionally to suckle. In C, which comprises species with large pouches and a single young, the young remain in the pouch, spending progressively longer periods out of the pouch until they are able to move independently. They then continue to follow the mother (initially riding on her back in the koala) for much of the time at least until weaning. This period between final pouch exit and weaning is termed the young-at-foot phase. Once young marsupials of pattern C reach the stage where their heads protrude regularly from the pouch, they have the opportunity to observe and learn about the physical and social environment and experience at close quarters their mother's reactions to environmental and social stimuli. Like placental mammals, the young of some marsupial species can be classified as 'followers', which maintain constant association with the mother up to weaning, while others are 'hiders' which are left in nests or dense cover apart from the mother for a large proportion of the time (Johnson 1987).

There are certainly major gaps in our knowledge of learning, social behaviour and ecology of marsupials compared with that on placental mammals (but see Table 5.1 for information relevant to opportunities for social learning across marsupial taxa). This reflects the concentration of this taxon into parts of the world with relatively few researchers and the fact that most marsupial species are small, cryptic and nocturnal. Thus, much background information required for studies of social learning in marsupials is not available, and knowledge of social learning is at preliminary stages. However, given the aims of this volume, it is pertinent to the general debate to consider the most likely taxa and circumstances for social learning within the group. As it happens there are good circumstantial observations of social learning in marsupials (see pp. 87–95) and, importantly, there are a number of hypotheses to stimulate research.

For example, among species that are relatively long-lived (Table 5.1), those in which young remain within their mothers' home ranges have a longer period in which they can learn socially from their mothers, particularly if mother and young continue to associate closely, than those in which the young disperse. Ashworth (1996) suggested for marsupials that regular association of adult female offspring with their mothers might enable daughters to learn from their mothers about food selectivity and distribution, as well as about the location of refuges and shelters and even mothering skills.

Similarly, pattern C species have a longer period in which they can socially learn from their mothers, as well as a closer association with them, than pattern A and B species. The degree to which young follow their mothers, associate closely with them in their activities, and engage in communication will further influence the potential for social learning. Although retaining the young within a pouch may have the primary function of protecting the young from predation (Lee and Ward 1989), it could have further adaptive value in allowing social learning to occur.

We further predict that body size will have an indirect influence on the opportunities for social learning among marsupials. Body size has a negative influence on litter size and a positive influence on duration of life-history phases (e.g. pouch life, time till weaning and sexual maturity, and longevity) (Lee and Ward 1989) and apparently on diurnal activity, opportunity for long-term social bonds, and intensity of maternal care of individual offspring (Russell 1982). Body size should thus be positively correlated with the availability of opportunities for social learning. Beyond the effects of body size, pouch life durations and age at weaning in macropodids and several arboreal species (generally parental care pattern C species) are greater than expected, while those of dasyurids and bandicoots (generally pattern A or B species) tend to be shorter than expected (Russell 1982).

Finally, the low sociality of most marsupial taxa indicates that there is

Table 5.1. *Factors affecting opportunities for social learning among marsupial families*

Classification follows Strahan (1991). The families included are those for which a moderate amount of information is available. No information available is indicated by a blank space; ? indicates that research has tentatively established this value.

Family	No. of species	Common name	Body size range (kg)	Range	Habitat	Substrate	Diet	Diurnality	Social unit	Litters/ year	Litter size	Parental care pattern	Age at weaning	Age of female at maturity	Longevity	Philopatry
Didelphidae	70	opossum	0.01–2	Am	F	T/A	I/O	N	A	1,2,3	3	A,B	1	1,2	1	P
Dasyuridae	52	dunmart, quoll (or genus)	0.03–8 (most < 0.1)	Aus PNG	M	T/S	I/O	N/C	A	1,2,3	3	A	1,2	1,2	1,2	P
Peramelidae	15	bandicoot	0.2–5	Aus PNG	M	F	I/O	N/C	A	3	2	B	1	1	1,2	D
Thylacomyidae	2	bilby	0.2–5	Aus	D	F	I/O	N	C?	3	2	B	1?	1	3?	
Petauridae	23	glider, ringtail possum	0.1–1.7	Aus PNG	F	A	I/B/P	N	A,C, D,E	2	2	B	2	2,3	2,3	P
Phalangeridae	11	brush-tailed possum, cuscus	1.0–5	Aus PNG	F	A	B	N	B,C	1	1,2	B	2	3	3	P
Burramyidae	7	pygmy possum, feathertail glider	0.007–0.045	Aus PNG	F/H	A	I/P	N	nests in groups?	3	3	A?	1	3		

Family	No. of species	Common name	Mass (kg)	Range	Habitat	Substrate	Diet	Diurnality	Social unit	Litter size	Litters/yr	Parental care	Age at weaning	Age at maturity	Longevity	Philopatry
Tarsipedidae	1	honey possum	0.007–0.020	Aus	H	S	P	C		3	2	B	1	1	1?	
Phascolarctidae	1	koala	6.5–13	Aus	F	A	B	N		1	1	C	1	3	3	P
Vombatidae	3	wombat	20–35	Aus	F/S/D	F	G	N/C		1	1	B?	1	3	3	
Potoroidae	7	bettong, potoroo	0.5–4	Aus, PNG	M	T/F	F/O	N	A,B,D	2,3		C	2	2	3	
Macropodidae	36	kangaroo, wallaby	1.3–66	Aus, PNG	M	T/(A)	G/B	D/C/N	A,B,C,D,E,F	1,2	1	C	2,3	2,3	3	P
References				1			1,4,5		1,6,7,8	8			7,8,9	1,8,9,13	1,7,10,11,12,14	1,15,16,17

Range: Am, North and South America; Aus, Australia; PNG, Papua New Guinea. *Habitat*: M, most terrestrial plant communities; F, forest; D, desert; H, heath; S, savannah. *Substrate*: T, terrestrial; A, arboreal; S, scansorial; F, semi-fossorial; D, partially diurnal (as well as nocturnal). *Diet*: G, grazer; B, browser; I, insectivore; O, omnivore; F, fungivore; P, plant part specialist, i.e. fruit, seed, nectar or exudate. *Diurnality*: N, nocturnal; C, crepuscular; D, partially diurnal (as well as nocturnal). *Social unit*: A, individual with young dispersing at weaning; B, individual with young dispersing at maturity or later; C, monogamous pair; D, harem; E, unstable group; F, stable group. *Litters/yr*: 1, ≤ 1; 2, > 1 and < 2; 3, > 2. *Litter size*: 1, ≤ 1; 2, > 1 and < 2; 3, > 2. *Parental care pattern*: A, young permanently exits pouch and left in nest as soon as releases nipple when relatively undeveloped; B, young first exits pouch and left in nest when relatively well developed and returns to pouch intermittently; C, young not left in nest and young leaves pouch for progressively longer periods generally following mother. *Age at weaning*: 1, ≤ 4 months; 2, > 4 and ≤ 10 months; 3, > 10 months. *Age of female at maturity*: 1, ≤ 8 months; 2, > 8 and ≤ 16 months; 3, > 16 months. *Longevity*: 1, ≤ 2 years; 2, > 2 and ≤ 4 years; 3, > 4 years. *Philopatry*: P, females relatively philopatric and males disperse; D, both sexes disperse.

References: 1 = Lee and Cockburn (1985), 2 = Russell (1984), 3 = Turner and McKay (1989), 4 = Lee et al. (1982), 5 = Seebeck and Rose (1989), 6 = Eisenberg (1981), 7 = Smith and Lee (1984), 8 = Russell (1982), 9 = K. Johnson (1989), 10 = Hunsaker and Sharpe (1977), 11 = Lee and Ward (1989), 12 = Martin and Handasyde (1990), 13 = Seebeck and Rose (1989), 14 = Wells (1989), 15 = Eisenberg pers. commun., 16 = Mitchell and Martin (1990), 17 = C. N. Johnson (1989).

generally little opportunity for social learning among adults. In general, we expect social learning by young from their mothers to be the main avenue by which such learning could occur.

Role of social learning in behavioural development

In general, there should be adaptive value in social learning of behaviours for which a flexible response to local or temporary conditions is needed, and for which individual learning would have high costs in terms of mortality risk or time/energy expended (Bandura 1977). Thus, in marsupial taxa for which there are appropriate opportunities for social learning, we predict that the following behaviours are most likely to develop through social learning:

- Aspects of predator recognition and avoidance in prey species (most marsupial taxa); e.g. knowledge of the taxonomic identity and habits of predators (particularly humans) of marsupials where these are highly variable in space and time.
- Aspects of predation behaviour in predatory species employing complex and flexible predation skills (mostly the larger Dasyuridae).
- Selection of food items, especially in species exhibiting high feeding selectivity (generally the smaller species).
- Use of the local habitat with respect to factors that are important to fitness in species/sexes with relatively stable home ranges (most species); e.g. knowledge of good feeding areas and escape routes.
- Relative dominance of conspecifics in the local area in species/sexes exhibiting dominance hierarchies (generally the relatively social species).
- Identities of conspecifics in the local area in species with social networks based on individual recognition (especially the Petauridae).
- Social behaviours involving complex motor patterns which are likely to improve with practice, most notably fighting behaviour by males in species with strong inter-male competition (e.g. the large Macropodidae).

Further, we predict that if the models or learning facilitators incur significant costs in fulfilling that role, then they are most likely to be close genetic kin of the learner(s).

Evidence of social learning in marsupials

There have been no studies of marsupials which experimentally assess the presence of social learning, nor which systematically investigate the relationship between the social experience of individuals and their subsequent behaviour patterns. The only available evidence of social learning is circumstancial, based mainly on detailed behavioural observations at the individual level.

There are two main overlapping aspects of behaviour for which there is the most substantial (though preliminary) evidence of social learning. Firstly, young macropodids apparently learn aspects of their physical and social environments, and appropriate responses to that environment, as a result of association with their mothers. Secondly, social play, which occurs in a range of taxa, but again has been most studied in macropodids (at least in the field), probably has a learning function. Detailed studies of marsupial behaviour, which are most likely to reveal evidence of social learning, have been carried out mostly in the macropodids. Therefore, it is not yet possible to determine whether the fact that evidence of social learning has been found mainly in macropodids reflects a taxonomic bias in research, or fulfilment of the predictions on p. 83 regarding the incidence of social learning among marsupial taxa.

Mother–young behaviour in *Macropus*

The relationships between mother kangaroos and wallabies and their young have been systematically studied in the wild in the whiptail wallaby, *Macropus parryi* (Kaufmann 1974), the red kangaroo, *M. rufus* (Croft 1980), the euro, *M. robustus* (Croft 1981), the eastern grey kangaroo, *M. giganteus* (Stuart-Dick 1987) and the red-necked wallaby, *M. rufogriseus* (Johnson 1987, Higginbottom 1991). Detailed descriptive studies of mother–young behaviour in captivity have also been carried out in tammar wallabies, *M. eugenii*, and red kangaroos (Russell 1973, 1989). These studies jointly form the basis of the aspects of the following description general to the genus *Macropus*.

The exploratory behaviour and apparent curiosity of *Macropus* pouch-young are conspicuous. Young with their heads out of the pouch frequently appear to be investigating their mothers' immediate environment, and are usually actively feeding, grooming, looking, or sniffing from side to side and at various objects, or investigating objects by sniffing or occasionally manipulation. They appear very alert to visual, acoustic and olfactory stimuli. Early in pouch life, the heads of young tammar wallabies are most often out of the pouch while their mothers are active as opposed to resting (Russell 1973). While their mothers are feeding, *Macropus* pouch-young – whose heads are brought in close contact with the ground in this position – often sniff or nibble at vegetation and various objects (Figure 5.1), and sometimes manipulate them using their forepaws. Initially, the food is rarely ingested, but well before the young leaves the pouch for the last time its feeding actions while in the pouch resemble those of adults. Both pouch-young and young-at-foot occasionally make nose-to-nose contact with their mother while she is foraging,

Figure 5.1. Red-necked wallaby (*Macropus rufogriseus*) young have the opportunity to learn about food selection while in the pouch.

and the young sometimes sniffs at her mouth, which may allow detection of the type of food she is eating. The young generally appears to initiate this contact. During the young-at-foot period, the mother tolerates her young feeding much closer to her than she does with other conspecifics, and they will often feed with their heads closely together on the vegetation, which may allow continued learning of appropriate food items.

When out of the pouch, the pouch-young spends much of its time investigating its environment and 'practising' locomotion, with its mother providing a secure base to which it can return whenever alarmed. For example, in red-necked wallabies, the pouch-young moves to and fro in radiating lines from its mother, usually stopping at the end of each 'radius run' to look, sniff or manipulate and then dash rapidly back to the safety of its mother (Higginbottom, unpublished data). The distance, duration and speed of these radius runs increase gradually with age. In eastern grey kangaroos and red-necked wallabies, the mother is more likely to let her young out of the pouch when she is away from other conspecifics and, in the case of red-necked wallabies, in a well-concealed location (Stuart-Dick 1987, Higginbottom, unpublished data). With younger pouch-young, she generally does so only after having surveyed the environment for some time, apparently to ensure there are no dangers lurking. She then remains much more vigilant than usual while the young is out of the pouch, and signals it to return to her using a vocalisation and a characteristic pouch-entry posture if she detects a disturbance (ibid).

Stuart-Dick (1987) has proposed that eastern grey kangaroo mothers actively train their newly emerging pouch-young to return to the pouch in response to the appropriate signals from the mother. This impression is based on the absence of an obvious immediate function, the sequential repetition of the routine, the progressive increase in the duration of the period during which the young is 'allowed' out of the pouch, and the great attentiveness of the mother to her young during the process. A similar, though less intensive process, appears to occur in red-necked wallabies (Higginbottom, unpublished data).

Late in pouch-life, the *Macropus* young appears to learn appropriate following behaviour. Although it is clearly predisposed to follow its mother, the mother often uses vocalisations to encourage her young to follow while waiting for it, then resumes hopping only once the young catches up (Stuart-Dick pers. commun.; Higginbottom, unpublished data). In eastern grey kangaroos, red kangaroos and tammar wallabies, this sometimes occurs in sequences where there is no obvious need for the mother to move, such that it appears she may be actively tutoring her young in following behaviour (Russell 1973, 1989; Stuart-Dick pers. commun.). Similarly, tutoring may be involved in teaching the young to negotiate obstacles (R. Stuart-Dick pers. commun.). Mother eastern grey kangaroos have frequently been observed hopping over an obstacle (such as a log) when an alternative and 'easier' route was readily available, then pausing and turning to call to their young to follow.

Once it has left the pouch for the last time, the *Macropus* young continues to follow its mother closely for much of the time until at least weaning. This is true even for some 'hider' species such as the red-necked wallaby (Higginbottom 1991). During this period the risk of mortality is at its highest, with about half of red-necked wallaby and eastern grey kangaroos dying during the young-at-foot period (Higginbottom 1991, Stuart-Dick 1987). In at least these species predation is the main proximate cause of mortality, so that it would be highly adaptive for the young to have learnt to avoid predators and to flee in response to appropriate cues from conspecifics or direct stimuli.

There is some evidence that *Macropus* young learn their mothers' responses to external stimuli, particularly those relating to potential predators. Both in the pouch and as young-at-foot, macropodids have extensive opportunities for observing the reactions of conspecifics, especially their mothers, to environmental stimuli, and associating these with the conspecifics' responses. As pouch-young, many young red-necked wallabies and eastern grey kangaroos initially display few fear or flight reactions to such stimuli; then later appear to learn to respond fearfully to a range of stimuli; and by the time they become adults have apparently learnt to respond selectively (Higginbottom, unpublished data; Stuart-Dick pers. commun.). There is anecdotal evidence in red-necked wallabies that some of this learning occurs vicariously through the mother (Higginbottom, unpublished data). A mother's sudden flight from a disturbance (such as a bird call, human intruder or gust of wind) is always immediately followed by the young fleeing; whereas if the mother showed no alarm but the young fled it usually goes only a short distance then returns. There would be significant adaptive value in learning under what circumstances the risks are sufficient to justify disruption of ongoing activities. If a young is in the pouch with its head out, its mother's sudden adoption of an upright and alert posture is usually followed immediately by the young also becoming alert, often simultaneously moving its head to look in the same direction as its mother (Figure 5.2). If a pouch-young is outside the pouch, the mother when alarmed usually signals for her young to return to the pouch. When disturbed during a suckling bout, a slight but sudden jerk of the mother's body or head, and occasionally, a sudden presssing with her forepaws on her young's back, is usually followed immediately by the young removing its head from the pouch and standing alert by its mother until some time has passed without further sign of agitation by the mother. Further, individual differences in flight distances of mother red-necked wallabies and eastern grey kangaroos to human intruders appear to be reflected in those of their young, even as adults no longer associating with their mothers (Higginbottom, unpublished data; Stuart-Dick pers. commun.). Thus, although many aspects of predator avoidance may be innate, and animals are probably predisposed to respond fearfully to certain types of stimuli (Mineka and Cook 1988), the fine tuning of these responses seems to be influenced by social learning.

Learning of the mother's behaviour may also occur in terms of activity state. In red-necked wallabies, the amount of time that an individual young-at-foot spent in each activity state (feeding, looking, resting, interacting, self-grooming or moving) was positively correlated with the time its mother spent in that activity, and their activity was the same for an estimated 70% of the time they spent together (Higginbottom 1991). A young usually commenced feeding or resting shortly after its mother did so, and looked immediately after its mother looked. Similarly, captive red kangaroo young-at-foot adopt vigilant behaviour in response to locomotion or vigilance of their mothers, and begin feeding if their mother is feeding (Witte 1993). Thus, young have the opportunity to associate particular activities with particular circumstances such as location, time of day and weather conditions.

In eastern grey kangaroos (Stuart-Dick 1987) and red-necked wallabies (Higginbottom 1991) the mother's activity state and spatial use of her home

Figure 5.2. Red-necked wallaby mother and pouch-young adopt alert postures in response to a sudden disturbance.

range are modified by association with her young, although in all species studied it is the young which is primarily responsible for maintaining proximity to its mother. Red-necked wallaby and eastern grey kangaroo females and probably other macropodids have highly individualised and consistent patterns of use of their home ranges, including favoured feeding areas, routes followed between feeding and resting areas, and in the case of wallabies, areas in which young are left alone for periods of time (Higginbottom, unpublished data; R. Stuart-Dick pers. commun.). A female's young-at-foot follows her repeatedly to these areas, and has the opportunity to associate certain areas with certain activities. Since daughters are to some extent philopatric (Higginbottom 1991, Stuart-Dick pers. commun.), they may benefit from learning appropriate areas for various activities. In the few cases where a daughter red-necked wallaby was observed intensively during its young-at-foot life and subsequent adulthood, she tended as an adult to use feeding areas and even areas for hiding young that were used by her mother more than would be expected on the basis of her home range location alone (Higginbottom, unpublished data). Both red-necked wallaby (Higginbottom, unpublished data) and eastern grey kangaroo mothers (Stuart-Dick pers. commun.) and their young were also observed on several occasions making exceptionally long 'tours' around their home ranges with no obvious function towards the end of the period of intense mother-young association, as if giving their young a final familiarisation with the area.

There can be marked individual variation between females in their maternal behaviour in red-necked wallabies (Higginbottom, 1991) and eastern grey kangaroos (Stuart-Dick 1987). In both cases, aspects of maternal behaviour were clustered according to the degree of 'active mothering' that they entailed. For example, wallaby mothers with high rates of interacting with their young also tended to spend a large proportion of their time with them, have frequent and long suckling bouts, and make relatively strong efforts to maintain proximity with their young. In eastern grey kangaroos some aspects of these 'maternal styles' apparently were reproduced in daughters once they also became mothers (R. Stuart-Dick pers. commun.).

Male and female *Macropus* typically have very different life histories and adult social experiences. These seem to be reflected to some extent in the different opportunities for social learning as young (see the next section, on social play, for additional information). For example, in red-necked wallabies, mothers spent a larger proportion of their time with sons, and interacted more frequently with them, while sons took more initiative than daughters in achieving reunions with their mothers (Higginbottom 1991). Thus, sons and daughters may have quantitatively different opportunities for social learning.

In conclusion, it seems likely that through close association with their mothers, *Macropus* young learn appropriate use of, and responses to, their environment. We would expect that this occurs mostly through a combination of local enhancement (where the young's attention is drawn to a stimulus by its mother's location or behaviour), and sometimes by vicarious associative learning (see Galef 1996a). It is also possible that some limited active tutoring may occur. However, without experimental studies it is not possible to determine what mechanisms for social learning, if any, exist.

The apparently intensive learning period in late pouch life would be expected to lead to a substantial increase in the young's chance of surviving the vulnerable young-at-foot phase. Mother macropodids probably incur significant costs in terms of time spent feeding and quality of food by modifying their behaviour when associating with their young; part of the adaptive reason for this may be to allow social learning to occur, so increasing the young's chances of surviving or reproducing.

In ungulates, which comprise the closest placental counterpart to the macropodids (Jarman 1991), a major function of maternal behaviour is the facilitation of learning processes in the infant, and traditions may be transmitted in this way (Lent 1971). Most importantly, the mother is thought to provide stimulation and a protective environment that will maximise the opportunities for the young to learn. Young are often aided in attaining contact with the teats and sometimes encouraged to suckle by the movements, vocalisations and postures of the mother. This seems to have parallels in the pouch orientation 'lessons' (whether by tutoring or not) of macropodids. Young of both groups also tend to have a strong innate tendency to follow their mothers, which gives rise to a similar range of opportunities for the young to learn by exposure to the mother's environment and by her response to environmental and social cues. However, the latter part of the pouch life of macropodids may provide an additional learning opportunity not available to ungulates. When carried in the pouch with the head out, the young has both a secure environment and ample time for learning, and is automatically exposed to its mother's immediate environment.

A different type of evidence that young *Macropus* employ social learning in developing appropriate adult behaviours can be derived by comparing the behaviour of animals reared in captivity and those reared in the wild, and from observing the effects of introducing the captive animals to experienced con-specifics. A study of the effects of hand-rearing on eastern grey kangaroos (Campbell 1994) showed that hand-reared young were less likely to cross unfamiliar fences (a common obstacle in the environments into which animals are normally released) or to select the most palatable grass species than expected for their wild counterparts. Further, females which had been hand-reared showed more variable and sometimes abnormal maternal behaviour. Thus, learning seems to play a role in the adaptive development of these behaviours, though it is not clear to what extent intraspecific social interactions are important for this learning to occur. Similar indirect evidence of the adaptive importance of social learning can be derived from reintroductions of captive-reared marsupials into the wild. Many such reintroductions of threatened species have been unsuccessful because the captive-bred animals succumb to predators (Serena 1995). Many of the researchers involved feel that this is often at least partly due to lack of appropriate predator recognition or avoidance behaviour, although this has not been experimentally assessed (but see Soderquist 1995 for circumstancial evidence of this in brush-tailed phas-cogales, *Phascogale tapoatafa*). Following unsuccessful reintroduction attempts of rufous hare wallabies, *Lagorchestes hirsutus*, due to predation mortality, McLean *et al.* (1995) 'trained' their captive-reared animals to recognise pred-ators by using an artificial fox or cat and simultaneously playing hare wallaby alarm calls. It was assumed that the animals could learn by observation and that cultural transmission occurs, though this has not yet been demonstrated in marsupials. The study found that after training, hare wallabies were more likely to hide from the models, although the effect did not persist eight months later. This provides some evidence that social learning (in this case based on conspecifics' alarm calls) assists predator recognition.

Social play

Play can be defined as activities that are structurally similar to those seen in functional contexts, but are performed outside those contexts and lack their consummatory goals (Bekoff and Byers 1981). Engaging in social play may facilitate animals developing their motor, cognitive and social skills and if so can be seen as a form of social learning. Demonstrating that this occurs, however, is notoriously difficult.

While some authors consider play to be less prevalent in marsupials than eutherians (e.g. Hunsaker and Shupe 1977, Gansloßber 1989), such conclusions may be premature given the paucity of observational data and difficulties in defining play (Watson and Croft 1993). Various forms of play, including social play between mothers and young, are extensive in several marsupial taxa, including the eastern grey kangaroo (Stuart-Dick 1987) and the larger dasyurids (Croft 1982). In macropodoids, play is comparable to that of placental herbivores, while in insectivorous and carnivorous marsupials it is comparable to that of placental carnivores (Lissowsky 1996). Fagen (1981) concluded that play is a conspicuous part of the behaviour of the young of many marsupial species and has been recorded in representatives of most families, though it may be low in frequency.

Play has been most studied in the macropodids, followed by the dasyurids, and only macropodid play has been systematically observed in the wild (Lissowsky 1996). Social play is common, usually involving the mother as the principal play-partner in species bearing a single young, and involving litter-mates as well as the mother for species bearing multiple young (Fagen 1981, Lissowsky 1996). Initiation is usually by the young, and termination usually by the mother. However, this pattern is sometimes reversed. For example, in *Dasyurus maculatus* the mother has a particular call that apparently leads to initiation of play with the young (Lissowsky 1996).

Hypotheses about the functional significance of play in marsupials are speculative. Social play in marsupials includes behaviours involving elements of agonistic, sexual, and predatory behaviour (in predatory species) as well as play chasing (Lissowsky 1996). For example, the most common components of sexual play in marsupials are mounting and tail-grasping (Lissowsky 1996).

Play-fighting is the most frequently observed form of social play in macropodids (Lissowsky 1996), although it can be difficult to distinguish from 'true' fighting. If defined as fighting which is not associated with direct conflict over resources and involving a different interaction pattern (Watson and Croft

1993), then play-fighting also commonly occurs in adult male macropodids (Croft and Snaith 1991). Watson (1993) presented evidence that the principal function of play-fighting in most age classes of red-necked wallabies was motor training. However, play-fights between adult male macropodids are thought to function mainly to provide assessment of relative fighting abilities in a low risk context (Croft and Snaith 1991, Watson and Croft 1993).

In eastern grey kangaroos, red kangaroos and red-necked wallabies, play-fighting between mothers and young is more common in at least some age groups of male young than female young (Lissowsky 1996, Watson and Croft 1993). The most likely adaptive explanation is that the development of fighting abilities and assessment of potential competitors in these species is more important to male reproductive success since it affects mating access to females (Watson and Croft 1993). On the other hand, in eastern grey kangaroos, female pouch-young initiated more affiliative interactions with conspecifics other than the mother than did male pouch-young. This may reflect benefits for females, which continue to associate in affiliative ways with conspecifics from the local area, as compared with males, which disperse (Stuart-Dick 1987).

In the dasyurids, which are predatory animals, the predominant form of social play is play-fighting (Croft 1982). Juvenile littermates chase and wrestle with each other in long bouts of play (Figure 5.3). It is not known why such play is more prominent in larger species. One possibility is that it is related to the greater need for developing predatory skills required for capturing and handling relatively large prey (ibid).

Conclusions regarding behaviours influenced by social learning

There is preliminary evidence that, within macropodids, social learning occurs with respect to some of the behaviours for which we predicted that it would be most adaptive on p. 86, that is: predator recognition and avoidance, food selection, and habitat use. Further, as predicted, social learning appears to occur with respect to development of fighting abilities in at least some marsupial species with strong inter-male competition, and perhaps with respect to predatory abilities in certain species employing complex predation skills.

Social learning: implications for management of marsupials

There has been a marked decline in the diversity and geographic range of most of the smaller Australasian marsupials in the last two centuries (Tyler 1979), although some of the larger kangaroos have increased in abundance and they

Figure 5.3. Spotted-tailed quolls (*Dasyurus maculatus*) play wrestling at 92 days old.

are managed as pests in pastoral regions (Cunningham 1981). Thus, many populations require intensive management for their maintenance or control, and captive breeding and release programmes are being employed to restore rare and endangered species to their former ranges (Serena 1995). An understanding of the incidence, function and requirements for social learning in marsupials may have significant implications for the success of these programmes.

For example, if we attempt to reintroduce species back into the wild then we should determine in what ways and to what extent social learning (as well as individual learning) is important for the development of behaviours essential to survival. This information could then be used for the design of appropriate training, or the introduction of wild-caught models, prior to reintroductions of captive-reared and threatened species (see e.g. McLean *et al.* 1995). On the other hand, where a population is harvested or culled then a knowledge of the role of various age/sex classes in facilitating social learning would allow prediction of the behavioural effects on the remaining animals of any selective removal of certain classes (especially of mothers).

Finally, species under scientific study are often maintained and handled in captive groups. Since marsupials may be quite susceptible to stress in captivity (see e.g. Williams 1990) or when caught (Shepherd 1990), knowledge of the extent and way in which individuals socially learn how to react to potentially threatening stimuli may help in reducing such damaging responses. For example, exposing wild-caught animals to a captive conspecific that does not flee when humans approach may help reduce their stress levels.

Future directions for research on social learning in marsupials

We suggest that knowledge of social learning in marsupials will be most efficiently advanced by initially focusing on:

- taxa and individuals in which social learning is most likely to occur;
- taxa and populations for which observation and experimentation is most practicable;
- taxa for which there is good background information on natural social behaviour.

Thus, the Macropodidae have relatively high opportunities for social learning, especially by young from their mothers (Table 5.1). Some of the Petauridae (especially the sugar glider, *Petaurus breviceps*), by virtue of their tendency to form cohesive and stable social groups, occasional monogamy, and life history pattern, may also have substantial opportunities. Given their social networks, which in certain species are probably more developed than those of the Macropodidae, the Petauridae may have the greatest opportunities for social learning between conspecifics other than mother and young. There may also be significant opportunities for social learning by young from their mothers among the Phalangeridae, Vombatidae, Potoroidae and Phascolarctidae.

Behaviours selected for study should be relatively easy to define operationally and to classify, and should have clear functional significance. Thus, the role of social learning in the development of food preferences and of responses to potentially threatening environmental stimuli seem to be good candidates.

Research should initially concentrate on controlled studies in captivity, to determine whether social learning occurs in various contexts and to explore mechanisms. Such studies should use large samples of animals exposed to different social learning opportunities and compare their subsequent behaviour. Studies which replicate as closely as possible (with appropriate modifications to ensure relevance for the species concerned) the most rigorous experimental studies of social learning in placental mammals would be particularly useful to the evolutionary debate (see especially Laland, chapter 10, this volume, and Galef 1996b regarding foraging preferences in rats). The extensive practice of hand-rearing of marsupials also provides abundant opportunity for exploring the effects of social deprivation. Long-term field studies of known individuals can be used to analyse correlations between aspects of individuals' social learning experiences and their subsequent behaviour. Although it will not be possible to deduce cause–effect relationships from such studies, it will provide hypotheses about functional importance that can then be further examined under controlled conditions.

Acknowledgements

The Cooperative Research Centre for Conservation and Management of Marsupials funded KH's contribution to this chapter, which forms part of their research programme. Robyn Stuart-Dick, John Eisenberg and Udo Gansloßber provided information from their unpublished research; U. Gansloßber and Jan Aldenhoven commented on the manuscript. We thank Hilary Box and the referees for their sound advice in improving the final draft.

References

Ashworth, D. L. (1996). Strategies of maternal investment in marsupials. In *Comparison of Marsupial and Placental Behaviour*, ed. D. B. Croft and U. Gansloßber, pp. 187–225. Fürth: Filander Verlag GmbH.

Bandura, A. (1977). *Social Learning Theory*. Englewood Cliffs, New Jersey: Prentice-Hall Inc.

Bekoff, M. and Byers, J. A. (1981). A critical reanalysis of the ontogeny and phylogeny of mammalian social play: an ethological hornet's nest. In *Behavioural Development in Animals and Man*, ed. K. Immelmann, G. Barlow, M. Main and L. Petrinovich, pp. 298–337. New York: Cambridge University Press.

Biggins, J. G. (1984). Communication in possums: a review. In *Possums and Gliders*, ed. A.

Smith and I. Hume, pp. 35–57. Sydney: Surrey Beatty and Sons.

Campbell, L. (1994). Effects of hand-rearing on eastern grey kangaroos (*Macropus giganteus* Shaw. Honours thesis. Sydney: University of New South Wales.

Charles-Dominique, P. (1983). Ecology and social adaptations in didelphid marsupials: comparisons with eutherians of similar ecology. In *Advances in the Study of Mammalian Behaviour*, ed. J. F. Eisenberg and D. G. Kleiman, pp. 395–422. The American Society of Mammalogists special publication no. 7.

Croft, D. B. (1980). Behaviour of red kangaroos, *Macropus rufus* (Desmarest, 1822) in northwestern New South Wales,

Australia. *Austr. Mammal.*, **4**, 5–58.

Croft, D. B. (1981). Social behaviour of the euro, *Macropus robustus*, in the Australian arid zone. *Austr. Wildl. Res.*, **8**, 13–49.

Croft, D. B. (1982). Communication in the Dasyuridae (Marsupialia): a review. In *Carnivorous Marsupials*, ed. M. Archer, pp. 291–309. Sydney: Royal Zoological Society of New South Wales.

Croft, D. B. (1989). Social organization of the Macropodoidea. In *Kangaroos, Wallabies and Rat-Kangaroos*, Vol. 2, ed. G. Grigg, P. Jarman and I. Hume, pp. 505–25. Sydney: Surrey Beatty and Sons.

Croft, D.B. and Snaith, F. (1991). Boxing in red kangaroos, *Macropus rufus*: aggression or play?

Int. J. Comp. Psychol., **4**, 221–36.

Cunningham, D. K. (1981). Arid zone kangaroos – pest or resource? In *The Ecology of Pests: Some Australian Case Histories*, ed. R. L. Kitching and R. E. Jones, pp. 19–54. Melbourne: CSIRO Australia.

Dawson, T. J. (1983). *Monotremes and Marsupials: The Other Mammals.* London: Edward Arnold.

Eisenberg, J. R. (1981). *The Mammalian Radiations.* Chicago: University of Chicago Press.

Fagen, R. (1981). *Animal Play Behaviour.* Oxford: Oxford University Press.

Galef, B. G., Jr (1996a). Introduction. In *Social Learning in Animals: The Roots of Culture*, ed. C. M. Heyes and B. G. Galef, pp. 3–15. San Diego: Academic Press.

Galef, B. G., Jr (1996b). Communication of information concerning distant diets in a social, central-place foraging species: *Rattus norvegicus*. In *Social Learning in Animals: The Roots of Culture*, ed. C. M. Heyes and B. G. Galef, pp. 119–41. San Diego: Academic Press.

Gansloßer, U. (1989). Agonistic behaviour in macropodoids – a review. In *Kangaroos, Wallabies and Rat-Kangaroos*, Vol. 2, ed. G. Grigg, P. Jarman and I. Hume, pp. 475–503. Sydney: Surrey Beatty and Sons.

Higginbottom, K. (1991). Reproductive success and reproductive tactics in female red-necked wallabies. Unpublished PhD thesis. Armidale: University of New England.

Hunsaker, D. (1977). Ecology of New World Marsupials. In *The Biology of Marsupials*, ed. D. Hunsaker, pp. 95–158. New York: Academic Press.

Hunsaker, D. and Shupe, D. (1977). Behaviour of New World marsupials. In *The Biology of Marsupials*, ed. D. Hunsaker, pp. 279–347. New York: Academic Press.

Jarman, P. J. (1991). Social behaviour and organization in the macropodoidea. *Adv. Stud. Behav.*, **20**, 1–50.

Jarman, P. J and Kruuk, H. (1996). Phylogeny and spatial organisation in mammals. In *Comparison of Marsupial and Placental Behaviour*, ed. D. B. Croft and U. Gansloßber, pp. 80–101. Fürth: Filander Verlag GmbH.

Johnson, C. N. (1987). Relationships between mother and infant red-necked wallabies (*Macropus rufogriseus banksianus*). *Ethology*, **74**, 1–20.

Johnson, C. N. (1989). Dispersal and philopatry in the macropodoids. In *Kangaroos, Wallabies and Rat-Kangaroos*, Vol. 2, ed. G. Grigg, P. Jarman and I. Hume, pp. 593–601. Sydney: Surrey Beatty and Sons.

Johnson, K. (1989). Thylacomyidae. In *Fauna of Australia*, Vol. 1B, *Mammalia*, ed. Australian Biological Resources Study, pp. 625–35. Canberra: Australian Government Publishing Service.

Kaufmann, J. H. (1974). Social ethology of the whiptail wallaby, *Macropus parryi*, in northeastern New South Wales. *Anim. Behav.*, **22**, 281–

369.

Kirsch, J. (1984). Marsupial origins: taxanomic and biological considerations. In *Vertebrate Zoogeography and Evolution in Australasia*, ed. M. Archer and G. Clayton, pp. 627–32. Perth: Hesperian Press.

Lee, A. K. and Cockburn, A. (1985). *Evolutionary Ecology of Marsupials.* Cambridge: Cambridge University Press.

Lee, A. K. and Ward, S. J. (1989). Life histories of macropodoid marsupials. In *Kangaroos, Wallabies and Rat-Kangaroos*, Vol. 2, ed. G. Grigg, P. Jarman and I. Hume, pp. 105–15. Sydney: Surrey Beatty and Sons.

Lee, A. K., Woolley, P. and Braithwaite, R. W. (1982). Life history strategies of dasyurid marsupials. In *Carnivorous Marsupials*, ed. M. Archer, pp.1–11. Mossman: Royal Zoological Society of New South Wales.

Lent, P. C. (1971). Mother–infant relationships in ungulates. In *The Behaviour of Ungulates and its Relation to Management*, ed. V. Geist and F. Walther, pp.14–55. Morges, Switzerland: IUCN.

Lissowsky, M. (1996). The occurrence of play behaviour in marsupials. In *Comparison of Marsupial and Placental Behaviour*, ed. D. B. Croft and U. Gansloßber, pp. 187–207. Fürth: Filander Verlag GmbH.

Martin, R. and Handasyde, K. (1990). Population dynamics of the koala *Phascolarctos cinereus* in southeastern Australia. In *The Biology of the Koala*, ed. A. K. Lee, K. A. Handasyde and G. D. Sanson, pp. 75–84. Sydney: Surrey Beatty and Sons.

McLean, I. G., Lundie-Jenkins G. and Jarman, P. J. (1995). Training captive rufous hare-wallabies to recognize predators. In *Reintroduction Biology of Australian and New Zealand Fauna*, ed. M. Serena, pp. 177–82. Chipping Norton: Surrey Beatty and Sons.

Mineka, S. and Cook, M. (1988). Social learning and the acquisition of snake fear in monkeys. In *Social Learning in Animals: The Roots of Culture*, ed. C. M. Heyes and B. G. Galef, pp. 51–74. San Diego: Academic Press.

Mitchell, P. and R.Martin (1990). The structure and dynamics of koala populations – French Island in perspective. In *The Biology of the Koala*, ed. A. K. Lee, K. A. Handasyde and G. D. Sanson, pp. 193–202. Sydney: Surrey Beatty and Sons.

Russell, E. M. (1973). Mother-young relations and early behavioural development in the marsupials *Macropus eugenii* and *Megaleia rufa*. *Zeitschr. Tierpsychol.*, **33**, 163–203.

Russell, E. M. (1982). Patterns of parental care and parental investment in marsupials. *Biol. Rev.*, **57**, 423–86.

Russell, E. M. (1984). Social behaviour and social organization of marsupials. *Mammal Rev.*, **14**, 101–54.

Russell, E. M. (1985). The metatherians:order Marsupialia. In *Social Odours in Mammals*, Vol. 1, ed. R. E. Brown and D. W. MacDonald, pp. 45–104. Oxford: Clarendon Press.

Russell, E. M. (1989). Maternal behaviour in the Macropodoidea. In *Kangaroos, Wallabies and Rat-Kangaroos*, Vol. 2, ed. G. Grigg, P. Jarman and I. Hume, pp. 549–69. Sydney: Surrey Beatty and Sons.

Salamon, M. (1996). Olfactory communication in Australian marsupials. In *Comparison of Marsupial and Placental Behaviour*, ed. D. B. Croft and U. Gansloßber, pp. 46–79. Fürth: Filander Verlag GmbH.

Schultze-Westrum, T. (1965). Innerartliche Verständigung durch Dufte beim Gleitbeutler *Petaurus breviceps papuanus* Thomas (Marsupialia: Phalangeridae). *Zeitschr. vergleich. Physiol.*, **50**, 151–220.

Seebeck, J. H. and Rose, R. W. (1989). Potoroidae. In *Fauna of Australia*, Vol. 1B, *Mammalia*, ed. Australian Biological Resources Study, pp. 716–39. Canberra: Australian Government Publishing Service.

Serena, M. (ed.) (1995). *Reintroduction Biology of Australian and New Zealand Fauna*. Chipping Norton: Surrey Beatty and Sons.

Shepherd, N. (1990). Capture myopathy. In *Care and Handling of Australian Native Mammals*, ed. S. J. Hand, pp. 143–7. Sydney: Surrey Beatty and Sons.

Smith, A. and Lee, A. (1984). The evolution of strategies for survival and reproduction in possums and gliders. In *Possums and Gliders*, ed. A. Smith and I. Hume, pp. 17–33. Sydney: Surrey Beatty and Sons.

Soderquist, T. R. (1995). The importance of hypothesis testing in reintroduction biology: examples from the reintroduction of the carnivorous marsupial *Phascogale tapoatafa*. In *Reintroduction Biology of Australian and New Zealand Fauna*, ed. M. Serena, pp. 159–64. Chipping Norton: Surrey Beatty and Sons.

Strahan, R. (ed.) (1991). *Complete Book of Australian Mammals*. North Ryde: Cornstalk Press.

Stuart-Dick, R. I. (1987). Parental Investment and Rearing Schedules in the Eastern Grey Kangaroo. Unpublished PhD thesis. Armidale: University of New England.

Turner, V. and McKay, G. M. (1989). Burramyidae. In *Fauna of Australia*, Vol. 1B, *Mammalia*, ed. Australian Biological Resources Study, pp. 652–64. Canberra: Australian Government Publishing Service.

Tyler, M. J. (ed.) (1979). *The Status of Endangered Australasian Wildlife*. Adelaide: Royal Zoological Society of South Australia.

Watson, D. M. (1993). The play associations of red-necked wallabies (*Macropus rufogriseus banksianus*) and relation to other social contexts. *Ethology*, **94**, 1–20.

Watson, D. M. and Croft, D. B. (1993). Playfighting in captive red-necked wallabies, *Macropus rufogriseus banksianus*. *Behaviour*, **126**, 219–45.

Wells, R. T. (1989). Vombatidae. In *Fauna of Australia*, Vol. 1B, *Mammalia*, ed. Australian Biological Resources Study, pp. 755–68. Canberra: Australian

Government Publishing Service.

Williams, R. (1990). Kangaroos in captivity. In *Care and Handling of Australian Native Mammals,* ed. S. J. Hand, pp. 109–21. Sydney: Surrey Beatty and Sons.

Winter, J. W. (1996). Australasian possums and Madagascan lemurs. In *Comparison of Marsupial and Placental Behaviour,* ed. D. B. Croft and U. Gansloßber, pp. 262–92. Fürth: Filander Verlag GmbH.

Witte, I. (1993). The temporal patterning of the behaviour of red kangaroos (*Macropus rufus*). Unpublished honours thesis. Sydney: University of New South Wales.

6

The social context for learning and behavioural development among wild African elephants

Phyllis C. Lee and Cynthia J. Moss

Introduction

Elephants are among some of the most socially complex mammals as well as being the largest extant land mammal (Spinage 1994). The Asian elephant (*Elephus maximus*) is currently thought to contain three subspecies, one in dense forest patches of India and Southeast Asia, one on Sri Lanka and one on Sumatra (Sukumar 1989). The African elephant (*Loxodonta africana*) currently has two recognised subspecies. The forest elephant (*L.a. cyclotis*) is a smaller, small-eared form with straight, downward-pointing tusks. It is found in the rainforests of central Africa and in forest fragments of west Africa. The larger savannah elephant is distributed throughout woodlands, bush and grasslands of sub-Saharan Africa in populations with some genetic substructuring (Georgiadis *et al.* 1994).

Our knowledge of elephant social behaviour in the wild has come from a number of long-term studies, primarily in east and southern Africa, on Sri Lanka and in south India (Moss 1988, Sukumar 1989). African forest elephants tend to live in small family groups but they can gather into larger groups in clearings where they come to obtain minerals (Turkalo and Fay 1995). The family units of savannah elephants are larger, and they frequently aggregate in large groups. Asian elephants also live in stable families, and these families may congregate in open areas of grassland or near water (Sukumar 1989). This family structure, with the potential to associate with other families in large aggregation, is the basis for the fission–fusion society of elephant, which has implications for learning in the social context discussed below.

Elephants have been tamed and used by humans for at least the last 4000 years, and feature in art, mythology, religion, logging and transport. What little we know of elephant learning is primarily derived from experience with these captive or 'domesticated' elephants. One striking feature of the process of elephant taming is the use of already tamed elephants as monitors and facilitators, and young captive-born animals are preferred for taming (de Alwis 1991). It can be suggested that the calm response of the monitor elephants initiates the process of learning that the human handler is to be obeyed and respected. Other evidence for the importance of the social context in learning comes from orphaned Tsavo elephants, where contacts with both humans and other older elephants provided information to translocated orphans about the new foods in the area, where and when to forage, and how to avoid dangers (McKnight 1992). Although anecdotal, in the absence of recent experimentation on elephant learning abilities these examples suggest that it is the social context of development that should enable us to understand, or at least speculate on, the conditions for the acquisition of knowledge – about the physical, social and emotional world – amongst this extraordinary, long-lived mammal.

In the wild, African elephants (*Loxodonta africana*) face complex problems during their developmental period. They have a period of nutritional dependence of 2–4 years, a period of social dependence of about 10–16 years, and a lifespan of over 60 years. Elephant brains are large and complex, with marked development of the cerebrum, cerebellum and temporal lobes (Shoshani 1991). The size of a female elephant's brain relative to body weight is over twice that expected for a typical mammal and their neuroanatomy suggests that memory, communication, and coordination are all highly developed in elephants. Thus, learning and social communication are likely to form the basis for their behaviour. The large body size of adult elephants, combined with the specialised foraging organs of trunks and tusks, allows for the exploitation of a wide range of foods, as well as the capacity for considerable selectivity. The complex coordination of the trunk must be practised by infants, and how to apply that organ to procuring appropriate foods from a specific environment must be learned (Moss 1988). These are challenges faced by young elephants where the required skills for foraging are probably far greater than for most grazing ungulates. One interesting issue is whether the social context contributes to the efficiency of food learning and in particular to this selectivity.

Elephant learning can be seen as the outcome of an interaction between the intrinsic behaviour of an individual, its social experience, and the accumulated social experiences of its mother and family, in addition to the physical and biotic environment in which it must forage and survive. The fission–fusion nature of savannah elephant sociality provides a heterogeneous social context within an ecologically variable environment. Calves mature in this rich and varied social world, and although males and females share the same family environment, they differ markedly in their behaviour and make different decisions as they develop. In addition to obvious physiological differences

(Short 1966, Short *et al.* 1967), the sexes differ in growth (Hanks 1972, Laws *et al.* 1975, Lee and Moss, 1995), early social development (Lee 1986, Lee and Moss 1986), life histories (Moss 1988, 1994), foraging and reproduction (Moss and Poole 1983, Moss 1988, Poole 1994). Sex differences in survival and growth thus provide the physical context for sexual differentiation in behaviour during development. The aim of this chapter is to summarise sex differences in development as a basis for examining the social context for learning, specifically in relation to the acquisition of foraging skills, maternal care patterns and decisions about dispersal among free-living African elephants.

The social context of African elephants

African savannah elephants, male and female, grow up within a family unit consisting of 2–35 individuals. Families typically are extended social networks of mothers and daughters, sisters and other relatives. Family units associate with other such units in larger groups on a temporary basis. Some families maintain consistent associations, affiliative behaviour and coordinated ranging (Moss 1988), called bond groups. Some bond groups represent an extended family unit which has undergone social fissioning as its size increased (Moss 1988, long-term records). Family units and bond groups are embedded in the further social network of other families; these join and separate in discrete ranges and across populations as a whole. From the perspective of the females and immature males within families, the social world varies temporally over 24 hours and over the year (Lee 1991). There are seasonal trends, with larger groups of females and calves forming during and after rainy periods when food is particularly abundant (Moss 1988, Lindsay 1994). Smaller aggregations of families, and indeed the fragmentation of families into individual mother–calf units, are common during severely dry periods. The social world of elephant calves thus consists of familiar (family unit members) individuals who engage in a range of affiliative and nurturing relationships with the calves, as well as others who may be more or less unfamiliar depending on the family unit's social network. Complex systems of communication (Poole *et al.* 1988, Poole and Moss 1989), which may be learned, revolve around this dynamic social structure.

Young males disperse from their natal families between 9–18 years of age and join the floating community of adult, reproductive males. Individual males associate with family units on a relatively temporary basis, either when pursuing an oestrus female for mating, or when older males are in musth. When not with females in aggregations, males are either found on their own or foraging in small male groups (Moss and Poole 1983, Poole 1989a, Poole and Moss 1989).

In terms of social development, two patterns can be clearly identified. Females establish relationships within a family unit, and reside in association with these kin females throughout their lives. Reproduction typically begins with first oestrus at about 12 years of age and first birth at 14 (Moss 1994). Within the family, females (young and old) contribute to the protection and survival of all calves (Lee 1987). Males also show a major transition at about this age when they leave the family, establish a male ranging area, integrate within the male dominance hierarchy, and eventually seek mating opportunities. Their access to oestrus females depends on their reaching about 25–30 years of age, whereupon they start having annual musth cycles (Poole 1987, Poole 1989b). Reproductive success increases with age, and thus longevity is crucial to male reproductive strategies (Moss 1983, Poole 1989a). An understanding of mechanisms underlying sex-specific social strategies should provide us with a clearer picture of the importance of the social context in the extended period of behavioural development among elephants.

Methods

The study population

The subjects of this study are a cohort of 381 calves born between 1976 and 1985 into the Amboseli elephant population. The population presently consists of close to 1000 individually recognised elephants living in and around the 390 km^2 Amboseli National Park in southern Kenya. The approximately 2000 km^2 area used by the elephants is relatively dry acacia bushland, with a series of permanent swamps and watercourses within the protected park. Rainfall averages 300–350 mm per year, with higher rainfall on the slopes of Mt. Kilimanjaro and lower rainfall to the north in the swamp basin.

The population has been intensively studied since 1972 by Moss and co-workers (Moss 1988). All births and deaths, as well as family unit size, composition and associations, are monitored. Other events, such as musth cycles among adult males and location within the habitat, are also continuously recorded. Recognition is based on photographic files of individual ear shape, size and characteristics such as vein patterns, as well as distinctive features of tusks or bodies.

Cohort sample

The 1976–85 cohort was chosen for these analyses for four reasons. Firstly, these individuals have been followed over a 20 year period (1976–1995) for demographic transitions: deaths, dispersal from the natal family and first

Table 6.1. *Sample sizes for the 1976–85 birth cohort, along with deaths, repro-ductive onset and dispersals over the period of 1976–1995*

Year of birth	Females			Males		
	Number born	Dead	First birth	Number born	Dead	Dispersed
76	9	4	3	19	15	4
77	1	0	1	1	1	0
78	1	0	1	4	1	2
79	29	6	15	28	10	12
80	30	6	16	22	11	5
81	15	3	5	9	3	0
82	17	4	3	16	6	4
83	43	16	4	37	18	1
84	18	9	0	22	7	0
85	33	9	2	27	10	1
Total	196	57	50	185	82	29

reproduction. Secondly, the ten year span of births occurred over a range of ecological conditions, from severe droughts to relatively wet. In addition, intensive behavioural observations were made on individuals in this cohort between birth and six years of age in 1982–84. Thus, data are available on foraging, suckling patterns, social interaction and associations for 129 calves in this cohort. Finally, this is the primary cohort that has been the focus of longitudinal growth monitoring.

The cohort consists of 196 females and 185 males born between 1976 and 1985, of which 139 have died during the 20 years considered here (Table 6.1). Of the 242 survivors, 50 females reproduced for the first time by 1995 and 29 males dispersed. In analyses of dispersal, a larger group of 136 males was used since the majority of the study cohort was still too young to disperse. The larger sample included all known family males between 10–18 years of age where the age of independence and time spent with the family was recorded.

It should be noted that this cohort analysis has limitations. While the sample of calves is large, and the calves experienced a range of rainfall and food availability which could be expected to influence survival and growth, it still may be an atypical sample and extrapolations to other populations or time periods should be made with care. The births were not evenly distributed across the 10 years (Table 6.1), and a high degree of clumping was observed.

Over the 20 years analysed, a number of ecological and demographic changes have occurred. The population size has increased from about 550 in 1976 to almost 1000 by 1997. There have been changes within the protected area associated with acacia tree loss, and an increase in swamp area and waterflow. While the number of family units has remained at about 50, the average size of a family has increased from 8 in 1976, to 15 in 1995. Thus, the social context experienced by these calves has not been constant. Variable opportunities for aggregation, for association with a wider range of individuals, and a change in the number of immature animals within families are likely to contribute to variance in the patterns and processes of social development. With this caveat, we aim to use the demographic variation as one factor explaining variation in developmental patterns.

Analyses

Rainfall was calculated over a 12 month period from November (onset of the short rains) through to October (end of the long dry) rather than a calendar year (Lindsay 1994). Mean rainfall in this cycle between 1974 and 1990 was 282 mm (Lindsay 1994). Drought years ($n = 3$) were defined as < 200 mm rainfall, moderate years ($n = 3$) as 200–320 mm, and wet years ($n = 4$) as > 320 mm, representing the lowest 25%, middle, and highest 25% of rainfall respectively. Calves were assigned to a rainfall category of dry, moderate or wet on the basis of the rainfall in the 12 month cycle which included their month of birth.

Survival analysis was carried out using the Cox hazard regression model, which takes into account truncated observations. The Cox's model is particularly relevant to modelling trends for individuals who have yet to die, reproduce or disperse.

Growth curves were constructed from data collected between 1976 and 1995, using measures of hind foot length from footprint impressions and photometric techniques to assess shoulder height (see Lee and Moss 1995, for further details). Curves were fit separately for males and females using a nonlinear asymptotic equation, and individuals partitioned into rapid, moderate and slow growing categories using residuals from the calculated curve.

A total of 129 calves aged between birth and five years was observed on a focal animal basis for 272 hours. These data are used in analyses of rates of suckling and social interaction. Suckling data are presented as independent bouts of suckling per hour (see Lee and Moss 1986 for details). *Suckling effort* is defined as (No. of attempts to make nipple contact/No. of rejections from nipple contact) × No. of successful bouts of suckling. This measure indicates the extent to which a calf is successful in obtaining milk in relation to attempts and conflict with the mother over access to the nipple. Suckling frequencies and suckling effort were natural log transformed for normality. Frequency was regressed against calf age. Residuals from the regression were then used to explore the effects of sex and maternal experience, controlling for calf age.

Scan samples ($n = 3348$) of proximity, play and maintenance behaviour were made on all calves between birth and six years old (see Lee 1986). Scan data are presented as a proportion of observations for an age–sex class in different activities or at specific distances to neighbours. We examined the effect of age on calf behaviour in two ways. Absolute age from known birth dates was used to assess changes over the first five years of life. In the exploration of suckling, we hypothesised that absolute age may have had less of an effect on suckling rates and effort than did the calf's age relative to maternal reconception, and again relative to it being weaned by the birth of a new sibling. Age relative to maternal conception and weaning was calculated in relation to the birth of a calf's sibling, and was derived from known inter-birth intervals and a gestation length of 22 months. These were expressed in categories as: calf was $<$ 36 months, 35–24 months, 23–12 months, and 11 months prior to the next conception, 1–12 months of gestation, and 13 months of gestation to new birth.

Analyses of natal male dispersal were based on independent sightings of males and families where the presence or absence of a male could be confirmed. Time away was defined as the proportion of sightings of a male away from the family/total sightings of male + family. A logistic curve was fitted using the nonparametric curve-fitting procedure in SPSS (PC). Residuals from the logistic curve were calculated for each male at each sample age, and these were used to assign a relative 'speed' for his independence process. These were categorised as *slow* (on average one quartile below the mean residual), *average* or *quick* (one quartile above the mean residual). Age of independence was determined as the year of age when $>$ 90% of sightings were away from the family, and when time away was maintained at this level.

Results

Background profiles to calf development

The cohort analysis of survivorship provides additional support for our earlier finding (Lee and Moss 1986) that male calves suffer higher mortality than do females (Figure 6.1). The effects of a drought or high food availability are clearly shown, with a significantly reduced survivorship in drought years (Sex effect: Wald = 8.15, df = 1, $p < 0.005$. Rainfall: Wald = 11.51, df = 2, $p = 0.003$). The effects of both sex and the environment on survival appear to be most

A) Males

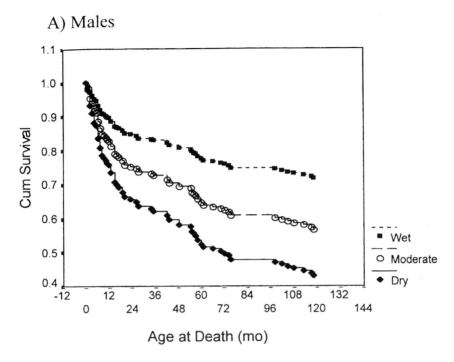

Age at Death (mo)

B) Females

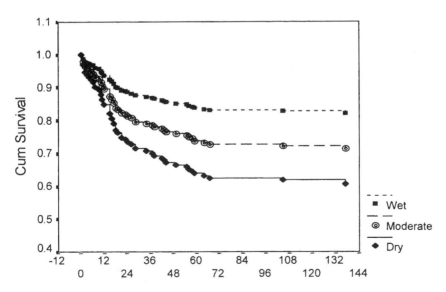

Figure 6.1. Survival (from Cox's regression) for male (A) and female (B) calves born between 1976 and 1985, plotted separately as a function of rainfall for calves' year of birth. Cum=cumulative survival.

marked in the first two years of life (Figure 6.1).

Growth curves for this cohort show less variation with respect to severity of environmental conditions. For males, no significant differences in height residuals were found, but there was a trend towards larger size and greater footlength among males born in wet years (ANOVA, $F_{2,105} = 2.52$, $p = 0.086$). More females were tall in moderate or wet years than in dry years ($F_{2,81} = 3.50$, $p = 0.035$). We have argued that slow growth could be associated with increased mortality among male elephants (Lee and Moss 1995). Since longitudinal growth monitoring can only take place on individuals who do survive, we are still unable to relate directly slow growth and poor survival.

The general pattern of maturation, as indicated by the age of first reproduction for females and age of dispersal from the family for males, differs significantly between the sexes (Wald = 27.25, df = 1, $p < 0.001$; Figure 6.2). Adolescent transitions are occurring earlier for females. The mean age of independence for 88 young males who have completed the transition is 14.2 years (± 1.80), but as many males have yet to disperse, this is likely to be biased by early completed dispersal within this sample. Indeed, the likelihood of dispersal before age 15 is only a quarter of the likelihood of first reproduction for females by this age (Figure 6.2).

The acquisition of foraging knowledge

Calves begin sampling their potential food environment around 1–2 months of age, well before actual ingestion. Typically, sampling consists of plucking food items with the trunk, rolling them in the trunk, placing them in the mouth and possibly chewing at the items. Calves also explore other elephants' foods, placing their trunk in the other's mouth and pulling stems, twigs or leaves from the mouth of the other elephant (see Figure 6.3). Of social contacts (greet, rub, explore food and invite play; $n = 1136$) between calves and other elephants, food exploration consisted of 15.8% of the total. Exploration of foods represented 13.6% of all mother–calf contacts in the first five years of life, while over a quarter (26.2%) of all contacts with nonmothers were explorations of food. Interestingly, there was no change in this behaviour with age (Table 6.2), possibly since calves were continuously being exposed to novel food items throughout this period, and were sampling both for novelty and consistency. The higher proportion of food explorations with nonmothers also represents interest received by calves when they were sampling foods (33.9% of food explorations were initiated by others vs. 0.02% initiated by mothers). No sex differences were found in food contacts with mothers ($\chi^2 = 1.92$, df = 2, NS) or nonmothers ($\chi^2 = 0.92$, NS).

Relatively little active time was devoted to food ingestion before 10 months

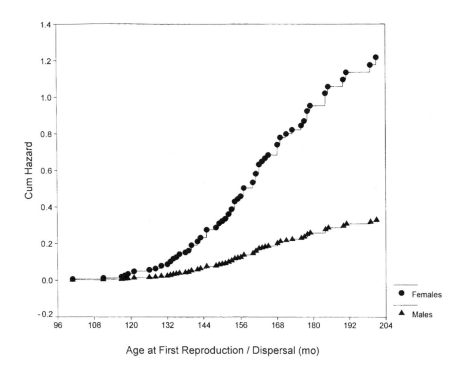

Figure 6.2. Hazard function (from Cox's regression) for the likelihood of first reproduction (females) or dispersal (males) by age in months.

(Figure 6.4). The amount of time spent feeding – ingesting rather than simply handling foods – increased between 12 and 24 months, with a variable but consistent increase over the next two years. No marked sex differences were observed in feeding time. Indeed, differences in seasonally available food types were probably more influential in determining overall feeding time than was sex.

While adult Amboseli elephants ate primarily grass from a total of 22 species, they incorporated as much as 20% browse from 29 species in seasonal diets, and the overall diet contained 91 plant species (Lindsay 1994). The highest dry matter intake was attained from swamp edge grasses ($n = 5$ species), although these had relatively low protein and digestible energy yields in comparison to browse (Lindsay 1994). Calves tended to feed on the predominant swamp edge species, with few observations of browsing among calves under five years. Physical constraints appear to limit calves to maximising intake on easily obtained, low nutrient resources.

Figure 6.3. A young female explores the food of another elephant by touching its trunk and mouth.

Table 6.2. *Food investigation as an hourly rate and as a percentage of friendly contacts by calf age*

Age (months)	With mother			With other		
	Hourly rate	%	N	Hourly rate	%	N
0–12	0.19	8.8	193	0.42	14.0	264
13–24	0.28	14.7	75	0.57	22.1	136
25–36	0.19	11.6	86	0.49	20.8	120
37–48	0.18	13.5	59	0.31	12.1	116
49–60	0.23	16.2	37	0.44	20.0	50
Total	0.20	11.6	450	0.44	16.9	686

Learning and mother–infant relationships

We have suggested that there is greater maternal investment in sons than that in daughters (Lee and Moss 1986, Lee 1986). Sex differences are observed in a number of social interactions: suckling, maintenance of proximity, and friendly contacts between mothers and calves. The question examined here is whether these sex differences are the product of differences in the behaviour of the two sexes of calves, or whether males and females are treated differently by

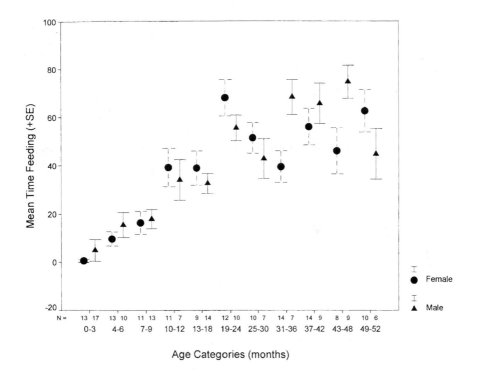

Figure 6.4. Mean time spent feeding for male and female calves at each age.

mothers and others, producing a sex-specific social response by calves. How do calves learn appropriate social responses for their sex and age, integrating these into the simultaneous physical changes in body size with age? Do size and age changes underlie behavioural differentiation or is it the nature of the interaction itself?

Over the first six years of life, males and females spend similar amounts of time very close (< 2 m) to their mothers (Figure 6.5). However, male calves are somewhat more likely to be found far (> 5 m) from the mother when not by her side ($t = 1.95$, df $= 5$, $p = 0.056$, one tailed). Male calves are also more likely to be found far (> 5 m) from other family members ($t = 2.5$, df $= 5$, $p = 0.026$, one tailed). Male calves appear to be more exploratory and less constrained to remain close to family members than are females. The maintenance of close proximity between mothers and sons may reflect suckling patterns. Male calves have consistently higher rates and durations of suckling contacts than do females (Lee and Moss 1986).

Suckling frequencies were compared here between the sexes as a function of

Figure 6.5. Young calves play in the security of a close association with their mothers.

absolute age of the calf (Figure 6.6A), showing the consistent male bias overall ($F_{1,223} = 9.64$, $p = 0.002$), and by calf age relative to maternal reconception and weaning (Figure 6.6B). Sex differences in rates of suckling were most marked in the 2–3 years prior to reconception ($F_{1,146} = 5.37$, $p = 0.02$), and became less marked once the mother had reconceived and independent feeding provided the majority of the calf's nutritional intake ($F_{1,76} = 0.82$, NS).

These observed sex differences in intake are due to males' more persistent and frequent attempts to suckle (Lee and Moss 1986). More attempts are initiated by sons and these are tolerated by mothers. Daughters, for the most part, appear to be less demanding as well as less persistent, indicated through the measure of suckling effort. Significant sex differences in suckling effort were found overall (ANOVA ln effort by sex, $F_{1,228} = 12.37$, $p = 0.001$) (Figure 6.7), but maternal experience also plays a role in the effort calves put into obtaining milk. Comparisons within categories of maternal parity found significant sex differences only between the calves of multiparous mothers (Mann–Whitney U, $Z = 2.80$, $p = 0.005$). Neither males nor females showed any significant change in effort with increasing parity (males: Kruskall–Wallis ANOVA, $\chi^2 = 0.3$, NS; females $\chi^2 = 2.12$, NS). Sons appeared to be working harder than daughters, and their success in attaining extra milk through this effort was greatest when they had experienced mothers.

Our original suggestion that males have higher suckling frequencies also appears to be partially a function of maternal experience. Residuals from the

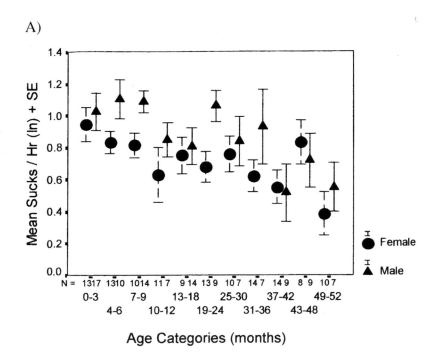

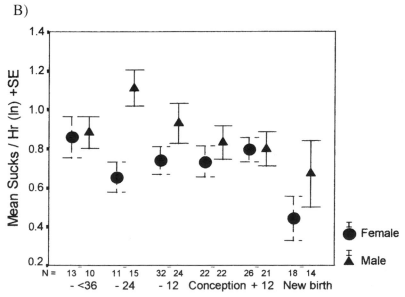

Figure 6.6. Suckling frequencies (ln) for male and female calves plotted against absolute age in months (A) and age relative to conception and the subsequent birth of a sibling (B), using the known inter-birth interval and a gestation length of 22 months.

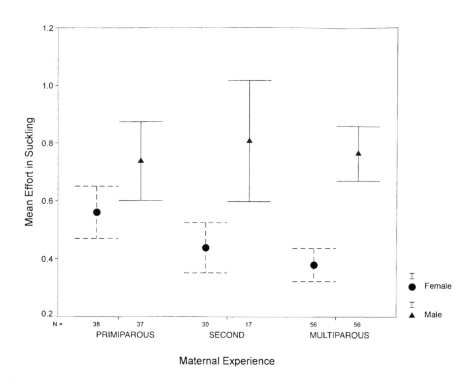

Figure 6.7. Suckling effort for male and female calves as a function of maternal parity, between multiparous and primiparous mothers. Effort is defined as the proportion of nipple contacts rejected for each successful suckling event.

regression of suckling frequency and age were compared between sons and daughters of primiparous mothers, mothers with second infants, and multiparous mothers (Figure 6.8). Within the sexes, there was no significant effect of maternal experience on suckling frequency (males ANOVA, $F_{2,107} = 0.23$, NS; females $F_{2,122} = 0.68$, NS). However, in comparisons between the sexes, differences were significant for multiparous mothers (ANOVA, $F_{1,109} = 6.357$, $p = 0.013$) with the highest age-specific residuals for their sons.

The social context for later development

The consequences of individual decisions during the critical period of adolescence can be explored in detail amongst young elephants in this sample by examining the process of independence from the natal family. Dispersal from the family unit is typically observed among males in savannah elephants. Males as young as five years of age begin to spend some time in groups of elephants that do not contain their own family. Such occasions are relatively rare, but

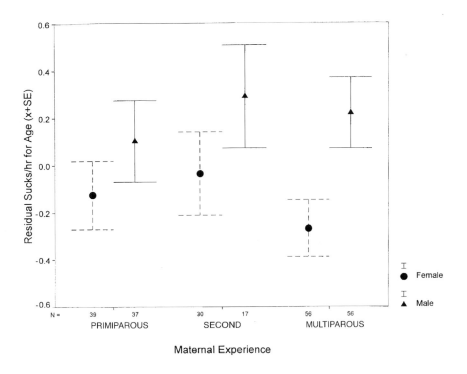

Figure 6.8. Mean age-specific residuals of suckling frequencies for male and female calves by maternal parity.

they increase with frequency over time (Figure 6.9A). By the age of 19, males will be associating with their family only on those rare occasions when their day ranges coincide, or when groups are extremely large and draw in many families and independent males from across the population as a whole.

Observations of time spent away from the family by age were fitted to a logistic curve (Figure 6.9B). While the mean observed and the predicted time away match closely ($r^2 = 0.557$, $p < 0.001$), there is enormous individual variance in how males accomplish the transition to independence, and how long this process takes (Table 6.3). In this sample, at least three trends could be distinguished: 1. Males who leave simply at the average age. The process appears to be relatively slow for such males, suggesting considerable experience of time away from the family prior to the final departure. 2. Males who leave late, but who tend to go relatively quickly. 3. Males who start the process early and take several years to complete it but do so at a relatively young age. However, the complexity of the process is such that we have yet to find

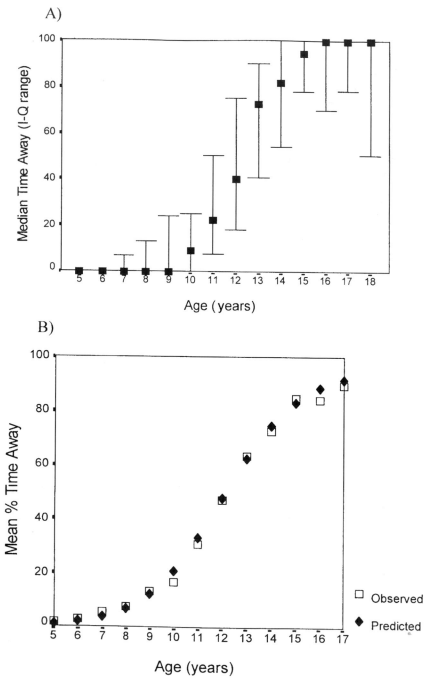

Figure 6.9. Time spent away from the family unit for males between 5 and 19 years. (A) Median and interquartile range for age. (B) Mean observed and predicted from the logistic curve for each age.

Figure 6.10. Two calves engage in trunk-wrestling play as a means of exploring each other's strength and identity.

explanatory factors for the individual variance, or common elements within the families of males who appear to share similar strategies.

Discussion

Social development and social learning appear to have a tiered structure among elephants. Elephant behavioural development thus can be viewed as a process involving an interaction of the intrinsic attributes of the individual, such as its sex and birth size, with the social and physical world which that individual experiences (see Figure 6.10). The results presented above suggest that while both survival and growth are a function of sex, the environment and possibly maternal condition, other aspects of behavioural development are unlikely to be driven solely by sex-specific physical processes but are likely also to reflect individual differences in behaviour within this social context. The physical processes of growth and reproductive maturation differ between the sexes, with consequences for the context of learning – for opportunities to learn social and environmental skills and for how males and females meet their different needs. Despite having apparently similar social environments during early development, sex differences arise through males and females altering

Table 6.3. *The frequencies of young males who dispersed from the family unit at three categories of speed (determined from residuals of the logistic curve), and who left early, average or late (determined from the quartiles for age at independence)*

	Rate of departure		
Age at complete independence (years)	Quick	Moderate	Slow
Early (< 13)	0	5	11
Average (13–15)	10	19	21
Late (> 15)	12	5	5

that social environment as a consequence of their behaviour, which interacts with the experiences of mothers and others in their social contacts with each sex.

Environmental, or specifically food learning, shows little early sexual differentiation. Social contacts may be especially important in providing opportunities for food learning among elephants, since opportunities to learn through individual trial and error are constrained by the size, strength and competence of young animals. In relation to the acquisition of foraging skills, the problems of size constraints are especially marked for immatures of species such as elephants. Constraints on acquisition due to the individual's stature and the food species' height above the ground, on processing foods before ingestion by breaking branches, removing thorns or stripping bark, and mechanical constraints on chewing woody vegetation probably limit calves to the grass-rich diets when small, relatively weak, lacking tusks, and with small molars in wear. Opportunities to learn about and sample diets thus depend less on direct experience with food types, but are more a function of sampling and observation within a social context. This may explain why so much calf social contact revolves around investigations of food, since the only means for experiencing the normative range of foods while young is through the actions of others.

During foraging, there appears to be an exchange of information from mother to calf, which is initiated by the calves, and a more reciprocal interest in foods between calves and nonmothers. This broader social context for sampling potential food items appears to be an important element of food learning among elephants, in that they can thus sample the foods of a number of individuals of different ages, sizes and sex. Other modes of sampling, which are less interactive, also exist. Calves eat the fresh dung of adults and juveniles, which may both provide appropriate digestive bacteria and extend the sampling of food types through their smell in the dung.

The major behavioural sex differences appear in interactions between calves and mothers, as well as in the social networks of older calves. Early in development, rates of nutrient intake through milk are a function of sex. Faster growth rates among males may underlie a higher suckling frequency. Although absolute size differences are small, growth rates among males are consistently faster, promoting greater metabolic requirements for their age and these metabolic requirements of males for rapid early growth are associated with higher suckling frequencies and potentially greater intake. In order to acquire this intake, male calves are more persistent and demanding – behaviour driven by underlying needs. Their success in these attempts is, however, a function of the mother's responsiveness, which depends, at least partially, on her previous experience with calves. More experienced mothers appear more sensitive to calf demands, with consequences for growth and survival. However, more experienced mothers are also larger and older, and more likely to have body reserves or foraging skills to cope with their sons' demands. It is unlikely that experienced mothers of sons and experienced mothers of daughters differ markedly in the quality or quantity of their milk, and thus the observed behavioural differences are not simply a consequence of differences in maternal ability to sustain suckling. Experienced mothers of sons appear to be adjusting their behaviour to respond to the demands of their sons. There is thus modification of both mother and calf behaviour in the context of suckling as a result of experience, and there is an interaction between the drive and persistence of sons in suckling contacts, and maternal responsiveness, which is acquired over a number of calves. The successful meshing of mother and calf in suckling depends both on characteristics of the calf, such as sex-specific growth rates, and on maternal experience and condition. Again, these results highlight the interaction between the physiological status of the animal and its behaviour, with consequences for the outcome of interactions between individuals, both during development and throughout the lifespan.

While female calves appear to take a more passive role in suckling, they are very active interactants with other family members, concentrating their social behaviour within the family. Female maturation is both physically and socially advanced over that of the males; they begin reproduction younger and thus function within the family as autonomous, but integrated, individuals at a younger age. In contrast to young males, females under the age of five show no marked tendency to concentrate their play on same-sex peers; rather, they tend to play more with infants of either sex, and especially with younger calves from the family or extended bond group. Juvenile females, in particular, play a critical role in the care of young calves and their interest in and interaction with infants makes a contribution to survival (Lee 1987). The caretaking role of juvenile female calves provides them with an array of caretaking experiences that persist until they give birth for the first time. Interestingly, despite 10 years of such social competence, first-born infants tend to have high mortality (Moss 1994). In part, increased calf mortality among primiparous females may be due to size constraints; young females are still actively growing and the trade-off between energy allocated to reproduction and that to individual growth may lead to smaller, more vulnerable neonates. Experience gained over a succession of births, however, may play a major role in successful calf rearing. Other factors such as family size and the degree of competition between females may also influence calf survival for younger, less experienced, mothers.

There are other elements of female reproductive behaviour which may require a social context for learning. In particular, female responses to males during oestrus change with individual experience and both the acquisition of oestrus behaviours and the choice of mates appear to be facilitated by the presence and behaviour of mothers of these young females during their oestrus (Moss 1983, 1988). Young females are frequently and persistently chased during their first few oestrus periods by males of all ages and only appear to learn to maintain proximity to large musth males with experience. Mothers of young oestrus females have been observed exhibiting oestrus postures, approaching and avoiding males, and running with their daughters during long chases. They also made post-copulatory calls after their daughters were mated, even when not in oestrus themselves. The first parturition is another event where the presence and behaviour of experienced females provides assistance to inexperienced mothers in coping with the physiological event and the demands of the newborn calf (Moss 1988).

Young males face fewer trade-offs between reproduction and growth early in life, but similar needs for acquiring social experience. During the period when young females are maintaining close caretaking contacts with the family, juvenile males are seeking out novel social partners, and engaging in high rates of play with these partners. We found no significant sex differences in overall rates of play between males and females. However, the choice of play partners did differ (Lee 1986). As males age, they tend to concentrate their play with same-sex peers, and a greater proportion of these peers are novel partners from outside the family. Male exploratory and play behaviour thus occurs in the context of seeking novelty outside the family, while females appear to consolidate their relationships within the family.

Departure from the natal family is typically a male strategy, which can be linked to the seeking of social novelty observed among young male calves. As a process, it appears to have few time or age constraints; males can depart at 9 or at 18 years. The transition can be accomplished within one year, or take as long as 3–4 years. The lack of a clear age-specific sequence to this process suggests that individual decisions within the context of the family's structure and character underlie a gradual and interacting process of exploration, play and independence among males. While age is obviously an important variable, it explains only just over half the variance in rates and timing. The remainder of the variance is due to individual characteristics of the males; their social history, their maternal experiences, their family's general sociability and so on. We have yet to find any common parameter between males who depart early or late, and those who leave quickly or gradually. In our opinion, we are observing the outcome of complex decision-making on the part of these young males. Are they big enough to cope away from the family; do they have peers who have already dispersed or are still within the family; are they the offspring of subordinate mothers and thus leaving to avoid aggression from family females; do they have sufficient social experience of other males, of male areas of residence and of the habitat to survive on their own? These decisions have potential survival costs, as male mortality is higher than that of females between 10 and 20 years of age (Moss, long-term data). We hope, with further analysis of these data, to be able to elucidate the decisions made by males as well as their survival and reproductive consequences.

Sex differences in interactions may be the result of individual decisions which result in discrimination between potential partners that are present equally in the early social environments of both males and females. There will, of course, be limits to the opportunities for engaging in such interactions as a function of family size and of its tendency to aggregate with 'strangers' in groups where calves can cross family boundaries in order to play. Furthermore, there may be differences between calves born to families with small or no bond groups, as opposed to those with large bond groups. Such individual differences in social experience will interact with sexually determined preferences for social partners, and produce individual differentiation in social experience which may ultimately be related to the marked individual characters or 'personalities' we observe among the adults (Moss pers. obs.).

Opportunities for social experience and learning among elephants are not, however, constrained to the developmental period. Throughout the 60 + year lifespan, both unpredictable environmental events and a continuously changing social world produce the need for constant behavioural modification. In the context of elephant social complexity, learning, and specifically learning about and within the social context, appears to be essential to survival and reproduction over the lifespan. Decision-making as an indicator of cognitive ability is influential for both sexes; however, the interaction between the social context and physical development results in different outcomes in terms of behavioural strategies for males and females throughout their lives.

Acknowledgements

We thank the Office of the President, the National Research Council, Republic of Kenya, for research permission, and the wardens of Amboseli National Park and the Kenya Wildlife Service for their support of this project. Financial support was provided by African Wildlife Foundation, The National Geographic Society, Eliefriends and many private donors. We gratefully acknowledge their contribution over the duration of the project. We also thank Soila Sayialel, Norah Njiraini and Katito Sayialel, who carry out the long-term monitoring with patience and dedication, and our colleagues Kadzo Kangwana Keith Lindsay, Karen McComb, Hamisi Mutinda, and Joyce Poole, whose collaboration has been essential to the project's success. We also thank G. Bentley, C. Ellington, R. Foley, M. Hauser, and especially K. McComb for helpful suggestions on analysis and drafts.

References

de Alwis, L. (1991). Working elephants. In *The Illustrated Encyclopaedia of Elephants*, ed. S. K. Eltringham, pp. 116–29. London: Salamander Books.

Georgiadis, N., Bischof, L., Templeton, A., Patton, J., Karesh, W. and Western, D. (1994). Structure and history of African elephant populations 1. Eastern and southern Africa. *J. Hered.*, **85**, 100–4.

Hanks, J. (1972). Growth of the African elephant. *East Afr. Wildl. J.*, **10**, 251–72.

Laws, R. M., Parker, I. S. C. and Johnstone, R. C. B. (1975). *Elephants and their Habitats*. Oxford: Clarendon Press.

Lee, P. C. (1986). Early social development among African elephant calves. *Nat. Geogr. Res.*, **2**, 388–401.

Lee, P. C. (1987). Allomothering among African elephants.

Anim. Behav., **35**, 278–91.

Lee, P. C. (1991). Social life. In *The Illustrated Encyclopaedia of Elephants*, ed. S. K. Eltringham, pp. 48–63. London: Salamander Books.

Lee, P. C. and Moss, C. J. (1986). Early maternal investment in male and female African elephant calves. *Behav. Ecol. Sociobiol.*, **18**, 352–61.

Lee, P. C. and Moss, C. J. (1995). Statural growth in known-age African elephants (*Loxodonta africana*). *J. Zool., Lond.*, **236**, 29–41.

Lindsay, W. K. (1994). Feeding ecology and population demography of African elephants in Amboseli, Kenya. PhD thesis: University of Cambridge.

McKnight, B. L. (1992). Behavioural ecology of orphan African elephants, *Loxodonta africana*, in Tsavo East National

Park, Kenya. M.Phil. thesis: University of Oxford.

Moss, C. J. (1983). Oestrous behaviour and female choice in the African elephants. *Behaviour*, **86**, 167–96.

Moss, C. J. (1988). *Elephant Memories*. London: Elm Tree Books.

Moss, C. J. (1994). Some reproductive parameters in a population of African elephants, *Loxodonta africana*. In *Advances in Reproductive Research in Man and Animals*, ed. C. S. Bambra. Nairobi: National Museums of Kenya.

Moss, C. J. and Poole, J. H. (1983). Relationships and social structure among African elephants. In *Primate Social Relationships*, ed. R. A. Hinde, pp. 315–25. Oxford: Blackwells.

Poole, J. H. (1987). Rutting behaviour in African elephants:

the phenomenon of musth. *Behaviour*, **102**, 283–316.

Poole, J. H. (1989a). Announcing intent: the aggressive state of musth in African elephants. *Anim. Behav.*, **37**, 140–52.

Poole, J. H. (1989b). Mate guarding, reproductive success and female choice in African elephants. *Anim. Behav.*, **37**, 842–9.

Poole, J. H. (1994). Sex differences in the behaviour of African elephants. In *The Differences Between the Sexes*, ed. R. V. Short and E. Balaban, pp. 331–46. Cambridge: Cambridge University Press.

Poole, J. H. and Moss, C. J. (1989). Elephant mate searching: group dynamics and vocal and olfactory communication. *Symp. Zool. Soc. Lond.*, **61**, 111–25.

Poole, J. H., Payne, K., Langbauer, W. R. and Moss, C. J. (1988). The social context of some very low frequency calls of African elephants. *Behav. Ecol. Sociobiol.*, **22**, 385–92.

Short, R. V. (1966). Oestrous behaviour, ovulation and the formation of the corpus luteum in the African elephant, *Loxodonta africana*. *East Afr. Wildl. J.*, **4**, 56–68.

Short, R. V., Mann, T. and Hay, M. F. (1967). Male reproductive organs of the African elephant. *J. Reprod. Fert.*, **13**, 517–36.

Shoshani, J. (1991). Anatomy and physiology. In *The Illustrated Encyclopaedia of Elephants*, ed. S. K. Eltringham, pp. 30–47. London: Salamander Books.

Spinage, C. (1994). *Elephants.* London: Poyser.

Sukumar, R. (1989). *The Asian Elephant.* Cambridge: Cambridge University Press.

Turkalo, A. and Fay, J. M. (1995). Studying forest elephants by direct observation: preliminary results from the Dzanga Clearing, Central African Republic. *Pachyderm*, **20**, 45–54.

7

Comparative social learning among arctic herbivores: the caribou, muskox and arctic hare

David R. Klein

Introduction

There are three medium-size to large mammalian terrestrial herbivores in the Arctic. These are the caribou (*Rangifer tarandus*) (caribou refers here to the wild form of the species and reindeer to the domestic form), the muskox (*Ovibos moschatus*) and the arctic hare (*Lepus arcticus*, *L. othus*, and *L. timidus* are included in the circumpolar distribution). Sociality among these species appears to have evolved primarily in response to predation pressure character-istic of other large herbivores living in open terrain (Hamilton 1971), but there are large differences among them in their morphology, locomotive efficiency and feeding patterns (Klein 1991, Klein and Bay 1994). Moreover, sociality presents opportunities for group members to acquire information within groups in all three species, although there are pronounced differences among the species in their requirements and opportunities for social learning. The distinction between inherent patterns of behaviour and those learned through social interaction is not readily apparent among these arctic mammals. Rudi-ments of many behaviours characteristic of the adults of a species appear to be present at birth, but require refinement through interaction with cohorts and adults, or through copying of adult behaviour (Klopfer 1970). Thus, the complex of adult behaviour patterns characteristic of caribou, muskoxen and arctic hares is a blend of patterns of behaviour that may originate in the individual from its genome, but are honed to completion through social experience, often via play. However, few attempts have been made to assess the role of social learning among arctic mammals. Our knowledge so far is primarily drawn from observations in the field under natural conditions that offer opportunities for comparative analysis. We may first consider a variety of characteristics of the species under consideration.

Figure 7.1. Caribou, shown here in northern Alaska, and reindeer are unique among the Cervidae in that both sexes grow antlers. The antlers are used primarily in social interactions that maintain hierarchical structures within the group.

The arctic ungulates

Caribou and muskoxen are both members of the Artyodactyla, but have divergent evolutionary histories in the Cervidae and Bovidae respectively. Caribou, with deciduous antlers used primarily in social interactions (Figure 7.1), and slender body proportions characteristic of most Cervidae, differ from muskoxen, which possess perennial horns and have the more compact and massive body characteristic of bovids (Klein 1991). Antlers of female caribou are too small to be effective weapons against wolves (*Canis lupus*) and those of adult males are only retained, upon completion of growth, until shortly after the rut in autumn. Thus, caribou are of insufficient body size and without adequate weaponry to defend themselves from attack by wolves, their primary predator. Speed of locomotion provides them with their most effective mode of avoiding predation (Table 7.1). Muskoxen, by contrast, have a more massive body and weigh more than twice that of caribou, and the sharp tipped, up-curved and persistent horns project to equal distances from the head in both sexes. These characteristics allow them to defend themselves against wolves by standing their ground and facing their attackers (Figure 7.2).

Morphological differences between caribou and muskoxen account for

Table 7.1. *Advantages gained from sociality by the three social mammals of the Arctic. Increasing advantage is indicated by the increasing number of plus signs*

	Predation avoidance				Other values	
Species	Increased surveillance	Flight response	Aggregation confusion	Group defence	Migration	Food selection
Muskox	++	+	+	+++	+	+++
Caribou	+++	+++	+++		+++	+++
Arctic hare	+++	+++	+++		?	+

Figure 7.2. Eskimo sled dogs have been used as surrogate wolves to hold muskoxen in their typical predator defence formation during a study of muskox group integrity in eastern Greenland (Clausen *et al.* 1984).

their divergent strategies to avoid predation; however, both strategies require sociality. Among many taxa of mammals and birds living in open terrain, individuals within groups share in the increased vigilance of the group over that of a single individual (Elgar 1989). Early response of the group to the approach of predators enables the keen-sighted and fleet-footed caribou to distance themselves from predators before they are close enough for a successful attack. For muskoxen, vigilance of the group also enables an early response to the threat of predation; they gain time to gather together in a stationary group in defence of themselves and their young.

Morphological differences between these species, however, also account for marked differences in their locomotive efficiency, thermoregulation, dependence on body reserves in winter, dietary selection, and associated behaviour patterns (Klein 1991). Muskoxen are among the most energetically conservative of the ungulates. Their short legs and small hooves, although contributing to their efficiency in retaining body heat in winter, equip them poorly for running and for travelling or foraging through deep snow. The slender bodied caribou are among the most locomotively efficient ungulates (Klein *et al.* 1987). They are strong runners, and are well adapted to travel and forage in snow, and to undertake long migrations. These are energetically demanding activities. Caribou, however, have a less insulative pelage than muskoxen; they are more vulnerable to insect harassment, are more limited in the proportion of body fat they can accumulate, and have a much greater surface to body mass ratio than muskoxen. Thus, caribou, in contrast to muskoxen, lead an energetically costly life (Klein 1991).

As a consequence of these differences in morphology and energetic expenditure, muskoxen are bulk foragers (Hofmann 1983). They move little and are capable of a rapid intake of low-quality plant material that is usually dominated by graminoids (Klein 1991). By contrast, caribou must forage selectively for higher quality forage (Klein 1991); this necessitates longer daily feeding bouts over greater distances (Roby 1978). In addition, most caribou expend energy digging through snow in winter to obtain lichens, a forage of high energy content (Klein 1991) that helps them to meet the energy costs of their long seasonal migrations and active lifestyle.

The arctic hare

The arctic hare, like the two arctic ungulates, is also well adapted for life in the Arctic. This is perhaps surprising for a mammal that is relatively small, remains active above the snow surface in winter and, in contrast to the ungulates, has a high surface area to body mass ratio. Arctic hares remain white throughout the entire year in the High Arctic (Figure 7.3), but they molt to a cryptic grey–brown in summer in the Low Arctic (Hall and Kelson 1959). In the High Arctic, where their densities are sufficient, they form into large groups in winter; this affords advantages in detecting predators, and in creating confusion in predators when hares run as an aggregation. In summer, in the High Arctic, the white adult hares form small groups during foraging bouts, but

Figure 7.3. The arctic hare is white in both winter and summer in the High Arctic. It joins social groups during daily foraging bouts presumably as a defence against predation.

return daily to solitary hides, usually on rocky slopes, where they have good visibility of the land below them and where they are relatively inconspicuous (Klein and Bay 1994). The precocial and cryptically coloured leverets do not accompany foraging adults but remain hidden at rendezvous sites where they are visited by their dam to be nursed. By early winter young hares in the High Arctic form aggregations and join later with increasing numbers of adults as winter progresses (Parker 1977).

In the Low Arctic, when hares undergo a molt to cryptic grey–brown in summer, they have not been observed to form aggregations. In winter in the Low Arctic, large aggregations of the white hares have only been observed in late winter, possibly in association with courtship behaviour and initiation of the breeding season. The less frequent aggregation of hares in the Low Arctic may be associated with the greater presence of shrub cover there, which is used to hide from predators.

Arctic hares, like caribou, are selective foragers with efficient locomotive ability, but their long daily resting periods and relatively short foraging time more closely fits the energetically conservative pattern of muskoxen. Although they have a higher surface to body mass ratio than either muskoxen or caribou, they are well insulated by their pelage and greatly reduce their exposed surface area when resting or feeding, especially under cold winter conditions. Their long legs in relation to their body size results in highly efficient locomotion and swift running ability, which are important adaptations for avoiding predators and for wide-ranging selective foraging (Klein and Bay 1994). Their large feet and low body weight enable them to readily travel over the wind-hardened snow of the arctic landscape. Their small body size in contrast to muskoxen and caribou, their smaller bite size, and their relatively smaller stomach than the rumen of these ungulates enable hares to selectively exploit smaller patches of vegetation within the patchy landscape of the High Arctic than is the case for either muskoxen or caribou. Thus, their small daily dietary intake and high locomotive efficiency enables arctic hares, even in groups, to exploit portions of the landscape with insufficient forage biomass to be of value to foraging muskoxen or caribou (Klein and Bay 1994).

The role of social learning

Potentially, socially acquired information may be advantageous to individual survival and fitness in a variety of ecological situations. Given the different lifestyles involved, however, we can expect differences between species in the opportunities for social learning, as well as in the potential ways in which social learning may occur. For example, migration in caribou requires coordination in timing, navigational ability, and group cohesion if optimal use of seasonal components of their total rangelands is to be achieved. Although initiation of the migratory urge in pregnant female caribou in spring is a function of approaching parturition, individual females presumably derive benefit from aggregating with like individuals. The potential exists for them to benefit from mutual stimulation for directional movement, collective navigational ability and terrain familiarity, shared effort in leadership and trail breaking through snow, and the advantage of predator avoidance within a group. Females experiencing their first pregnancy, and other young animals accompanying the older females have the potential to learn characteristics associated with migration that favour their fitness. The strength of the female–young bond in preparing the young for social living is expected to be greater in caribou, because of the more frequent inter-group exchanges, than in muskoxen, and of little importance in arctic hares where young remain isolated from social foraging groups during most of summer. Group structure and female–young associations are likely to be disrupted more frequently in caribou than muskoxen due to their more extensive movements over obstacles in their environment (Duquette 1984), such as swift-flowing rivers, and their likelihood of being scattered by predators. The strong female–young bond facilitates their re-uniting after separation. Selective foraging is more important in determining dietary composition of caribou than muskoxen and therefore young caribou may benefit more than young muskoxen in learning to select forage while in close association with others in the group.

Among muskoxen, which are dependent upon group fidelity for predator avoidance, we can hypothesise that young animals experience strong pressures from other group members to learn patterns of social behaviour that facilitate the success of all individuals in avoiding predation. This may also result from a greater likelihood of extended kinship within groups of muskoxen in contrast to caribou. Muskox groups generally consist of females and young accompanied by a dominant or 'herd' bull. The herd bull will often tolerate a second bull in the group. Although some exchange of animals between groups may occur, group composition remains relatively stable throughout the year. Other bulls may be present as single animals or in small all-male groups, but may challenge herd bulls during the rut. In contrast, most bull caribou remain in groups isolated from females and young except during the autumn migration to the wintering grounds, which coincides with the rut. As a consequence of this difference in the relations between the sexes in the two species, muskox males play a greater role in influencing patterns of intra-group behaviour, including social learning, than is the case among caribou (Gray 1987).

Artctic hares, although social while foraging during summer in the High Arctic, do not involve young animals in this sociality. Social learning opportunities for young hares in association with adults are limited to the brief encounters with females during the single daily suckling events. Although Hudson *et al.* (chapter 8, this volume) have shown that maternal contact with young in the European rabbit (*Orycotolagus cuniculus*) resembles that of the arctic hare, with contact limited to one brief nursing bout daily, learning from the doe is shown to include olfactory identification by the young of the does belly, teat location, and components of her diet. Little is known of the behaviour of arctic hares during early winter when young hares, as well as adults, form large aggregations and mixing of age groups begins to occur. From studies of the closely related Eurasian species, *L. timidus* and *L. europaeus*, sociality appears to be an adaptation for predator avoidance and is largely restricted to foraging periods when the hares are potentially exposed to predators (Monaghan and Metcalfe 1985, Tapper 1987). Dominance hierarchies develop among these foraging groups associated with controlling access to favoured feeding sites (Lindlof 1978).

The caribou
Social learning among caribou and reindeer has not received the extensive attention directed to some primates and carnivores for example, or even to

some other ungulates (Geist and Walther 1974) at lower latitudes and especially in Africa (Jarman 1974, Estes 1974). Nevertheless, an understanding of the
sociality of the species has benefited from the long human association with
semidomestic reindeer. Traditional knowledge of reindeer behaviour derived
from generations of experience of reindeer herders has often provided the basis
for interpretation of behavioural patterns in the light of evolutionary and
ecological theory. Baskin (1970) drew upon this experience while living and
working with reindeer herders in eastern Siberia, and as a result was able to
describe the dominance relationships that exist within groups of reindeer.

In the context of the present chapter it is important that caribou and
reindeer have long been hypothesised to learn specific migration routes
through their social interaction with more experienced individuals. Moreover,
the calving grounds are the one fixed location in the annual migratory movements of caribou, and pregnant females have a high fidelity for use of the same
birthing area from year to year. Females that are breeding for the first time
travel to these calving grounds in the company of older experienced females.
Adult males are less precise in the timing of their spring movements, and tend
to follow the availability of high quality, new growth forage associated with the
advancing plant phenology northward; they arrive in a more dispersed distribution on the tundra of the summer ranges (Calef 1981). Further, movements of caribou at other times of the year have less well defined routes and
locations and are subject to weather conditions that determine levels of insect
harassment in summer, the timing of the arrival of winter snows, and the
regional variation in the depth of snow.

Interestingly, experience with reindeer in Sweden that, due to early winter
snows, were trucked several hundred kilometres from their summer to winter
ranges in the 1960s led to many animals being disoriented in their attempts to
return to the summer ranges in spring. They were not able to follow their
traditional migration routes (Espmark 1970). This suggested that initially the
learned navigation necessary for movements between the seasonal ranges
required experiene through annual travel between seasonal ranges. Familiarity
with features of the terrain appeared to be important to enable the reindeer to
follow specific routes. Alternatively, experience with caribou in Alaska supports another hypothesis, that caribou do not learn specific migration routes to
the calving grounds based on features of the terrain. Rather, they navigate
during migration to a fixed point, with local variation imposed by environmental features that restrict or channel movements. This occurred following
the heavy snow winter of 1981–82 when many caribou of the Porcupine Herd
continued their autumn migration to the south and west, well beyond their
traditional wintering areas (Russell et al. 1993). The new wintering area

overlapped with that of a smaller caribou herd. Some animals in both herds were radio-collared, and their movements could be tracked. When spring arrived the Porcupine Herd animals headed north over completely unfamiliar terrain, and returned directly to their traditional calving grounds. What navigational cues are used by migrating caribou is not known. However, in this case familiarity with the terrain, at least over the first portions of the route travelled, was certainly not a factor. The important point is that the ability to navigate is presumably innate, but its use by the caribou appears to be a learned response to their environment, and young animals learn to use navigational cues in the company of adults.

Social learning has been suggested also as a means whereby young animals learn to forage selectively. Among caribou and reindeer, the close bonding of female and young results in their close proximity as the young are beginning to forage. In winter, the female shares the feeding crater she has dug through the snow with her offspring, but not with other animals in the group (Shea 1979). The young animal benefits from the adult female's experience in selection of specific feeding sites and may learn in this way to select the more favourable foraging sites and palatable plants on which to forage. Association with the group may assist in learning to identify broader feeding areas. Selection for individual plant species and parts, however, may also be tied to the nutritional–physiological requirements of the developing animal, as well as being an inherent attribute. In experimental work in Alaska, however, with a tame reindeer orphaned at birth, we observed that its selection of forage at over one year of age was nearly identical to that of free-ranging reindeer present in the same area, but with which it had not associated. I have also observed that caribou calves raised in captivity in isolation from their dams show a high preference for lichens when first offered to them, which is another example of inherent responses for selection of forage. It is reasonable to hypothesise that the role of social learning in development of foraging efficiency among caribou and reindeer is most likely to be tied to learning to recognise favourable microhabitats while in association with their dams, and at a larger habitat scale with the foraging group, rather than in regard to information about specific food material.

Social learning by individual caribou is also strongly associated with the hierarchical structure within social groups. Although group structure does not remain fixed, with the exception of female–young bonds, individuals must learn their own social status in relation to other individuals in the group. As individuals are exchanged between groups there is a short-term testing and learning of the constantly changing social hierarchies. Although body size, antler size, and seasonal sexual drives are important factors in determining hierarchical structures, learning the behaviours associated with establishing an individual's status with relation to others, and learning to interpret the behaviours of others in the group, are of critical importance (Espmark 1964).

Further, play provides a social context for learning social behaviour. Chasing, mutual threat displays, kicking and mounting are common among young caribou during the first summer after birth (Lent 1974). Play involves learning to use inherent patterns of behaviour employed by adults as well as replications of oberved adult behaviour (Fagen 1974). Skogland (1989), in investigations in Norway, observed less play among young caribou in a predator-free environment and suggested that this supported the hypothesis that the risk of predation is a factor promoting social living. The threat of predation, while a stimulus to social living, is a factor promoting aspects of play behaviour that yield benefits for predator avoidance. Alarm responses of caribou to the approach of predators or other strange animals include an alert response in which the animal that has made the observation looks directly at the source of disturbance and spreads it hind legs, usually also urinating (Pruitt 1960). Other animals in the group are then alerted to the disturbance and the group responds accordingly. The first animal in the group to initiate flight often does an 'excitation' leap from its hind legs before running off. Wolves target animals within the group that delay in response to this signal and thus show aberrant behaviour. This exerts a selection for surviving young to adopt adult behaviour. Even though behaviour such as the excitation leap may be innate, young animals are stimulated to employ the behaviour through observation of the context within which they are used by adults.

The muskox

Compared with caribou it may seem that there is considerably less opportunity among muskoxen for social learning. For example, parturient muskox females do not normally isolate themselves from their social group when giving birth, as do caribou. In caribou, this isolation from the group enables the cow and calf to identify accurately the unique characteristics of the odour and vocalisations of one another, and is the basis for a strong female–young bond. Thus, in muskoxen there is less opportunity for development of such a strong bond between the mother and her offspring. Whereas female muskoxen and their calves readily identify one another at times of nursing, calves generally socialise with other calves in the group more than with their own mother or other muskoxen in the group (Gray 1987). Allo-nursing and adoption of young, which have been reported among muskoxen in the wild (Tiplady 1990), would be more likely to occur under this behavioural regime than among caribou, and it has not been reported among caribou. Female and young muskoxen and some of the adult males live in cohesive social groups, and learning of hierarchical structures within the group is a dominant aspect of social learning within the species.

Muskoxen are not normally migratory; they have well-defined home ranges, but may move moderate distances between seasonal ranges. Fidelity to home ranges is stronger in females than males. In newly established populations of muskoxen in Alaska resulting from translocations into unoccupied habitats (Klein 1988), male groups tend to be dispersers from core areas of habitat. They then frequently locate in new areas of suitable habitat (Reynolds 1998). These males may return during the rutting period to traditional ranges to challenge herd bulls in female groups and, if successful, may lead groups of females and young to the newly found habitat. Hence, we may hypothesise that the home ranges of these female groups are either altered or expanded through learning from association with pioneering males.

The stationary group defence of muskoxen against wolves requires that these less visually alert animals maintain closer grouping in their daily activities than caribou. Young muskoxen are much more vulnerable to predation, without the protection offered by adults, than are caribou. Wolves are more frequently successful in their attacks on muskoxen when they can induce a flight response rather than when confronted by the stationary defence. In such cases calves are the most likely victims (Mech 1988). As a consequence, adult muskoxen, both males and females, butt calves and occasionally may hook them with their horns, if they are slow to move into the centre of the protection offered by the adults when under attack by wolves. Adult muskoxen, therefore, play a more direct role in the social learning of the predator avoidance behaviour of their young than do caribou. This behaviour has been observed in Greenland when using sled dogs as surrogate wolves to facilitate darting the adult muskoxen to attach identifying markers (see Figure 7.2) (Clausen *et al.* 1984).

The sharp, upturned horns of adult muskoxen are effective weapons against predators, but they pose a higher risk of serious injury in aggressive intraspecific interactions than do antlers. Young animals must learn to avoid close proximity to the heads of adults to avoid the head swing annoyance response and possible injury from adult horns. The avoidance of such injury continues to be an important element in the within-group behaviour of animals throughout their lives, and may account for the less frequently challenged dominance–subordinance relationships within muskox groups than in caribou (pers. obs.).

Play behaviour of muskox calves, as is the case with caribou, includes testing of cohort members and refining adult behaviours. In muskoxen these include head butting, chasing, mounting, 'king of the castle', gland rubbing, and flanked running or standing. There is a greater variety in the play behaviours of young muskoxen than caribou, which is probably a reflection of the stronger fidelity of individuals to the group in muskoxen in contrast to caribou. One of the factors adding to the complexity of behaviour in the muskox group is the presence of a dominant herd bull, and often a second subdominant bull in groups consisting primarily of cows and young animals. These structures are much more fixed than among caribou groups, although females and young move between groups and dominant bulls may be challenged by other bulls, but this is primarily during the rut (Gray 1987).

The arctic hare

Sociality among arctic hares differs substantially in a variety of ways from that of caribou and muskoxen. For example, the young nursing animals do not initially join the adult group structure. Additionally, unlike ruminants that forage, ruminate and rest sequentially throughout a 24 hour day, hares, having small stomachs, forage selectively and intensively until the stomach is full and then rest for an extended period while digestion takes place. The feeding and resting cycles of hares are usually synchronised, and in summer show a diurnal pattern of a single feeding bout at night when the arctic sun is low. The animals then remain largely inactive during the remainder of the day.

White, adult arctic hares in the High Arctic in summer employ two distinct strategies for predation avoidance. They aggregate at night when the sun is low into small groups to forage where plant productivity and species composition are favourable. After several hours of selective foraging these groups disband and the hares return to individual hides on rocky hill slopes. Adult hares, in the high arctic summer, therefore, seek to avoid predation by using both social and solitary behaviour. Females bear their young in seclusion and leave the young in hiding for extended periods between nursing bouts.

The opportunity for social learning by young hares is through interaction with litter mates or their dams, but little is known of these interactions. Young hares do not join the adult aggregations at the end of summer, but rather initially form their own juvenile aggregations (Parker 1977). Presumably social learning is involved in development of hierarchical structures within the group. Mixing of adult and juvenile groups occurs later as winter progresses (Parker 1982). Separation of adult and young hares during most of their first summer, which largely eliminates opportunities for young to learn from adults, presumably is necessitated by the limited running ability of the young hares and the large litter size (Parker 1977), which limits opportunity for female–young bonds within a social context. The behavioural strategy of hiding the young away from the female is obviously functional. In contrast to young caribou and muskoxen, the young hares reach adult body size by the beginning of winter and do not then require the association and protection offered by continued female–young bonding that characterises the early development of both caribou and muskoxen.

There is an increasing degree of sociality among arctic hares in winter from their southern distribution into the High Arctic, and this appears to be a function of decreasing availability of vegetative cover from south to north. In this regard, they mirror the trend among other Leporids (Tapper 1987), as well as among ungulates (Hirth 1977), to be solitary where cover is dense and to be social in open habitats where cover is lacking, but where visibility is good .

Sociality among hares in the High Arctic yields benefits in predator avoidance but there are associated costs that include constraints on optimal foraging and the time and energy involved in learning and maintaining hierarchical structures within the group. The presence of conspecifics nearby, increases rates of interactions, and animals 'test' one another as they attempt to move up the hierarchy. Nonetheless, we may hypothesise that in hares as in ungulates, the cost of group living is minimised through a well defined hierarchical structure, and rapid learning by individuals of their place within the structure.

Conclusions

Sociality in the open terrain of the Arctic appears to have been adopted by caribou, muskoxen and arctic hares as a means of minimising risks of predation. We may hypothesise that individuals of all three species must learn group hierarchical structures and associated behaviours. However, differences exist between these species in their requirements and opportunities for social learning. In caribou, social learning facilitates migration, location of specific calving areas, and intra-group responses to predators. The larger bodied, more sedentary muskoxen, which possess sharp upturned horns as weaponry, employ a group defence with counter attack against the wolf, their primary predator. This lifestyle results in strong group integrity and necessitates that adults aggressively encourage the young to take advantage of the protection offered by adults, which, along with intra-specific dominance interactions, requires constraint in the use of horns to avoid injury to group members. Arctic hares are social in winter throughout most of their distribution, as well as in summer in the High Arctic. In the High Arctic in summer, the conspicuous, white adult hares form small social groups during foraging bouts, but disperse to individual hides when inactive. The foraging group strategy for predator avoidance requires social learning of hierarchical status, and alarm and escape responses. The young hares, although precocial, do not accompany foraging adults that respond appropriately to predators, for example; the leverets remain hidden at a rendezvous site where they are visited by their dam to be nursed. This behaviour is, nevertheless, functional because of the rapid growth of the young to adult body size in their first summer, in contrast to caribou and muskoxen.

References

Baskin, L. M. (1970). Reindeer ecology and behavior. *Nauk,* Moscow, USSR Academy of Science, 149 pp. (English translation).

Calef, G. (1981). *Caribou and the Barren-lands.* Canadian Arctic Resources Committee. Toronto: Firefly Books Ltd, 176 pp.

Clausen, B., Hjort, P., Strandgaard, H. and Sorensen, P. L. (1984). Immobilization and tagging of muskoxen (*Ovibos moschatus*) in Jameson Land, Northeast Greenland. *J. Wildl. Dis.,* **20,** 141–5.

Duquette, L. S. (1984). Patterns of activity and their implications to the energy budget of migrating caribou. M.Sc. thesis, University of Alaska, Fairbanks, 95 pp.

Elgar, M. A. (1989). Predator vigilance and group size in mammals and birds: a critical review of the emperical evidence. *Biol. Rev.,* **64,** 13–33.

Espmark, Y. (1964). Studies in dominance-subordination relationship in a group of semidomestic reindeer. *Rangifer tarandus* L. *Anim. Behav.,* **12,** 420–6.

Espmark, Y. (1970). Abnormal migratory behavior in Swedish reindeer. *Arctic,* **23,** 199–200.

Estes, R. D. (1974). Social organization of the African Bovidae. In *The Behaviour of Ungulates and its Relation to Management,* ed. V. Geist and F. Walther, pp. 166–205. Morges, Switzerland: International Union for the Conservation of Nature.

Fagen, R. M. (1974). Selective and evolutionary aspects of animal play. *Am. Nat.,* **108,** 850–8.

Geist, V. and Walther, F. (eds.) (1974). *The Behaviour of Ungulates and its Relation to Management.* Morges, Switzerland: Papers of an International Symposium, International Union for the Conservation of Nature.

Gray, D. R. (1987). *The Muskoxen of Polar Bear Pass.* Ontario, Canada: Fitzhenry and Whiteside, 191 pp.

Hamilton, W. D. (1971). Geometry for the selfish herd. *J. Theor. Biol.,* **31,** 295–311.

Hirth, D. H. (1977). Social behavior of white-tailed deer in relation to habitat. *Wildl. Monogr.,* **53,** 55 pp.

Hofmann, R. R. (1983). Adaptive changes of gastric and intestinal morphology in response to different fibre content in ruminant diets. In *Dietary Fibre in Human and Animal Nutrition,* ed. L. Bell and G. Wallace, pp. 51–58. Wellington: Royal Society New Zealand.

Jarman, P. J. (1974). The social organization of antelopes in relation to their ecology. *Behaviour,* **58,** 215–67.

Klein, D. R. (1988). The establishment of muskox populations by translocation. In *Translocation of Wild Animals,* ed. L. Nielsen and R. D. Brown, pp. 298–318. Kingsville, TX: The Wisconsin Humane Society and The Caesar Klegerg Wildlife Research Institute.

Klein, D. R. (1991). Comparative ecological and behavioral adaptations of *Ovibos moschatus* and *Rangifer tarandus. Rangifer,* **12,** 47–55.

Klein, D. R. and Bay, C. (1994). Resource partitioning by mammalian herbivores in the high Arctic. *Oecologia,* **97** 439–50.

Klein, D. R., Meldgaard, M. and Fancy, S. G. (1987). Factors determining leg length in *Rangifer tarandus. J. Mammal,* **68**, 642–55.

Klopfer, P. H. (1970). Sensory physiology and esthetics. *Am. Sci.,* **58**, 399–403.

Lent, P. C. (1974). Mother–infant relationship in ungulates. In *The Behavior of Ungulates and its Relationship to Management,* ed. V. Geist and F. Walther, pp. 14–54. IUCN Publ. New Ser. 24.

Lindlof, B. (1978). Aggressive dominance rank in relation to feeding by the European hare. *Viltrevy,* **10**, 146–57.

Mech, L. D. (1988). *The Arctic Wolf: Living With the Pack.* Stillwater, MN: Voyageur Press.

Monaghan, P. and Metcalfe, N. B. (1985). Group foraging in wild brown hares: effects of resource distribution and social status. *Anim. Behav.,* **33**, 993–9.

Parker, G. R. (1977). Morphology, reproduction, diet, and behavior of the arctic hare (*Lepus arcticus monstrabilis*) on Axel Heiberg Island, Northwest Territories. *Can. Field Nat.,* **91**, 8–18.

Parker, G. R. (1982). Hordes of hopping hares: an Arctic enigma. *Can. Geogr.,* **102**, 30–3.

Pruitt, W. O., Jr. (1960). Behavior of the barren-ground caribou. *Biol. Pap. Univ. Alaska,* **3**, 1–43.

Reynolds, P. C. B. (1998). Ecology of a reestablished population of muskoxen in northeastern Alaska. PhD thesis, University of Alaska, Fairbanks.

Roby, D. D. (1978). Behavioral patterns of barren-ground caribou of the Central Arctic herd adjacent to the Trans-Alaska oil pipeline. MSc thesis, University of Alaska, Fairbanks.

Russell, D. E., Martell, A. M. and Nixon, W. A. C. (1993). Range ecology of the Porcupine Caribou Herd in Canada. *Rangifer Special Issue,* No. 8, 167 pp.

Shea, J. C. (1979). Social behavior of wintering caribou in northwestern Alaska. MSc thesis, University of Alaska, Fairbanks, 112 pp.

Skogland, T. (1989) Comparative social organization of wild reindeer in relation to food, mates and predator avoidance. *Adv. Ecol. Ethol.,* No. 29, 74 pp.

Tapper, S. (1987). *The Brown Hare. Shire Natural History Series,* No. 20. Aylesbury, UK: Shire Publications Ltd. 24 pp.

Tiplady, B. A. (1990). Multiple nursing in free-living muskoxen, *Ovibos moschatus. Can. Field Nat.,* **104**, 450–4.

8

Transmission of olfactory information from mother to young in the European rabbit

Robyn Hudson, Benoist Schaal and Ágnes Bilkó

The European rabbit (*Oryctolagus cuniculus*) belongs to the order Lagomorpha, which comprises two families, the pikas or rock rabbits (Ochotonidae, 1 genus, *c.* 16 species) and the hares and rabbits (Leporidae, 11 genera, *c.* 46 species) (Fox 1994). The natural range of the European rabbit is essentially the Iberian Peninsula and northern Africa, although through translocation by humans and escape of domestic stock it has established itself in many countries, and reached plague proportions in Australia, New Zealand and parts of Britain (Thompson and King 1994). It is a herbivore which eats mainly grasses and forbes. However, it is not particularly selective, and depending on locality and season can adjust its diet to include a wide range of plant material (Rogers *et al.* 1994).

The European rabbit is the only Lagomorph known to burrow and to form stable social groups. Two to 20 animals may share a common underground burrow system which provides protection from harsh environments and from predators. Although the rabbit's lifestyle of communal grazing suggests opportunities for social learning of strategies optimising feeding, this may be restricted to transmission from mother to young (reviewed below), since adults of both sexes demonstrate high levels of intrasexual aggression (Cowan and Bell 1986).

Olfactory stimuli play a significant part in regulating most aspects of the lives of terrestrial mammals (Stoddart 1980, Vandenbergh 1983, Albone 1984, Brown and MacDonald 1985, Doty 1986). This is particularly true during early development. Not only are the chemical senses among the first to develop (Gottlieb 1971, Smotherman and Robinson 1990, Schaal and Orgeur 1992, Schaal *et al.* 1995a), but in the limited sensory world of the uterus or nest, they may have particular salience (Alberts 1981, Mair 1986, Schaal and Orgeur 1992, Lecanuet *et al.* 1995). The European rabbit is one of the best documented examples, with a large number of studies both in the wild and laboratory attesting to the importance of chemical signals in the social life of this gregarious species (Mykytowycz 1970, Mykytowycz *et al.* 1984, Bell 1980, 1985), and to the critical role of olfactory signals emanating from the mother for early survival (Schley 1977, 1981, Hudson and Distel 1983, 1995, Distel and Hudson 1985).

In this chapter we review recent findings suggesting that young rabbits may acquire a range of olfactory information from their mother by weaning age, and consider the possible functional significance of this in terms of their natural ecology. The chapter is divided into four main parts: 1. an introductory description of the mother–young relationship and general developmental context, 2. a consideration of pups' early olfactory competence, 3. a review of findings that they may learn a variety of odours associated with their mother and even develop enhanced perceptual sensitivity to them, and 4. a consideration of the biological relevance of these findings. A central theme is that the flow of olfactory information from mother to young can be best understood in the context of the rabbit's developmental ecology, the key feature of which is the unusually limited but highly efficient pattern of maternal care. Since to our knowledge there is little difference between wild and domestic rabbits in the behaviours discussed here, no distinction will be made between them (Bell 1984).

The developmental context

Maternal behaviour
Maternal behaviour in the rabbit represents a well organised series of events under tight circadian and hormonal control (Hudson and Distel 1982, 1989, González-Mariscal *et al.* 1994a, Jilge 1993, 1995, Hudson 1995, 1998, Hudson *et al.* 1995, 1996a,b, 1997). Towards the end of the approximately 31–day pregnancy, the doe normally digs a short nursery burrow which she lines with dried grass and fur pulled from her ventrum. Parturition is extremely rapid, usually lasting not more than 10 to 15 minutes for the birth of 10 or more pups (Fuchs and Dawood 1980, Hudson and Distel 1982, Hudson *et al.* 1995, in press). Immediately after giving birth the doe leaves the pups, closes the burrow entrance, and only returns to reopen it and nurse once approximately every 24 hours (Niethammer 1937, Deutsch 1957, Venge 1963, Zarrow *et al.* 1965, Lloyd and McCowan 1968, Kraft 1979, Hudson and Distel 1982, 1989; Broekhuizen and Mulder 1983, Jilge 1993, 1995, Hudson *et al.* 1996a).

The nursing visit is extremely short, lasting only about 3 to 4 minutes. On entering the nest the doe simply positions herself over the litter without giving the pups any direct behavioural assistance to suckle. Towards the end of nursing she deposits a few hard faecal pellets in the nest then jumps abruptly away from the pups, leaving them alone until the following day (Deutsch 1957,

Lincoln 1974, Kraft 1979, Hudson and Distel 1982, 1983, 1989, González-Mariscal *et al.* 1994a, Hudson *et al.* 1996a). She does not brood them, cleans them little if at all, and does not even retrieve pups which stray from the nest (reviewed in Hudson and Distel 1982, 1989).

Female rabbits typically enter postpartum oestrus and can be mated again almost immediately after giving birth (Lincoln 1974, Hudson *et al.* 1995, 1996a). This means, at least in the wild, that they are often both pregnant and lactating (Brambell 1944, Stephens 1953). While this enables does to raise several litters in a season, the sudden weaning of the young at about day 26 associated with preparation for the arrival of the next litter 4 to 5 days later means that pups have to make the transition to independent life with little or no direct assistance from their mother (Lincoln 1974, Hudson and Altbäcker 1994, Hudson *et al.* 1996b).

This extreme pattern of maternal care may be explained by the fact that rabbits are fugitive animals and heavily preyed upon (Rogers 1978, Delibes and Hiraldo 1981, Vitale 1989). Their main protection is to flee into the communal warren, which, with its many openings, offers some chance of escaping even predators able to pursue them underground. In such a situation the new-born young would be unable to escape, and it is probably for this reason, as well as to protect the pups from attack by other females (cf. Künkele 1992), that the doe constructs a separate and well disguised nursery burrow, the single opening of which she closes after each visit. Presumably to reduce the risk of predators locating the nest and trapping her and the pups there, the time spent with the young is kept to a minimum (Zarrow *et al.* 1965, Lincoln 1974, Hudson and Distel 1982, 1989).

Behaviour of the pups

For the pups, however, such limited care poses a range of problems, particularly as they are rather immature at birth. They are born naked, with eyes and outer ears sealed, and with poor motor coordination. By day 7 they are capable of limited orienting responses to auditory stimuli, and may also perceive light changes, although they only begin to open their eyes on day 9 or 10. They start to leave the nest when 13 to 18 days old, by which time they are able to maintain a stable body temperature and have much improved motor coordination (reviewed in Gottlieb 1971, Hudson and Distel 1982, 1989).

During the first and for most of the second week of postnatal life the pups depend entirely on the mother's milk for their nutritional needs. However, during the second week they start eating the faecal pellets deposited by her in the nest and also start nibbling the nest material. At first they simply bite through the long stalks, reducing the hay to a kind of rough chaff but by the middle of the third week clearly measurable amounts are consumed, and by the end of the fourth week complete independence is possible (Hudson and Distel 1982, Hudson and Altbäcker 1994, Hudson *et al.* 1996a,b, 1997).

Survival of the young under conditions in which they receive such limited care as nestlings and little direct assistance in making the transition to independent life at weaning is facilitated by what recent studies are showing to be an extensive range of olfactory cues provided by the mother.

Early olfactory function

Anatomically and physiologically new-born rabbits are well equipped to perceive and process olfactory information. Although, as in other mammals (Breipohl 1986, Brunjes and Frazier 1986, Meisami 1989), the olfactory system undergoes substantial postnatal development (Meisami *et al.* 1990, Stahl *et al.* 1990), the sensory epithelia and primary processing areas of both the accessory and main olfactory systems are well developed at birth (Hudson and Distel 1986, 1987, Meisami *et al.* 1990, Stahl *et al.* 1990). This, together with reports from other altricial species that foetal receptor neurons establish connections with the brain several days before term (Gesteland *et al.* 1982, Brunjes and Frazier 1986), suggest that the olfactory system is also functional *in utero* (cf. Schaal 1988a, Smotherman and Robinson 1990, Schaal and Orgeur 1992, Schaal *et al.* 1995a). The behavioural studies reported below confirm such early functional competence and in fact demonstrate that new-born rabbits depend completely for their survival on the ability to perceive and respond appropriately to olfactory cues provided by the mother.

The nipple-search pheromone
Rabbit pups are able to drink up to 25% of their body weight in the short time available each day (Lincoln 1974, Hudson and Distel 1982, Hudson *et al.* 1996a). Once the doe has settled to nurse, pups take only a few seconds to attach to the nipples (Lincoln 1974, Hudson and Distel 1983). Their search behaviour is highly stereotyped and is shown in response to any lactating doe at any time of day (Hudson and Distel 1983, 1984). While making rapid probing movements with the muzzle deep into the fur, pups move across the doe's belly with a sewing machine-like action until a nipple is reached. Surprisingly, they do not remain on one nipple but change them frequently, repeating the whole search sequence several times even though this reduces the actual time they spend on nipples to an average of only about 110 seconds per nursing (Drewett *et al.* 1982, Hudson and Distel 1983).

By investigating the cues governing this effective orienting behaviour it has been shown that an odour on the doe's belly, the so-called nipple-search pheromone, is essential for the release and maintenance of searching, for rapidly guiding pups to nipples and for nipple attachment (Schley 1977, 1981, Hudson and Distel 1983, Distel and Hudson 1985). Production of the pheromone is under hormonal control, with the sequential administration of estradiol, progesterone and prolactin to non-pheromone producing does stimulating emission within a few days (Hudson *et al.* 1990, González-Mariscal *et al.* 1994b, Hudson and Distel 1995). The pups are very sensitive to these cues, which are present not only on the doe's ventrum but also in the milk (Keil *et al.* 1990). Testing the reaction of pups to fresh milk presented on a glass rod, it was found that even milk diluted 10 000-fold elicits significantly more searching and grasping responses than cow's milk or other odorants. This dependence on maternal olfactory cues explains why new-born rabbits are so difficult to raise by hand (Appel *et al.* 1971, Schley 1981) and why they are completely unable to suckle and will starve if made anosmic (Schley 1977, 1981, Distel and Hudson 1985, Hudson and Distel 1986).

Since the reaction of mammals to pheromones or pheromone-like substances is usually at least partly dependent on experience (Beauchamp *et al.* 1976), it is notable that rabbit pups are able to respond appropriately to their mother at the very first nursing. Even pups delivered by Caesarean section one day before term respond to a lactating doe with normal search and suckling behaviour (Hudson 1985). However, this does not exclude the possibility that the response is dependent on prenatal experience of chemical characteristics of the uterine environment. In fact, this might even be considered likely given the steep rise in pheromone emission in late pregnancy (Hudson and Distel 1984), and given reports that rabbit pups are able to learn prenatally odour cues associated with their mother's diet (Bilkó *et al.* 1994, Semke *et al.* 1995, Coureaud *et al.* 1997; see below).

Early odour learning

The nipple-search pheromone – vital for the survival of the pups – is only one of a range of odours pups may encounter via their mother.

In the nest

Although the pups do not need to learn postnatally the cues governing nipple-search behaviour, they rapidly learn to associate other odours present on the mother with suckling. Pups nursed by a doe whose ventrum has been scented with an artificial odorant quickly learn to respond to this scent as to the pheromone itself. After only one such three-minute pairing, pups show the full search and suckling sequence when placed on the belly of a non-pheromone producing doe or even a female cat scented with the experimental odour. In contrast, control pups with no experience of the scent respond by crawling around or resting, and cannot be induced to grasp nipples (Hudson 1985, Hudson and Distel 1987, Kindermann *et al.* 1994). Perception of both the pheromone and the learned odour cues is mediated by the main rather than by the accessory olfactory system (Hudson and Distel 1986, 1987).

That this learning represents a form of olfactory imprinting (Hudson 1993a) is suggested by the finding that responsiveness to learned suckling odours is retained for some time without further experience (Hudson 1985). Just as pups raised without experience of the pheromone respond to it when tested on day 5, so pups conditioned on day 1 but either bottle fed or nursed by an unscented foster doe still show vigorous nipple-search behaviour in response to the learned odour when tested on day 5. This one-trial learning is restricted to a sensitive period and is only possible during the first three days postpartum. However, conditionability is maintained to at least day 5 in pups deprived of normal suckling experience by bottle feeding, although not in hand-raised pups allowed to search daily on a lactating doe (Kindermann *et al.* 1994).

Rabbit pups also learn odours arising from their mother's diet. During pregnancy and lactation the lab chow diet of does was supplemented with either juniper berries or thyme, both plants which form part of the diet of a population of wild rabbits under study in the Kiskunság National Park, Hungary (Kertész *et al.* 1993, Mátrai *et al.* 1998). At weaning on day 28, pups were caged individually and presented with a 'cafeteria', a food trough divided into three equal parts and containing equal volumes of lab chow, juniper berries or thyme arranged in random order (Figure 8.1). First food chosen and amount of each eaten in repeated test sessions during the first week post weaning demonstrated food choice by the pups to have been influenced by the diet of their mother. Pups raised by juniper-fed does ate significantly more juniper than pups from thyme-fed or control does, whereas pups from thyme-fed does ate significantly more thyme than the other two groups (Altbäcker *et al.* 1995). This learning is very robust and a current study in the Hungarian laboratory shows that pups raised from weaning without further experience of the experimental foods still demonstrate a preference for them when tested as adults six months later (Bilkó unpublished data).

In a series of cross-fostering experiments in which pups were either raised in control nests but nursed by juniper-fed mothers or raised in nests containing faecal pellets from juniper-fed mothers but nursed by control dams, the pups

Figure 8.1. 'Cafeteria' used for testing food choice in weanling rabbits; a food trough with three pots containing equal volumes of lab chow, dried thyme or juniper berries arranged in random order. On postnatal day 28, pups were separated from the mother, placed in individual cages and presented with the cafeteria at regular intervals. First food eaten and relative amounts of each food consumed in these test sessions were recorded during the first week post weaning (photograph by V. Altbäcker).

showed essentially the same degree of preference for juniper at weaning as pups from the first experiment (Bilkó *et al.* 1994). Even ambient juniper odour produced by placing the berries beneath a mesh floor in the nest is sufficient to induce the preference, demonstrating the learning to be mediated, at least in part, by olfaction (Altbäcker unpublished data).

In utero

Consistent with findings in rats and sheep (Hepper 1988, Schaal and Orgeur 1992, Schaal *et al.* 1995b), intra-uterine learning of odours associated with the mother's diet has also been demonstrated in rabbits. Pups exposed only *in utero* to juniper, and immediately after birth cross-fostered to control does, show as strong a preference for juniper as the pups described above receiving various forms of postnatal experience (Bilkó *et al.* 1994). Pups are able to express this learning soon after birth and when given a simple side preference test on day 1, choose the side scented with the odour of their mother's diet (Semke *et al.* 1995, Schaal and Coureaud unpublished observations). However, pups do not respond to experimental odours experienced *in utero* or in the nest with nipple-search behaviour although they can be conditioned to do so if these are paired with suckling (Hudson 1993b, Hudson unpublished data).

Experience-induced enhancement of olfactory sensitivity

The significance of early odour experience is reinforced by the finding that exposure *in utero* to odours associated with the mother's diet results in stimulus-specific enhancement in the sensitivity of the olfactory receptors. Electrical recordings were made from isolated olfactory epithelia of prenatally juniper-exposed and control pups during the application of juniper odour or a novel control odour, isoamyl acetate. Whereas no difference between epithelia of juniper-exposed and control pups was found in the response to isoamyl acetate, the response to juniper odour was significantly greater in the juniper-exposed pups than in controls (Semke *et al.* 1995, Hudson and Distel 1998).

Biological relevance

There can be little doubt as to the functional significance of the nipple-search pheromone – the rabbit's whole strategy of limited maternal care depends upon it, and possibly in the wild even more than in captivity. But what of the other odour cues pups encounter and learn via their mother? Are these simply inconsequential by-products of laboratory manipulations or do they reflect processes operating under natural conditions and representing a transfer of information from mother to young of real survival value?

Suckling

Although the nipple-search pheromone does not need to be learned post-natally, learning of odour cues in the context of suckling might be of functional significance in several respects. The ability of pups to learn odours encountered prenatally raises the possibility that the pheromone, or components of it, is also learned before birth and that odour learning during the early suckling period represents an extension of this process, enabling pups to gain rapid confirmation of the olfactory signals governing suckling in the final functional context (Kindermann *et al.* 1994).

Indeed, support for the idea that such chemosensory continuity between the pre- and postnatal environments might be of advantage for neonates (Schaal 1988a,b, Schaal and Orgeur 1992, Marlier *et al.* 1997, 1998) has

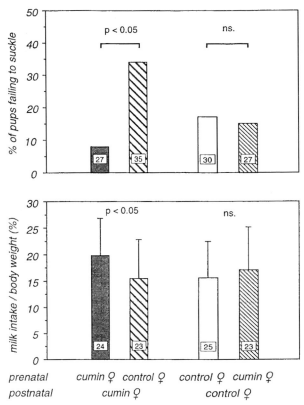

Figure 8.2. Suckling performance on day 1. Pups were divided into four groups: those subjected to a perinatal odour match, i.e. born to and raised by either a cumin-fed or a control doe, and those subjected to a perinatal odour mismatch, i.e. born to a control doe but nursed by a cumin-fed doe, or the reverse. The percentage of pups failing to obtain milk is shown above, and milk intake of those that succeeded in suckling is shown below. Pups were drawn from at least six litters per group. Means, standard deviations and numbers of pups are shown; groups were compared using the Cochran test.

recently been obtained for the rabbit. When presented with a choice between the odour of their amniotic/placental membranes and the odour of their mother's colostrum in a side preference test, new-born pups without postnatal ingestive experience do not orient differentially to the two odours (Coureaud *et al.* unpublished observations). Thus, both these substrates of the perinatal environment emit volatiles that are treated similarly by the pups on a chemosensory and/or motivational basis. Such an overlap may represent a means by which the mother transmits predictable cues to the young in preparation for the otherwise unpredictable postnatal environment.

This chemosensory continuity hypothesis has been further investigated by manipulating the predictability of the postnatal niche (Coureaud *et al.* 1997). New-born pups from does that had had cumin added to their diet during pregnancy and from control does were tested for their suckling performance when given either to control foster does or to foster does which continued to be fed the cumin additive. Pups faced with a perinatal odour mismatch – born to a control doe but nursed by a cumin-fed doe or the reverse – obtained significantly less milk during the first critical suckling episode than pups experiencing odour continuity – pups born to control or cumin-fed does and nursed by control or cumin-fed does, respectively (Figure 8.2). Thus, information gained *in utero* about the olfactory characteristic of the mother may help prepare pups for one of the most important postnatal tasks, the highly competitive race for nipples.

Longer-term consequences of such early odour learning are suggested by reports that in various species odour exposure early in development may affect filial attachment, mate preference or, as mentioned above, food choice (Mainardi *et al.* 1965, Carter and Marr 1970, Fillion and Blass 1986, Galef 1990, Hepper 1987, 1991, Moore *et al.* 1996). This is also consistent with the report that rabbit pups learn the specific odour of their mother's anal glands during the first postnatal week, and that this information may help them recognise her territory and so avoid being attacked by strange females when they leave the nursery burrow (Mykytowycz and Ward 1971).

Food choice

Longer-term consequences of early odour learning are also suggested by the ability of pups to acquire information about the diet of their mother before leaving the nest, and to use this in selecting foods at weaning. Moreover, there seem to be several apparently equally effective means by which this can be achieved – *in utero*, during nursing or via the doe's faecal pellets. Such redundancy may help ensure that less aromatic substances or their components transmitted differentially by these various routes are learned, and that pups can acquire a preference for a variety of foods eaten by their mother at different times during the 2-month gestation and nursing period (Bilkó *et al.* 1994).

First evidence that similar mechanisms might also operate in the wild comes from a study currently in progress in Hungary (Bilkó unpublished data). The juniper preference of wild rabbits taken from a region where rabbits are known to eat juniper (Mátrai *et al.* 1998) was tested as in the previous study demonstrating that prenatal experience is sufficient to induce a clear preference (Bilkó *et al.* 1994). Pups born in the wild, brought to the lab and raised by a lab chow-fed foster mother showed a clear juniper preference at weaning in contrast to second generation pups from a wild female bred in the lab without juniper experience (Figure 8.3). Although the sample is still small and needs to be extended by testing the response of wild-born pups taken from a non-juniper region, the results suggest that pups from the juniper region were exposed to juniper odour *in utero*, and that the memory of this carried over to influence their food choice at weaning almost a month later.

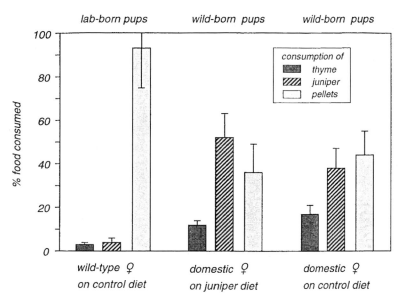

Figure 8.3. Food choice at weaning of three groups of wild-type rabbits: pups born in the laboratory and raised by a second generation, wild-type doe fed lab chow, pups born in Bugac juniper forest and raised by a domestic doe fed a juniper supplement, and pups born in Bugac juniper forest and raised by a domestic doe fed lab chow only. On postnatal day 28, pups were separated from the mother, placed in individual cages and presented at regular intervals with the cafeteria shown in Figure 8.1. The relative amounts of the three test foods eaten on the first day post weaning are shown (cf. Bilkó *et al.* 1994; $N = 5$ pups per group taken from three different litters).

Whether such information is actually of use to the rabbits is a further question. However, at the Hungarian study site the rabbits can choose from at least 40 different plant species (Kertész *et al.* 1993), many of which are of low nutritive value or are toxic to some degree (Tóth 1984). Thus, any information the young can acquire from their mother as to the most suitable foods in such a complex environment should help them survive the, in many respects, hazardous postweaning period (Hudson and Altbäcker 1994).

Some conclusions

There is increasing evidence that the learning of maternal odours is part of the natural repertoire of young rabbits and that, together with experience-induced enhancement of receptor sensitivity, this may function both to ensure sensory continuity across developmental stages and to detect biologically relevant odours. Furthermore, similar findings for a growing range of mammals indicate that such processes are relevant to an understanding of social ecology more generally, including our own. Thus, recent studies suggest that chemical stimuli present in the prenatal environment – for example, in the amniotic fluid – are perceived by the human foetus, and when encountered postnatally – for example, on the mother's skin or in her milk – elicit a positive response (Schaal 1988b, Schaal *et al.* 1995a,c, Marlier *et al.* 1997, 1998).

However, having documented such phenomena, two challenges remain – to demonstrate the relevance of experimental laboratory findings in ultimate, survival terms, and to understand the proximate mechanisms supporting them. With regard to the first point, bridging the gap between laboratory and field studies involves at least two steps – the demonstration that similar phenomena occur in animals living under natural conditions, and the demonstration that such phenomena have consequences for the survival of individuals or their progeny.

Regarding the second point, an understanding of the mechanisms underlying such phenomena is important for theoretical as well as basic neurobiological reasons. The ability of young, nonprimate species to acquire information in such a rapid and robust manner can help counterbalance more mentalistic accounts of mammalian social learning which so easily imply the operation of higher-order cognitive, conscious or even intentional processes. As demonstrated by the early learning of maternal odours in the rabbit, the potential for the transmission of information, even to very young nervous systems, may be considerable. However, here it is important to know what actually is learned. In the case of the rabbits, without testing the preferences of odour-exposed young in a variety of contexts – for example, in choosing resting sites or social partners – it is not possible to conclude that animals have acquired a preference relating specifically to foods. They may simply have come to regard stimuli associated with their mother or encountered in the nest as positive – or at least familiar – and when encountering these in a food-related context, express this as a food preference.

Prenatal odour learning is of particular interest with regard to learning mechanisms. As with the learning of ambient odours present in the nest, it is not yet clear what the nature of the reinforcer is supporting such learning – if indeed any is required. Related to this is the question whether the acquisition of responsiveness to ambient odours represents a form of incidental learning distinct from and supported by different mechanisms than the more familiar associative paradigms such as conditioning to suckling odours (Schaal and Orgeur 1992, Hudson 1993a,b).

Whatever the case, it would seem to make biological sense for young animals to be particularly receptive to stimuli associated with their mother.

Mothers are individuals who have succeeded in surviving to reproductive age and, having had time to learn features of their environment such as the most nutritious and least poisonous plants, represent a source of information which can be relied upon and taken advantage of, even by very immature nervous systems.

References

Alberts, J. R. (1981). Ontogeny of olfaction: reciprocal roles of sensation and behavior in the development of perception. In *Development of Perception: Psychobiological Perspectives*, ed. R. N. Aslin, J. R. Alberts and M. R. Petersen, pp. 321–54. New York: Academic Press.

Albone, E. S. (1984). *Mammalian Semiochemistry*. Chichester: Wiley.

Altbäcker, V., Hudson, R. and Bilkó, A. (1995). Rabbit mothers' diet influences pups' later food choice. *Ethology*, **99**, 107–16.

Appel, K., Busse, H., Schulz, K. and Wilk, W. (1971). Beitrag zur Handaufzucht von gnotobiotischen und SPF-Kaninchen. *Z. Versuchtierkd*, **13**, 282–309.

Beauchamp, G. K., Doty, R. L., Moulton, D. G. and Mugford, R. A. (1976). The pheromone concept in mammalian chemical communication: a critique. In *Mammalian Olfaction, Reproductive Processes and Behavior*, ed. R. L. Doty, pp. 143–60. New York: Academic Press.

Bell, D. J. (1980). Social olfaction in lagomorphs. *Symp. Zool. Soc. Lond.*, **45**, 141–64.

Bell, D. J. (1984). The behaviour of rabbits: implications for their laboratory management. In *The UFAW Handbook on Standards in Laboratory Animal Management*, pp. 151–62. Potters Bar.

Bell, D. J. (1985). The rabbits and hares: order Lagomorpha. In *Social Odours in Mammals*, ed. R. E. Brown and D. W. MacDonald, pp. 507–30. Oxford: Clarendon Press.

Bilkó, A., Altbäcker, V. and Hudson, R. (1994). Transmission of food preference in the rabbit: the means of information transfer. *Physiol. Behav.*, **56**, 907–12.

Brambell, F. W. R. (1944). The reproduction of the wild rabbit *Oryctolagus cuniculus* (L.). *Proc. Zool. Soc. Lond.*, **114–145**, 1–45.

Breipohl, W. (ed.) (1986). *Ontogeny of Olfaction*. Berlin: Springer Verlag.

Broekhuizen, S. and Mulder, J. L. (1983). Differences and similarities in nursing behaviour of hares and rabbits. *Acta Zool. Fennica*, **174**, 61–3.

Brown, R. E. and MacDonald, D. W. (eds.) (1985). *Social Odours in Mammals*. Oxford: Clarendon Press.

Brunjes, P. C. and Frazier, L. L. (1986). Maturation and plasticity in the olfactory system of vertebrates. *Brain Res. Rev.*, **11**, 1–45.

Carter, C. S. and Marr, J. N. (1970). Olfactory imprinting and age variables in the guinea pig. *Anim. Behav.*, **18**, 238–44.

Coureaud, G., Schaal, B., Orgeur, P., Hudson, R., Lebas, F. and Coudert, P. (1997). Perinatal odour disruption impairs neonatal milk intake in the rabbit. *Adv. Ethol.*, **32**, 134.

Cowan, D. P. and Bell, D. J. (1986). Leporid social behaviour and social organization. *Mammal Rev.*, **16**, 169–79.

Delibes, M. and Hiraldo, F. (1981). The rabbit as prey in the Iberian mediterranean ecosystem. In *Proceedings of the World Lagomorph Conference*, ed. K. Myers and C. D. McInnes, pp. 614–22. Guelph.

Deutsch, J. A. (1957). Nest building behaviour of domestic rabbits under semi-natural conditions. *Br. J. Anim. Behav.*, **5**, 53–4.

Distel, H. and Hudson, R. (1985). The contribution of the olfactory and tactile modalities to the performance of nipple-search behaviour in newborn rabbits. *J. Comp. Physiol. A*, **157**, 599–605.

Doty, R. L. (1986). Odor-guided behavior in mammals. *Experientia*, **42**, 257–71.

Drewett, R. F., Kendrick, K. M., Sanders, D. J. and Trew, A. M. (1982). A quantitative analysis of the feeding behavior of suckling rabbits. *Devel. Psychobiol.*, **15**, 25–32.

Fillion, T. J. and Blass, E. M. (1986). Infantile experience with suckling odors determines adult sexual behavior in male rats. *Science*, **231**, 729–31.

Fox, R. R. (1994). Taxonomy and genetics. In *The Biology of the Laboratory Rabbit*, 2nd edn., ed. P. J. Manning, D. H. Ringler and C. E. Newcomer, pp. 1–26. New York: Academic Press.

Fuchs, A. R. and Dawood, Y. (1980). Oxytocin release and uterine activation during parturition in rabbits. *Endocrinology*, **107**, 1117–26.

Galef, B. G. (1990). Necessary and sufficient conditions for communication of diet preferences by Norway rats. *Anim. Learn. Behav.*, **18**, 347–51.

Gesteland, R. C., Yancey, R. A. and Farbman, A. I. (1982). Development of olfactory receptor neuron selectivity in the rat fetus. *Neuroscience*, **7**, 3127–36.

González-Mariscal, G., Díaz-Sánchez, V., Melo, A. I., Beyer, C. and Rosenblatt, J. S. (1994a). Maternal behavior in New Zealand white rabbits: quantification of somatic events, motor patterns and steroid levels. *Physiol. Behav.*, **55**, 1081–9.

González-Mariscal, G., Chirino, R. and Hudson, R. (1994b). Prolactin stimulates emission

of nipple pheromone in ovariectomized New Zealand white rabbits. *Biol. Reprod.*, **50**, 373–6.

Gottlieb, G. 1971. Ontogenesis of sensory function in birds and mammals. In *The Biopsychology of Development*, ed. E. Tobach, L. R. Aronson and E. Shaw, pp. 67–128. New York: Academic Press.

Hepper, P. G. (1987). The amniotic fluid: an important priming role in kin recognition. *Anim. Behav.*, **35**, 1343–4.

Hepper, P. G. (1988). Adaptive fetal learning: prenatal exposure to garlic affects postnatal preferences. *Anim. Behav.*, **36**, 935–6.

Hepper, P. G. (1991). Recognising kin: ontogeny and classification. In *Kin Recognition*, ed. P. G. Hepper, pp. 259–88. Cambridge: Cambridge University Press.

Hudson, R. (1985). Do newborn rabbits learn the odor stimuli releasing nipple-search behavior? *Devel. Psychobiol.*, **18**, 575–85.

Hudson, R. (1993a). Olfactory imprinting. *Curr. Opinion. Neurobiol.*, **3**, 548–52.

Hudson, R. (1993b). Rapid odor learning in newborn rabbits: connecting sensory input to motor output. *German J. Psychol.*, **17**, 267–75.

Hudson, R. (1995). Chronoendocrinology of reproductive behavior in the female rabbit (*Oryctolagus cuniculus*). In *Bericht 9. Arbeitstagung über Haltung und Krankheiten der Kaninchen, Pelztiere und Heimtiere*, ed. S. Matthes, pp. 1–11.

Celle: Deutsche Veterinärmedizinische Gesellschaft.

Hudson, R. (1998). Potential of the newborn rabbit for circadian rhythms research. *Biol. Rhythms Res.*, **29**, 546–55.

Hudson, R. and Altbäcker, V. (1994). Development of feeding and food preference in the European rabbit: environmental and maturational determinants. In *Behavioral Aspects of Feeding: Basic and Applied Research in Mammals*, ed. B. G. Galef, M. Mainardi and P. Valsecchi, pp. 125–45. Chur: Harwood Academic Publishers.

Hudson, R. and Distel, H. (1982). The pattern of behaviour of rabbit pups in the nest. *Behaviour*, **79**, 255–71.

Hudson, R. and Distel, H. (1983). Nipple location by newborn rabbits: behavioural evidence for pheromonal guidance. *Behaviour*, **85**, 260–75.

Hudson, R. and Distel, H. (1984). Nipple-search pheromone in rabbits: dependence on season and reproductive state. *J. Comp. Physiol., A*, **155**, 13–17.

Hudson, R. and Distel, H. (1986). Pheromonal release of suckling in rabbits does not depend on the vomeronasal organ. *Physiol. Behav.*, **37**, 123–9.

Hudson, R. and Distel, H. (1987). Regional autonomy in the peripheral processing of odor signals in newborn rabbits. *Brain Res.*, **421**, 85–94.

Hudson, R. and Distel, H. (1989). The temporal pattern of suckling in rabbit pups: a model of circadian synchrony between mother and young. In *Development of Circadian Rhythmicity and Photoperiodism in Mammals*, ed. S. M. Reppert, pp. 83–102. Boston: Perinatology Press.

Hudson, R. and Distel, H. (1995). On the nature and action of the rabbit nipple-search pheromone: a review. In *Chemical Signals in Vertebrates VII*, ed. R. Apfelbach, D. Müller-Schwarze, K. Reuter and E. Weiler, pp. 223–32. Oxford: Elsevier Science Ltd.

Hudson, R. and Distel, H. (1998). Induced peripheral sensitivity in the developing vertebrate olfactory system. *Ann. New York Acad. Sci.*, **855**, 109–15.

Hudson, R., Cruz, Y., Lucio, R.A., Ninomiya, J. and Martínez-Gómez, M. (in press). Temporal and behavioral patterning of parturition in rabbits and rats. *Physiol. Behav.*

Hudson, R., González-Mariscal, G. and Beyer, C. (1990). Chin marking behavior, sexual receptivity, and pheromone emission in steroid-treated, ovariectomized rabbits. *Horm. Behav.*, **24**, 1–13.

Hudson, R., Müller, A. and Kennedy, G. (1995). Parturition in the rabbit is compromised by daytime nursing: the role of oxytocin. *Biol. Reprod.*, **53**, 519–24.

Hudson, R., Bilkó, A. and Altbäcker, V. (1996a). Nursing, weaning and the development of independent feeding in the rabbit. *Z. Säugetierkde*, **61**, 39–48.

Hudson, R., Schaal, B., Bilkó, A. and Altbäcker, V. (1996b). Just three minutes a day: the behaviour of young rabbits viewed in the context of limited maternal care. In *Proceedings of the 6th World Rabbit Congress, Vol. 2*, ed. F. Lebas, pp. 395–403. Toulouse: Lempedes.

Hudson, R., Schaal, B., Bilkó, A. and Altbäcker, V. (1997). 'Juste trois minutes par jour' ou des soins materels très restreints. *Cunuculture*, **24**, 253–60.

Jilge, B. (1993). The ontogeny of circadian rhythms in the rabbit. *J. Biol. Rhythms*, **8**, 247–60.

Jilge, B. (1995). The ontogeny of the rabbit's circadian rhythms without an external zeitgeber. *Physiol. Behav.*, **58**, 131–40.

Keil, W., Stralendorff, F. v and Hudson, R. (1990). A behavioral bioassay for analysis of rabbit nipple-search pheromone. *Physiol. Behav.*, **47**, 525–9.

Kertész, M., Szabó, J. and Altbäcker, V. (1993). The Bugac Rabbit Project. Part 1. Description of the study site and vegetation map. *Abstr. Bot.*, **17**, 187–96.

Kindermann, U. Hudson, R. and Distel, H. (1994). Learning of suckling odors by newborn rabbits declines with age and suckling experience. *Devel. Psychobiol.*, **27**, 111–22.

Kraft, R. (1979). Vergleichende Verhaltensstudien an Wild- und Hauskaninchen. I. Das Verhaltensinventar von Wild- und Hauskaninchen. *Z. Tierzüchtg. Züchtungsbiol.*, **95**, 140–62.

Künkele, J. (1992). Infanticide in wild rabbits (*Oryctolagus cuniculus*). *J. Mammal*, **73**, 317–20.

Lecanuet, J. P., Krasnegor, N. A., Fifer, W. and Smotherman, W. P. (eds.) (1995). *Prenatal Development: A Psychobiological Perspective*. Hillsdale: Lawrence Erlbaum.

Lincoln, D. W. (1974). Suckling: a time-constant in the nursing behavior of the rabbit. *Physiol. Behav.*, **13**, 711–14.

Lloyd, H. G. and McCowan, D. (1968). Some observations on the breeding burrows of the wild rabbit, *Oryctolagus cuniculus*, on the island of Skokholm. *J. Zool. (Lond.)*, **156**, 540–9.

Mainardi, D., Marsan, M. and Pasquali, A. (1965). Causation of sexual preference in the house mouse. The behavior of mice reared by parents whose odor was artificially altered. *Atti. Soc. Ital. Sci. Nat., Milano*, **104**, 325–38.

Mair, R. G. (1986). Ontogeny of the olfactory code. *Experientia*, **42**, 213–23.

Marlier, L., Schaal, B. and Soussignan, R. (1997). Orientation responses to biological odours in the human newborn: initial pattern and postnatal plasticity. *C. R. Acad. Sci., Paris/Life Sci.*, **320**, 999–1005.

Marlier, L., Schaal, B. and Soussignan, R. (1998). Neonatal responsiveness to the odor of amniotic and lacteal fluids: a test of perinatal chemosensory continuity. *Child. Devel.*, **69**, 611–23.

Mátrai, K., Altbäcker, V. and Hahn, I. (1998). Seasonal diets of rabbits and their browsing effect on juniper in Bugac Juniper Forest (Hungary). *Acta Theriol.*, **43**, 107–12.

Meisami, E. (1989). A proposed relationship between increases in the number of olfactory receptor neurons, convergence ratio and sensitivity in the developing rat. *Dev. Brain Res.*, **46**, 9–19.

Meisami, E., Louie, J., Hudson, R. and Distel, H. (1990). A morphometric comparison of the olfactory epithelium of newborn and weanling rabbits. *Cell Tissue Res.*, **262**, 89–97.

Moore, C. L., Jordan, L. and Wong, L. (1996). Early olfactory experience, novelty, and choice of sexual partner by male rats. *Physiol. Behav.*, **60**, 1361–7.

Mykytowycz, R. (1970). The role of skin glands in mammalian communication. In *Advances in Chemoreception. I. Communication by Chemical Senses*, ed. J. W. Johnson, D. G. Moulton and A. Turk, pp. 327–60. New York: Appleton-Century-Crofts.

Mykytowycz, R. and Ward, M. M. (1971). Some reactions of nestlings of the wild rabbit, *Oryctolagus cuniculus* (L.), when exposed to natural rabbit odours. *Forma Functio*, **4**, 137–48.

Mykytowycz, R., Goodrich, B. S. and Hesterman, E. R. (1984). Methodology employed in the studies of odour signals in wild rabbits, *Oryctolagus cuniculus*. *Acta Zool. Fenn.*, **171**, 71–5.

Niethammer, G. (1937). Ergebnisse von Markierungsversuchen an Wildkaninchen. *Z. Morphol. Ökol. Tiere*, **33**, 297–312.

Rogers, P. M., Arthur, C. P. and Soriguer, R. C. (1994). The rabbit in continental Europe. In *The European Rabbit: The History and Biology of a Successful Colonizer*, ed. H. V. Thompson and C. M. King, pp. 22–63. Oxford: Oxford University Press.

Rogers, R. M. (1978). Predator–prey relationship between rabbit and lynx in Southern Spain. *La Terre et la Vie*, **32**, 83–7.

Schaal, B. (1988a). Discontinuité natale et continuité chimiosensorielle: modèles animaux et hypothèses pour l'homme. *Ann. Biol.*, **27**, 1–41.

Schaal, B. (1988b). Olfaction in infants and children: developmental and functional perspectives. *Chem. Senses*, **13**, 145–90.

Schaal, B. and Orgeur, P. (1992). Olfaction in utero: can the rodent model be generalized? *Q. J. Exp. Psychol.*, **44**, 245–78.

Schaal, B. Orgeur, P. and Rognon, C. (1995a). Odor sensing in the human fetus: anatomical, functional, and chemoecological bases. In *Prenatal Development: A Psychobiological Perspective*, ed. J. P. Lecanuet, N. A. Krasnegor, W. Fifer and W. P. Smotherman, pp. 205–37. Hillsdale: Lawrence Erlbaum.

Schaal, B., Orgeur, P. and Arnould, C. (1995b). Olfactory preferences in newborn lambs: possible influence of prenatal experience. *Behaviour*, **132**, 351–65.

Schaal, B., Marlier, L. and Soussignan, R. (1995c). Responsiveness to the odour of amniotic fluid in the human neonate. *Biol. Neonate*, **67**, 397–406.

Schley, P. (1977). Die Ausschaltung des Geruchsvermögens und sein Einfluß auf das Saugverhalten von Jungkaninchen. *Berl. Münch. Tieräztl. Wschr.*, **90**, 382–5.

Schley, P. (1981). Geruchssinn und Saugverhalten bei Jungkaninchen. *Kleintier Praxis*, **26**, 261–3.

Semke, E., Distel, H. and Hudson, R. (1995). Specific enhancement of olfactory receptor sensitivity associated with foetal learning of food odors in the rabbit. *Naturwissenschaften*, **82**, 148–9.

Smotherman, W. P. and Robinson, S. R. (1990). Rat fetuses respond to chemical stimuli in gas phase. *Physiol. Behav.*, **47**, 863–8.

Stahl, B., Distel, H. and Hudson, R. (1990). Effects of reversible nare occlusion on the development of the olfactory epithelium in the rabbit nasal septum. *Cell Tissue Res.*, **259**, 275–81.

Stephens, M. N. (1953). Seasonal observations on the Wild Rabbit (*Oryctolagus cuniculus* L.) in West Wales. *Proc. Zool. Soc. Lond.*, **122**, 417–34.

Stoddart, D. M. (1980). *The Ecology of Vertebrate Olfaction*. London: Chapman and Hall.

Thompson, H. V. and King, C. M. (eds.) (1994). *The European Rabbit: The History and Biology of a Successful Colonizer*. Oxford: Oxford University Press.

Tóth, K. (ed.) (1984). *Tudományos Kutatások a Kiskunság Nemzeti Parkban, 1975–1984 (Research in Kiskunság National Park, 1975–1984)* (in Hungarian with English abstracts). Budapest: Hungexpo.

Vandenbergh, J. G. (ed.) (1983).

Pheromones and Reproduction in Mammals. New York: Academic Press.

Venge, O. (1963). The influence of nursing behaviour and milk production on early growth in rabbits. *Anim. Behav.,* **11**, 500–6.

Vitale, A. F. (1989). Changes in the anti-predator responses of wild rabbits, *Oryctolagus cuniculus* (L.) with age and experience. *Behaviour,* **110**, 47–60.

Zarrow, M. X., Denenberg, V. M. and Anderson, C. O. (1965). Rabbit: frequency of suckling in the pup. *Science,* **150**, 1835–6.

9

Social transfer of information in domestic animals

Donald M. Broom

Introduction

Most domestic animals are especially social species, indeed an ability to live socially and have an elaborate social structure is a prerequisite for satisfactory domestication. There are several questions concerning the fundamental likelihood that there is social transfer of information within groups of domestic animals such as cattle, sheep, pigs and dogs. Firstly, is information about the knowledge or actions of group members of value to an individual in the group? Secondly, have individuals the ability to detect and respond to actions or other cues from others, some of which may be very subtle? Thirdly, do individuals respond to group members and to what information? In some circumstances, in order to use information effectively it is necessary to recognise individuals and respond to them differentially so a fourth question is whether or not this is possible. Since it will often be difficult to evaluate potentially available information and to act on it appropriately, a fifth question is whether or not the animals have the cognitive ability to do so.

Information which might be obtained from conspecifics, which would normally be other members of the social group in domestic animals, can be of value in defence against predators, finding appropriate food, finding suitable places to nest, modifying the physical environment effectively, and selecting an appropriate mate or social companions. Some of these biologically important functions are often overlooked by the owners of domestic animals, but they are no less important to the animal because of this. Examples of such functions are given below as the answers to the other questions posed above are considered. Further examples are described and discussed by Nicol (1995).

Recognition of and response to other individuals

In most studies, individual recognition is evident because a behavioural response is shown, but there are also some physiological studies which provide information about the sensory and analytical basis for recognition. Unit recording studies by Kendrick and Baldwin (1987) and Kendrick (1992) show that sheep have cells in their temporal cortex whose response is specialised for the detection of sheep in general, sheep with horns, individual sheep, humans or dogs. The cell which responds to a familiar, individual sheep but not to unknown sheep could be responding to the set of familiar visual cues provided by that sheep, i.e. to familiarity, rather than to that sheep exclusively. However, in a behavioural discrimination task Kendrick *et al.* (1996) found that sheep could recognise the faces of individual familiar sheep and of individual unfamiliar sheep. The experimental studies of Alexander and Shillito (1977) show that ewes can recognise their own lambs using olfactory and visual cues.

When cows are living in a large group, they respond to different individuals in different ways and the responses are sufficiently rapid and sufficiently independent of the current behaviour of the cow which they are observing to make it clear that it is a consequence of individual recognition. For example, Potter and Broom (1987) reported that subordinate cows responded to the approach of a dominant individual by putting their heads into a feeding place in a feed rail or stepping into a cubicle. No such change in behaviour was shown when various other cows approached. In a study of the feral Chillingham cattle, Hall *et al.* (1988) reported that bulls responded by a behaviour change, including vocalisations, to individuals with whom their home range was shared. There were two types of lows, type 1 and type 2, and the bulls switched from one to the other when the dominant animal was detected. Young calves also respond differentially to individuals, for example spending much time over a period of more than a year associating with particular individuals (Broom and Leaver 1978) and such responses indicate that they still recognise one another after a period of separation lasting for 7–8 months (Bouissou and Hövels 1976). It could be said that a cow or calf responds in a particular way to another individual because it is categorised as being of a certain social type rather than because it is that particular individual. However, in some cases it is the only one of the individuals observed which elicits a certain reaction. An animal might: avoid contact including eye contact with one individual, defer to another when threatened, stand next to another when feeding and lie next to yet another. Hence, although it might be that the subject animal regards these animals as being members of four different categories, it seems more likely that it has a concept of them as individuals.

Studies of social interactions in other farm animal species also provide evidence for individual recognition. Subordinate sows in a group, fed with an electronic feeder which recognises the transponders worn by individual sows, will look out from a place where they are inconspicuous and approach the feeder only if certain potentially aggressive individuals are not in the vicinity of the feeder (Hunter *et al.* 1988). The subordinate sows may adopt either a fight-back or a submission and avoidance strategy (Mendl *et al.* 1992) but differential responses to individuals occur, probably according to the possibility of an attack and often before that individual has indicated that it recognises the presence of the subordinate.

Some examples of recognition with associated appropriate responses indicate impressive sensory ability, for example, dogs distinguishing individual humans, even identical twins, by odour after exposure to an item of clothing (Sommerville *et al.* 1990, Settle *et al.* 1994). The dog is able to retain information about the odour characteristics available on the item of clothing and to distinguish these same characteristics in one person out of a series of people sniffed. Hence this is recognition of an individual scent pattern with no necessity for any wider concept of that individual (see Johnston and Jernigan 1994), but the recognition of cohabiting humans by pet dogs is likely to involve elaborate concepts of those individual humans. The recognition and responses reported for a dog in a recent study by Sheldrake and Smart (1996), if correct, are particularly remarkable. A pet dog at home and its owner at work were video-recorded simultaneously. The owner worked at various distances from home and might return at any time during a three hour period but when the owner's behaviour indicated the intention of returning, the dog stood up and walked to the door. Since the frequency with which the dog showed this behaviour at other times was low, some form of communication between owner and dog is indicated.

Social transfer of information can occur accidentally with no cognitive ability required from the source animal and information processing in the recipient animal which, at its simplest, may involve associating some recognisable response to danger or food identifier with appreciation that a conspecific is carrying out an action which results in the cue. On the other hand, considerable cognitive ability is required when some information is transferred. Two examples, plus those described later in this chapter, should suffice to show that domestic animals have considerable cognitive ability and hence that social transfer of complex information could occur. Cattle and pigs are sometimes fed individually using systems in which the animal wears a transponder which activates a computer-controlled feed delivery. The animals learn rapidly to operate these systems and then, unless the design is very good, learn how to obtain a little extra food (Fraser and Broom 1990, p. 23). One sow, whose transponder was on a collar which triggered a once daily food drop in an electronic sow feeder, found a collar which had come off another sow and regularly obtained an extra meal by carrying it through the sow-feeder. It is likely that the sow initially carried the extra collar through the sow-feeder unit by accident but, on receiving extra food in the feeder, put down the collar, ate the food, picked up the collar again, left the feeder and put down the collar until it was required again later for an extra meal. A second example is of flocks of sheep studied in alpine pastures by Favre (1975). Many small areas of pasture separated by steep and hence energetically expensive paths were exploited by the sheep. Once a pasture area had been grazed, sheep seldom returned until regrowth had occurred about three weeks later, so they remembered where they had grazed and husbanded their food resources effectively without wasting energy.

Social facilitation

Social facilitation is defined by Fraser and Broom (1990) as behaviour which is initiated or increased in rate or frequency by the presence of another animal carrying out that behaviour. Galef (1988) suggests that contagious behaviour is a more precise name, but since dictionary definitions of 'contagious' refer to communication by contact as the original meaning or as one of the meanings, this use of the word is not more precise. The key point is that the behaviour performed by the second animal is the same as that performed by the first. However, there is a range of complexity of social facilitation. Indeed, as Nicol (1989) pointed out, it is possible for the frequency of an action in one individual to be increased by the mere proximity of another individual. She observed that preening by hens was stimulated by the presence of other hens which may or may not be preening. Hence this is not social facilitation according to the above definition unless the preening behaviour, rather than just the presence of the first hen, is necessary for the increase in preening in the second hen. This distinction was clear in the work of Tolman (1968) on social facilitation of pecking in chicks. The chicks pecked when an active companion was present but pecked most when that companion was seen or heard pecking.

An experimental study which demonstrated some of the consequences of social facilitation (Benham 1982) involved 31 cows whose behaviour was recorded throughout daylight hours for 15 days. After 5 days the herd was divided into a group of 16 with longer grazing duration and a group of 15 with shorter grazing durations; the long duration grazers were removed for 5 days. Hence the behaviour of the short duration grazers with and without the influence of the long duration grazers could be compared. In the presence of the long duration grazers, the short duration grazers spent less time lying, lay down later and stood up faster than they did in their absence. When the long duration grazers were still grazing, they had a social facilitatory effect on the other animals for these finished grazing and stood still or walked around, lying down only when the long duration grazers did so. Cows feeding in groups generally eat more than cows fed individually and heifers also show social facilitation of mounting behaviour (Phillips 1993, p. 129). Pigs also show strong social facilitation of feeding behaviour (Hsia and Wood-Gush 1984).

The practical importance of social facilitation is also evident in studies of calves by Barton (1983). Calves which were fed from teats connected to a milk reservoir were found to go to the teat more often when there was another calf sucking the teat. If several teats were present, close together but positioned so that young calves could suck simultaneously, the social facilitation effects resulted in greater uniformity of weight, more with five teats than with two and more with two teats than with one. In a further experimental study of social facilitation (Barton and Broom 1985), calves confined alone in open-sided pens drank 5.5 l when offered unlimited milk but if a hungry calf was then put into the next pen with milk provided, the first calf drank 7.5 l. If the calves were put in a pen together with a muzzled hungry calf which attempted to get to a teat which would normally supply it with milk, the first calf drank 9.2 l.

Social facilitation also appears to be important in predatory attacks by dogs. If several dogs are in the vicinity of a flock of sheep and one attacks the sheep, the other group members are more likely to attack. This also happens (Borchelt *et al.* 1983, Podberscek 1994) in cases where young humans are attacked by dogs. When one dog in a group of 15 bit a girl, the others also did so and in a case where two Rottweilers killed a girl, first one reacted to her falling by attacking and then the other did so.

Young mammals learning from their mothers about food

Food preferences in young mammals such as rats and sheep (Galef and Sherry, 1973, Nolte and Provenza 1992) can be affected by the content of the mother's milk, which is itself affected by the mother's diet. The food which mothers eat must be known to their offspring by olfactory and visual means and there are several examples of the food preferences of offspring being altered as a consequence. Wyrwicka (1978) trained a mother cat to eat banana and potato which would not normally be eaten. The kittens of this cat also ate these foods when tested alone at 9–27 weeks. Similarly, Lobato *et al.* (1980) found that lambs would eat more of blocks containing molasses and urea if their mothers had eaten them in their presence than if they were not exposed to these blocks until after weaning.

It seems that the effects on a young lamb's future food preferences of seeing

Table 9.1. *Effects of experience on consumption of strange foods by lambs*

	Consumption
Lamb never exposed to food or to any sheep eating it	x
Lamb exposed to ewe, not mother, eating food	$2x$
Lamb exposed to mother eating food	$4x$
Lamb exposed to mother eating food and eats the food itself	$8x$

Data from Thorhallsdottir *et al.* (1987).

x is the amount of a strange food consumed by a lamb which had never encountered it before.

the mother eat certain foods does not require that the lamb eat at the same time as the mother. Lambs kept at pasture do not normally accept whole grain wheat as food when it is offered to them. However, if lambs have been with their mothers during the suckling period and the mothers, but not the lambs, have eaten wheat, then those lambs accept wheat far more readily when it is offered to them many weeks later (Lynch *et al.* 1983). When lambs were with mothers which ate wheat for only five hours when the lambs were six weeks of age, in a consumption test at 34 months of age, the animals ate a mean of 357 g of wheat per day, but if they had been exposed to the wheat for the same period at the same age but without their mothers, consumption in the test was only 38 g per day. A comparison of the effects of various kinds of actual and observational experience of food by Thorhallsdottir *et al.* (1987) is shown in Table 9.1.

It seems likely that the kind of transmission of information about food which is indicated by these experimental studies of sheep also occurs in the grazing situation because Youssef *et al.* (1994) found that lambs avoided the pasture grass red fescue if the mothers avoided it.

Experiments involving demonstrator and observer animals

As part of the training of animals to perform tricks, they are often shown a sequence of actions carried out by a human or by a conspecific in the hope that they will copy the action. Experimental investigations of such procedures were carried out by Adler (1955) and by Adler and Adler (1977) using cats and dogs. Those cats or dogs which had observed a trained conspecific pull a cart by a ribbon and had been rewarded with food, performed the same action much faster than others which had not observed the trained animals.

Whenever the behaviour of an individual is altered by observing another

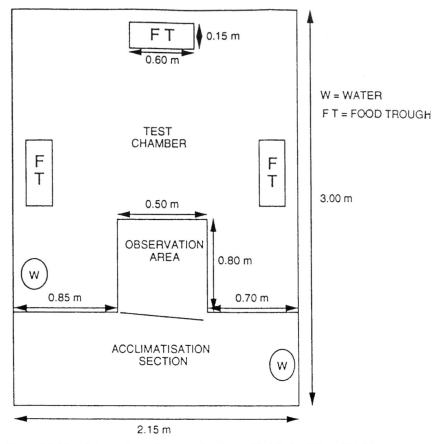

Figure 9.1. Plan of observation and test area for pigs used by Nicol and Pope (1994b).

animal carrying out an action, it is possible that any increased likelihood that a behaviour is shown is a result of reduced fear in the experimental situation rather than social transfer of information. Johnson *et al.* (1986) reported that chickens acquired a key peck response more readily if they spent some time observing another chicken pecking at a key when a light on the key was illuminated and then receiving food. In order to control for the possibility that this effect was due to fear reduction in the observer bird, Nicol and Pope (1992) repeated the study with a trained demonstrator hen and a control hen which was untrained. The observer hen acquired the key-pecking response much faster if the demonstrator hen was trained than if it was untrained. An enhancement of key-pecking in observer hens was also seen in a flock of eight hens which included a trained bird. The presence of an additional food source did not affect the result but the increase in key-pecking was greater if the demonstrator bird was a dominant individual (Nicol and Pope 1994a).

Similar studies, involving demonstrators and a food reward, have been conducted using pigs and horses. Nicol and Pope (1994b) used the observation and test situations shown in Figure 9.1. A young observer pig was put into the room at normal feeding time. There it observed a demonstrator eating standard food from a shallow trough. The demonstrator did not eat from the other troughs because they contained sawdust. In the test situation, the pigs were able to eat from any of the troughs, each of which contained normal food with a layer of sawdust over it. The observer pigs fed earlier and for longer from the demonstrated trough. However, in another experiment, there was no increase in consumption of blue mint-flavoured food if the demonstrator pig was observed to eat it. A similar result was obtained by Clarke *et al.* (1996), whose observer horses saw a demonstrator walk 13 m in an arena and select a black and white bucket but not a yellow bucket on 20 occasions. When compared with horses which had not had this experience, observers had a shorter approach latency (18 s versus 119 s) and a shorter latency to eat (35 s versus 181 s) but no preference for the black and white bucket.

Discussion

A range of studies indicate that domestic animals have various strategies for controlling their lives and that in order to have this control they use their considerable sensory and cognitive ability. Each of the species which has been investigated carefully has been found to be able to attend to the activities of others and to learn from them. It is likely that most of the behaviour in these very social species is affected by social transfer of information.

Sceptics ask, however, whether the animals are indeed aware of what is happening when their behaviour is changed and what is the proof that cows, sheep, pigs, dogs, etc. acquire information from other members of their social group. In answering this, some basic scientific attitudes may be questioned. Awareness is a state in which complex brain analysis is used to process sensory stimuli or constructs based on memory (Broom 1998) and there are different degrees of awareness (Sommerville and Broom 1998). Most of the examples described in this paper of domestic animal behaviour changes consequent upon observation of the behaviour of others are best explained as requiring 'assessment awareness' in the observer, i.e. that observer must have been able to assess and deduce the significance of a situation in relation to itself over a short time span. The sceptic's explanation might be that all behaviour changes by the observer animal are explicable in terms of a series of automatic responses.

The argument that whenever there are two explanations for an observed phenomenon, the simpler of the two is preferable is dangerous when complex systems such as those in the vertebrate brain are under consideration. The ready rejection of the less simple explanation on all occasions where there is some doubt has led to serious impairment of scientific progress in the area of brain functioning. As evidence accumulates concerning the complexity of animal awareness, it seems that on many occasions where a simple explanation is under consideration as an explanation for a behavioural phenomenon, this explanation is wrong and an explanation which assumes more elaborate brain processing is correct. These animals are so far from being automata that it is naive to assume that they are. It is likely that information is obtained frequently from other individuals and that much of the processing of that information is similar in domestic animals and in humans.

Many of the experimental studies of social transfer of information in domestic animals have concerned feeding because the results of transfer of information about food are relatively easy to recognise and the principal objective of the experimenters was to demonstrate whether or not such information transfer can occur. However, as should be apparent from some of the other observations which have been mentioned, it is likely that social learning is a significant aspect of the whole of the life style in many of these species. Herd-living herbivores like sheep, cattle, goats and horses are likely to learn from one another about: plants in pasture which should or should not be selected, locations of patches of high quality food and methods of husbanding these plant resources. Pigs would also learn about methods of obtaining food and carnivores such as dogs would learn about methods of hunting. In addition, individuals of each of these species would learn from one another about methods of avoiding predation and other dangers and about the social capabilities and general individual qualities of other group members. The abilities of individuals in the group to detect food or danger, avoid danger, fight, mate, or care for offspring might be evaluated, not only by direct observation of those individuals but also by interpreting the behaviour of other individuals. Individuals in groups of domestic animals may collaborate in achieving objectives and exact retribution for previous cheating or other misdemeanours so they must be obtaining sophisticated information from one another in order to do this.

References

Adler, H. E. (1955). Some factors of observational learning in cats. *J. Genet. Psychol.*, **86**, 159–77.

Adler, L. I. and Adler, H. E. (1977). Ontogeny of observational learning in the dog (*Canis familiaris*). *Devel. Psychobiol.*, **10**, 267–71.

Alexander, G. and Shillito, E. E. (1977). The importance of odour, appearance and voice in maternal recognition of the young in Merino sheep (*Ovis aries*). *Appl. Anim. Ethol.*, **3**, 127–35.

Barton, M. A. (1983). Behaviour of group-reared calves fed on acid-milk replacer. *Appl. Anim. Ethol.*, **11**, 77.

Barton, M. A. and Broom, D. M. (1985). Social factors affecting the performance of teat-fed calves. *Anim. Prod.*, **40**, 525.

Benham, P. F. J. (1982). Synchronisation of behaviour in grazing cattle. *Appl. Anim. Ethol.*, **8**, 403–4.

Borchelt, P. L., Lockwood, R., Beck, A. M. and Voith, V. L. (1983). Dog attack involving predation on humans. In *New Perspectives on our Lives with Companion Animals*, ed. A. H. Katcher and A. M. Beck, pp. 219–31. Philadelphia: University of Pennsylvania Press.

Bouissou, M-F. and Hövels, J. (1976). Effet d'un contact précoce sur quelques aspects du comportement social des bovins domestiques. *Biol. Behav.*, **1**, 17–36.

Broom, D. M. (1998). Welfare, stress and the evolution of feelings. *Adv. Stud. Behav.*, **27**, 371–403

Broom, D. M. and Leaver, J. D. (1978). The effects of group-housing or partial isolation on later social behaviour of calves. *Anim. Behav.*, **26**, 1255–63.

Clarke, J. V., Nicol, C. J., Jones, R. and McGreevy, P. D. (1996). Effects of observational learning on food selection in horses. *Appl. Anim. Behav. Sci.*, **50**, 177–84.

Favre, J. Y. (1975). *Comportement d'ovins gardés*. Ministére de L'Agriculture, Ecole Nationale Supérieure Agronomique de Montpellier.

Fraser, A. F. and Broom, D. M. (1990). *Farm Animal Behaviour and Welfare*. Wallingford: CAB International.

Galef, B. G., Jr (1988). Imitation in animals: history, definition, and interpretation of data from the psychological laboratory. In *Social Learning, Psychological and Biological Perspectives*, ed. T. R. Zentall and B. G. Galef, pp. 3–28. Hillsdale, NJ: Lawrence Erlbaum.

Galef, B. G., Jr and Sherry, D. F. (1973). Mother's milk: a medium for transmission of cues reflecting the flavor of mother's diet. *J. Comp. Physiol. Psychol.*, **4**, 432–9.

Hall, S. J. G., Vince, M. A., Shillito Walser, E. and Garson, P. J. (1988). Vocalisation of the Chillingham cattle. *Behaviour*, **104**, 78–104.

Hunter, E. J., Broom, D. M., Edwards, S. A. and Sibly, R. M. (1988). Social hierarchy and feeder access in a group of 20 sows using a computer controlled feeder. *Anim. Prod.*, **47**, 139–48.

Hsia, L. C. and Wood-Gush, D. G. M. (1984). Social facilitation in the feeding behaviour of pigs and the effect of rank. *Appl. Anim. Ethol.*, **11**, 265–70.

Johnson, S. B., Hamn, R. J. and Leahey, T. H. (1986). Observational learning in *Gallus gallus domesticus* with and without a conspecific model. *Bull. Psychon. Soc.*, **24**, 237–9.

Johnston, R. E. and Jernigan, P. (1994). Golden hamsters recognise individuals, not just individual scents. *Anim. Behav.*, **48**, 129–36.

Kendrick, K. M. (1992). Cognition. In *Farm Animals and the Environment*, ed. C. Phillips and D. Piggins, pp. 209–31. Wallingford: CAB International.

Kendrick, K. M. and Baldwin, B. A. (1987). Cells in temporal cortex of conscious sheep can respond preferentially to the sight of faces. *Science, NY*, **236**, 448–50.

Kendrick, K. M., Atkins, K., Hinton, M. R., Heavens, P. and Keverne, B. (1996). Are faces special for sheep. Evidence from facial and object discrimination learning tests showing effects of inversion and social familiarity. *Behav. Processes*, **38**, 19–35.

Lobato, J. F. P., Pearce, G. R. and

Beilharz, R. G. (1980). Effect of early familiarization with dietary supplements on the subsequent ingestion of molasses-urea blocks by sheep. *Appl. Anim. Ethol.*, **6**, 149–61.

Lynch, J. J., Keogh, R. G., Elwin, R. L., Green, G. C. and Mottershead, B. E. (1983). Effects of early experience on the post-weaning acceptance of whole grain wheat by fine-wool Merino lambs. *Anim. Prod.*, **36**, 175–83.

Mendl, M., Zanella, A. J. and Broom, D. M. (1992). Physiological and reproductive correlates of behavioural strategies in female domestic pigs. *Anim. Behav.*, **44**, 1107–21.

Nicol, C. J. (1989). Social influences on the comfort behaviour of laying hens. *Appl. Anim. Behav. Sci.*, **22**, 75–81.

Nicol, C. J. (1995). The social transmission of information and behaviour. *Appl. Anim. Behav. Sci.*, **44**, 79–98.

Nicol, C. J. and Pope, S. J. (1992). Effects of social learning on the acquisition of discriminatory keypecking in hens. *Bull. Psychon. Soc.*, **30**, 293–6.

Nicol, C. J. and Pope, S. J. (1994a). Social learning in small flocks of laying hens. *Anim. Behav.*, **47**, 1289–96.

Nicol, C. J. and Pope, S. J. (1994b). Social learning in sib-

ling pigs. *Appl. Anim. Behav. Sci.*, **40**, 31–43.

Nolte, D. L. and Provenza, F. D. (1992). Food preferences in lambs after exposure to flavours in milk. *Appl. Anim. Behav. Sci.*, **32**, 381–9.

Phillips, C. J. C. (1993). *Cattle Behaviour*. Ipswich: Farming Press.

Podberscek, A. L. (1994). Dog on a tightrope: the position of the dog in British society as influenced by press reports on dog attacks (1988 to 1992). *Anthrozoös*, **7**, 232–41.

Potter, M. J. and Broom, D. M. (1987). The behaviour and welfare of cows in relation to cubicle house design. In *Cattle Housing Systems, Lameness and Behaviour*, ed. H. K. Wierenga and D. J. Peterse, *Current Topics in Veterinary Medicine and Animal Science*, Vol. 40, pp. 129–47. Dordrecht: Martinus Nijhoff.

Settle, R. H., Sommerville, B. A., McCormick, J. and Broom, D. M. (1994). Human scent matching using specially trained dogs. *Anim. Behav.*, **48**, 1443–8.

Sheldrake, R. and Smart, P. (1996). Pets that know when their owners are coming home. In *Further Issues in Research in Companion Animal Studies*, ed. J. Nicholson and A. Podber-

scek, p. 67. Collander: Society for Companion Animal Studies.

Sommerville, B. A. and Broom, D. M. (1998). Olfactory awareness in domesticated animals. *Appl. Anim. Behav. Sci.*, **57**, 269–86.

Sommerville, B. A., Green, G. A. and Gee, D. J. (1990). Using chromatography and a dog to identify some of the compounds in human sweat which are under genetic influence. *Chemical Signals in Vertebrates*, Vol. 5, pp. 634–9. Oxford: Oxford University Press.

Thorhallsdottir, A. G., Provenza, F. D. and Balph, D. F. (1987). Food aversion learning in lambs with or without a mother: discrimination, novelty and persistence. *Appl. Anim. Behav. Sci.*, **18**, 324–40.

Tolman, C. W. (1968). The varieties of social stimulation in the feeding behaviour of domestic chicks. *Behaviour*, **30**, 275–86.

Wyrwicka, W. (1978). Imitation of mother's inappropriate food preference in weaning kittens. *Pavlovian J. Biol. Sci.*, **13**, 55–72.

Youssef, M. Y. I., Phillips, C. J. C. and Metwally, M. (1994). The effect of pre-weaning grazing experience and presence of adult ewes on the grazing behaviour of weaned lambs. *Appl. Anim. Behav. Sci.*, **44**, 281.

Part 3

Rats, bats, and naked mole rats: animals with information centres

Editors' comments

KRG and HOB

Norway rats and naked mole rats live in communal nesting sites containing adults of both sexes and immature animals. Similarly, many bats roost at communal sites during at least a portion of the year. Communal sites serve as potential information centres (Galef and Wigmore 1983) for the social acquisition and transmission of information in vertical, oblique and horizontal directions. Hence, communally nesting or roosting species can potentially build feeding and other local traditions as well as readily exchange information about transiently available resources.

The Norway or brown rat (*Rattus norvegicus*) lives in communal burrows that contain long tunnels, nests, and food storage sites and that may house up to 300 animals (Barnett 1975, Laland, chapter 10). Norway rats survive well in captivity because they are omnivorous and highly adaptable. Hence, for much of this century, they have been the experimental animal of choice for large numbers of psychologists. That local traditions for dietary preferences and foraging style have been found in populations of wild Norway rats suggests that they may also be a highly suitable species for experimental studies of social learning. Previous works by Galef and his associates (cited in Laland, chapter 10) have, in fact, demonstrated the existence of a number of social learning mechanisms in this species. In particular, Norway rats can acquire dietary cues from the breath and excretory products of conspecifics and also by following other rats and feeding in proximity with them.

Kevin Laland summarises a series of elegant laboratory experiments designed to determine whether foraging information and dietary preferences can become fixed in a laboratory population and persist even after the initially trained animals have been removed from the group. Most laboratory experiments in the field of social learning have followed a single paradigm in which one trained animal serves as a demonstrator and a second animal serves as an observer. In contrast, in the transmission chain studies discussed here, observer animals then act as demonstrator animals to new observers. Such studies provide information relevant to the naturalistic question of whether learned traditions can survive in animal populations even after the original animals have died or emigrated.

In the first experiment described by Laland, laboratory rats exhibited enhanced abilities to dig for carrots hidden under loose soil if they had observed other rats doing so, irrespective or whether the demonstrator rats had been trained in the digging task or had discovered it on their own. Three principles emerged from these experiments. (1) Information can be lost as well as gained over a transmission chain. (2) Innovation can be acquired through a series of small steps and does not necessarily require unusually clever individuals (see also Terkel 1996). (3) Populations may be able to maintain only an equilibrium amount of information. In his second set of transmission chain experiments, Laland demonstrated that the equilibrium amount of information that can be maintained in a population increases if animals are given opportunities to practice newly acquired behaviours, and if they are exposed to multiple, as opposed to single, dietary cues. These experiments also demonstrated that dietary cues received from the breath or excrement of other animals have their greatest effects when they reinforce, rather than conflict with, natural taste preferences (e.g. Laland's control rats preferred cinnamon to cocoa).

Overall, Laland's experiments demonstrate that in rats social learning capacities can potentially interact with natural taste preferences to allow them to develop communal dietary preferences and to transmit these preferences to succeeding generations. At the same time, rat social learning capacities may potentially facilitate the communal exploitation of transiently available food sources.

Many species of bats roost communally in caves, trees or dense foliage during at least part of the year. At these sites, as Wilkinson and Boughman (chapter 11) note, members of some species can potentially acquire information about food availability from faeces deposited within minutes of the original food ingestion, from pollen that accrues on fur of nectar feeding bats, and from food items that bats who feed on relatively large fruits, insects, or vertebrates may bring to the roosting site. Bats also have well developed following responses and often leave roosting sites in pairs or in communal groups. Those bats who follow other bats who have just fed appear to have greater feeding success where food is patchily distributed.

Additional potential sources of social information acquisition derive from the use by Microchiropteran bats of echolocation to locate food. This may enable some bats to locate food resources by eavesdropping on the echolocation calls of others. In bats, vocalisations may also be used to defend feeding territories and to coordinate foraging activities among groups of females. Given the importance of vocalisations in the behavioural repertoire of many bats, it would not be surprising if socially mediated vocal learning were found to exist in some bat species and/or if vocal communication were used to transmit foraging or other information. Indeed, as Wilkinson and Boughman note, evidence indicates that, depending on the species, young bats may learn to use their calls in appropriate contexts and may change the characteristics of their calls in response to playbacks of maternal calls. Adult bats may modify the frequency of their calls when foraging near conspecifics and may modify the acoustic structure of their calls in response to group composition. In greater spear nosed bats, the females may also use calls to coordinate foraging activity and to indicate and assess group identity.

The final section of the Wilkinson and Boughman chapter may long serve as a landmark example of naturalistic approaches to social learning. Here, they present and test a model designed to predict foraging conditions in which social versus individual learning will be most advantageous. Their model assumes that social learning is determined by four variables: the probability of social transmission; measures of environmental predictability (i.e. probability of patch disappearance); the probability that a naive forager finds food for itself; and the fitness advantage associated with adopting a knowledgeable conspecific's diet or feeding location. They test this model by examining food predictability and availability in populations of three species of bats, all of which exhibit following behaviour from a communal roost. According to the terms of the model, at no time during the year would all members of any of the populations studied be expected to acquire food primarily via social learning. However, in two populations, a combination of individual and social learning is predicted for the month of January. Given the contrast between the low predictability of social learning and the high frequency of following behaviour, the authors suggest that their model may be incomplete. One missing parameter is an estimation of possible energetic costs associated with individual versus social foraging. If bats who search for food as individuals incur greater energetic costs than social foragers, then social foraging might be more advantageous to more bats than is suggested by the current model.

In the final chapter in this section (chapter 12), Faulkes focuses on the behaviour and social learning capacities of naked mole rats – subterranean rodents who reside in arid regions of Kenya, Ethiopia and Somalia. Naked mole rats live in burrows consisting of communal nesting and toilet chambers, and of an extensive tunnel system from which the animals dig for roots, tubers and corms. Naked mole rats resemble termites, ants and bees in that they live in large colonies of genetically related individuals, and each colony contains a single breeding queen and one to three breeding males. Other colony members help rear the young, maintain the nests and burrows and forage for food.

The underground food resources exploited by these animals are patchily distributed. When individuals find a large patch of food, they return to the nest and recruit others to help with the foraging tasks. Foraging recruitment is aided by a specialised chirp vocalisation given by the returning naked mole rats and by the animals' ability to recognise individuals and follow their odour trails. Naked mole rats also possess a second specialised, more aggressive, vocalisation that is used when young animals approach digging sites of adults. When approached in this manner, adults often relinquish the sites to the young.

Animals who follow others to food resources save the energy that they would otherwise have expended searching for food themselves. Young animals that usurp another's site save digging energy and may also gain foraging information. It is also clear that parents who transmit foraging information to their own young, or who permit them to scrounge at their own feeding sites, may gain long-term reproductive advantages. What is less clear is what potential advantages may accrue to animals that donate information or feeding sites to individuals other than their own offspring. Aoki and Feldman (1987, cited and expanded in Sibly, chapter 4 this volume) have predicted that specialised donor behaviours, such as the food-recruitment chirps of naked mole rats, will only occur among populations of genetically related individuals or in other special circumstances. Naked mole rats live in colonies of closely related, highly inbred animals, and, hence, meet the expectations of genetic models. This common genetic heritage also explains why they may cooperate in rearing their young; i.e. in doing so, they are assuring the survival of their own genes.

To what extent colonies of rats or bats may also consist of closely related individuals in less clear. What is clear, however, is that Norway rats and many species of bats forage on foods that are patchily dispersed, and that occur in clumps sufficiently large to feed more than one animal. Hence, animals who share information about feeding sites may later benefit as others share information with them.

References

Aoki, K. and Feldman, M. W. (1987). Toward a theory for the evolution of cultural communication: Coevolution of signal transmission and reception. *Proc. Natl. Acad. Sci., USA*, **84**, 7164–8.

Barnett, S. A. (1975). *The Rat: A Study in Behaviour*. London: Methuen.

Galef B. G., Jr. and Wigmore, S. W. (1983). Transfer of information concerning distant foods: a laboratory investigation of the 'information centre' hypothesis. *Anim. Behav.*, **31**, 748–58.

Terkel, J. (1996). Cultural transmission of feeding behavior in the black rat (*Rattus rattus*). In *Social Learning in Animals: The Roots of Culture*, ed. C. Heyes and B. G. Galef, Jr., pp. 17–47. San Diego: Academic Press.

10

Exploring the dynamics of social transmission with rats

Kevin N. Laland

Introduction

Over the last century, the study of animal social learning has been carried out in two distinct traditions: field studies of natural populations of animals describing behaviours thought to be socially transmitted, and laboratory experiments exploring the psychological processes underlying social learning (Lefebvre and Palameta 1988). Both traditions have their advantages: field studies shed light on the natural context and social processes leading to social learning, while laboratory experiments introduce a scientific rigor and control of key variables that generates reliable and replicable findings. However, each tradition also has its disadvantages. In the field, it is difficult to establish through observation alone that the spread of a behaviour results from social learning, as opposed to asocial learning, or to investigate the processes underlying diffusion. Traditional laboratory experiments sacrifice validity for reliability, and tell us little about how, even whether, social learning operates in natural populations.

Yet in both cases the deficiencies of the approach reflect convention rather than obligate methodological restrictions. Ethologists and behavioural ecologists routinely carry out experimental studies of natural animal populations (Slater 1985, Krebs and Davies 1991). What is more, laboratory experiments exploring animal social learning have been extraordinarily conservative in the experimental approaches employed. Typically, each subject or 'observer' is paired with a single 'demonstrator' conspecific previously trained to exhibit a target behaviour and, after observing the demonstrator, the observer is tested alone in order to establish the nature of the information transmitted between the two animals. This design can be represented in abstract terms as 'A → B', since information passes from the demonstrator (animal A) to the observer (animal B). The vast majority of laboratory experiments on social learning fit this description, despite a plurality of valid, alternative experimental designs. For example, there have been very few social learning experiments employing designs with multiple demonstrators or multiple numbers of naive observers, opportunities for alternative forms of social interaction (e.g. scrounging), transmission of information through a population or along a chain of animals, investigating sex differences in social learning, or investigating the patterns of ecological resources favouring reliance on social cues. The enduring conventional experimental design reflects the narrow focus of those experimental psychologists and biologists interested in social learning. In the main part, these experiments have been conducted to establish whether animals can imitate, rather than to understand how social learning operates in the natural world.

Fortunately, the last 20 years have witnessed the emergence of more innovative experimental approaches. Field workers have begun to carry out experimental studies of social learning in free-living animal populations (Lefebvre 1986, Wilkinson 1992, Langen 1996). Investigators have translated field observations into elegant experiments in captivity (Terkel 1996). Other researchers have analysed the controlled diffusion of socially learned traits through captive groups of animal (Lefebvre 1986, Giraldeau and Lefebvre 1986, 1987). In addition, transmission chain studies have explored the passage of social information along chains of animals. Here I summarise the findings of transmission chain studies to investigate the dynamics of social transmission in populations of rats (Laland and Plotkin 1990, 1991, 1992, 1993, Galef and Allen 1995).

The transmission chain approach was pioneered by Curio et al. (1978), who carried out laboratory studies that demonstrated that mobbing of artificial stimuli could be socially transmitted along a chain of blackbirds. Observer birds learned to mob a novel nonraptorial bird, and even a plastic bottle, as a consequence of witnessing another transmitting bird exhibit mobbing behaviour. As a result of this learning, observers were themselves able to act as effective transmitters of the mobbing behaviour to other naive observers, allowing transmission to be effective along a chain of six birds. The distinguishing feature of the transmission chain design is that the 'observer' animals are given the opportunity to act as 'demonstrators' to other animals. This transmission chain design can be represented in the same abstract terms as 'A → B → C → ...', as each animal acts as both receiver and transmitter of information.

Transmission chain designs are not restricted to single demonstrator-to-observer interactions. For example, experiments employing small groups of individuals, with group membership partially changed each step in the transmission chain, have been successfully utilised to study social transmission in humans (Jacobs and Cambell 1961). Recently, similar designs have been used to explore animal social transmission (Galef and Allen 1995, Laland and Williams 1997). Using the same notation, such designs can be represented as

'ABCD → BCDE → CDEF...'. By manipulating the size of the group, and the degree of overlap between successive group membership, experimenters can explore how other processes of social interaction facilitate and interfere with social transmission. These chains allow transmission to be investigated in a controlled but semi-naturalistic context, and with subjects in social groups (Galef and Allen 1995).

Experimental studies of social transmission in the Norway rat

The Norway, or brown, rat (*Rattus norvegicus*) lives in colonies of up to 300 animals inhabiting a fixed burrow complex, with a home range of about 25–150 metres in diameter (Calhoun 1962). The underground burrows contain long, branching tunnels, several exits, and rooms for nests and food storage. Field and laboratory studies indicate a rich communication network in rat populations, which influence individual movements and feeding habits (Calhoun 1962, Telle 1966, Galef and Wigmore 1983). Rat burrows have been described as 'information centres' (Galef and Wigmore 1983), since they are used as a base from which rats forage for food, and since rats have been found to communicate information concerning distant foods and food sites, by a variety of mechanisms (Galef and Clarke 1971, Galef and Wigmore 1983, Galef and Beck 1985).

The Norway rat's natural history, particularly its colonial existence and its complex communication network, makes it an excellent subject for social learning research. Field studies have reported the social transmission of foraging behaviours and food preferences among Norway rats (Calhoun 1962, Steiniger 1950). Partly for these reasons, and partly because the rat is a traditional subject of psychological and learning theory experimentation, a large body of laboratory-based empirical research has accumulated on social learning in rats. Approximately half of these studies have been concerned with the social learning or enhancement of arbitrary tasks, such as maze/alley running or bar pressing for food rewards, by observing or interacting with conspecifics. Since 1970, studies have tended to be more concerned with the social transmission of food preferences and taste-aversion learning (e.g. Galef and Clarke 1971, Galef and Wigmore 1983, Galef 1996). The social transmission of information about foods might partly explain the rat's success as an omnivore.

Such studies provide compelling evidence that rats are capable of social learning, but do not establish whether behaviour patterns that have been changed by social learning have the potential for becoming fixed in a rat population. For this to occur, social learning must be extended into repeated transmission episodes such that the relevant information can spread through a population. The following sub-sections describe experiments that use social learning to study transmission, by allowing each observer to act as a demonstrator for the next observer. These experiments provide laboratory-based data on the conditions required for the fixation of behaviour patterns in social groups by way of information transmission between animals.

Social transmission of foraging information: digging for buried food

Rats have been reported digging for buried foods, or foraging in loose litter for food items (Barnett 1975). Laland and Plotkin (1990, 1992) carried out a series of experiments that explored the social learning and transmission of foraging information concerning buried food. Demonstrator rats were trained to dig for pieces of carrot buried underneath a 5-cm layer of lightly compressed peat, or soil, in a plastic walled enclosure measuring $52 \times 36 \times 18$ cm. Each experimental subject, or observer, was then placed in a similar enclosure containing buried food, in the presence of a single demonstrator separated from the observer by a wire mesh partition, for a 10-minute period. A preliminary experiment established that the foraging performance of observer rats was enhanced through social learning from a trained demonstrator conspecific, with control groups confirming that the elevated performance of observers could not be attributed to social facilitation or motivational factors (Laland and Plotkin 1990). This established, Laland and Plotkin embarked on a series of experiments with an 'A → B → C . . .' transmission chain design.

In the first such experiment the animals were assigned to one of three groups.[1] Two of these were social transmission groups in which each animal first had the role of observer of a demonstrator conspecific that had varying degrees of experience at foraging by digging. The observer then became the demonstrator for the next animal in the line of transmission. In the first, or 'standard transmission group', the first demonstrator had been trained to dig for carrots beneath the peat surface of the enclosures. Thereafter, each observer then served as a demonstrator for another animal. In the second group, the innovator group, the initial demonstrator was untrained, but again each observer became the demonstrator for the next animal. The third group was a control, in that no social transmission could occur, and so there was no transmission chain. Each observer was paired with an untrained demonstrator that had no carrots buried on its side of the enclosure, and thus each animal performed on the basis of the occurrence of its own individual learning. Having had one such foraging opportunity the control animals were removed from the experiment.

(a)

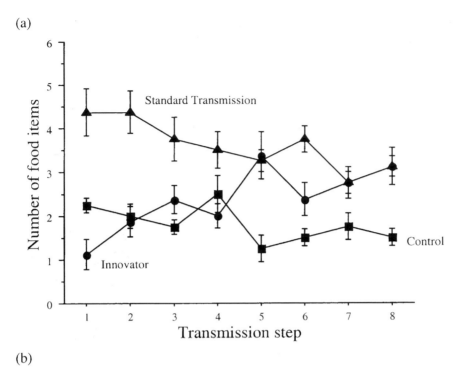

(b)

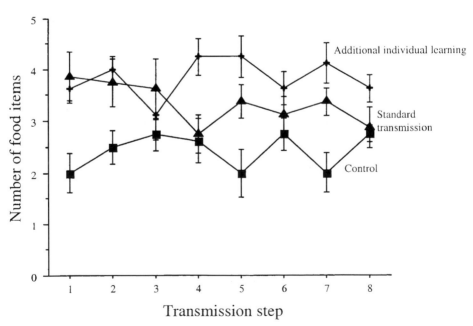

The results of this experiment are presented in Figure 10.1a. The animals in the standard transmission group exhibited elevated foraging performance relative to the control group, throughout the transmission chain. Animals in the standard transmission group were more active, began digging earlier, and dug up more pieces of carrot than did animals in the control group. It is clear that pertinent foraging information has been transmitted from the original trained demonstrators to other animals in the chain. Although this group shows some decline in performance in the early steps of the chain, this appears to flatten off at a level that is still significantly above that of the control group. This pattern is consistent with the interpretation of an initial loss of information at each transmission step, resulting in each successive animal being a less effective demonstrator for the next animal in the line of transmission.

The innovator group was so called because, since the transmission chain began with an untrained demonstrator, whatever information was being transmitted between animals was the result of cumulative innovation. Animals in this group also reached performance levels that were significantly better than those for the control group. There was an upward trend in performance from the earlier transmission steps, where performance of the innovator group was comparable to that of the controls, to the latter stages when it was significantly better than the behaviour of the controls and indistinguishable from the behaviour of the standard transmission group. Thus, for this group, there seemed to be an accumulation of information during the first few steps of the transmission chain, with performance asymptoting at a similar level to that found in the standard transmission group. The results from the innovator group suggests that the loss of information that occurs throughout the chain can be offset by the enhancement of the performance of successive observers by the sum of the demonstrator's social and individual learning. Towards the end of the transmission chains, both transmission groups appear to have reached an equilibrium level of performance where equivalent amounts of information are gained and lost.

The performance curves of subjects in the two transmission groups predicate certain distinctive properties of social transmission. First, information can

Figure 10.1. *(opposite)* Social transmission of foraging information. (*a*) Rats in the two transmission chain groups (standard transmission and innovator) that benefited from socially learned foraging information recovered more buried food items than those in the control group, which received no social cues. The standard transmission and innovator groups exhibit decreasing and increasing trends, respectively (Laland and Plotkin 1990). (*b*) Rats in a transmission chain group that were allowed to practise their socially acquired foraging skills (additional individual learning) recovered more food items than rats in the standard transmission group that had no extra foraging experience (Laland and Plotkin 1992).

be gained as well as lost throughout transmission. Second, innovation does not necessarily require creative or clever individuals, but can accrue through the accumulated activities of many individuals. Third, there may be an equilibrium amount of socially transmitted information that can be stably transmitted through a population. Where performance levels are higher than the equilibrium, information is likely to be lost, while where performance levels are below the equilibrium information may accrue.

If the third of these postulates is correct it should be possible to shift the level of this equilibrium, by enhancing opportunities to acquire or lose information throughout the chain. Laland and Plotkin (1992) went on to test this hypothesis in a further experiment by introducing another transmission group, labelled the 'additional-individual-learning' group, and characterised by the opportunity to reinforce any foraging information acquired as an observer with a period of lone foraging, prior to the animals acting as demonstrators. As each transmission step was separated by 24 hours, it was possible to replace subjects in the additional-individual-learning group alone in a fresh enclosure containing buried food for a further 10 minute period, 6 hours after their experience as observers, and 18 hours prior to their acting as demonstrators.

The results of this experiment are presented in Figure 10.1b. Animals in the additional-individual-learning group unearthed significantly more carrot pieces than those in the standard transmission group, and these groups dug up more carrot pieces than the controls. Figure 10.1b suggests that the opportunity to reinforce socially learned foraging information through individual experience allowed the additional-individual-learning group to transmit more information, thereby maintaining an elevated equilibrium level of foraging, and preventing any decay in performance along the chain.

Social transmission of dietary preferences

Several laboratory studies have established that rats can acquire dietary information by interacting with conspecifics (Galef and Clarke 1971, Galef and Wigmore 1983, Galef 1996). Galef and Heiber (1976) established that weanling rats preferred to eat from a food site in a section of the enclosure marked by the excretory deposits of a mature female conspecific rather than from a food site in an unmarked section. Galef and Beck (1985) found that rats can mark feeding sites, making them more attractive to weanling conspecifics than unmarked sites, and concluded that the communication of food site preferences could be mediated by olfactory cues surrounding particular feeding sites. Laland and Plotkin (1991) found that urine marking on and around the food site combines with faecal deposits to establish a stimulus complex that renders food sites attractive to rats, and can result in the social learning of dietary preferences. These experiments show that food site marking may be an effective mechanism for the passage of food-related information among rats. If, as field reports suggest (Steiniger 1950), this process operates in natural populations, it is possible that food preferences may be socially transmitted by these means. Rats that feed at a site marked by conspecifics' excretory deposits may themselves mark the site with their own products.

Laland and Plotkin (1993) investigated whether dietary preferences could be socially transmitted among rats through excretory marking. The first such experiment had a no-transmission control group, and two transmission chain groups, that investigated whether preferences for cinnamon- and cocoa-flavoured diets could be socially transmitted. In the transmission groups, four rats were placed in an enclosure for 48 hours and allowed to eat from a single food bowl containing powdered food flavoured with cinnamon (the cinnamon demonstration group) or cocoa (the cocoa demonstration group). These demonstrators were then removed, and a second food bowl containing the alternative flavoured diet, together with a single naive 'observer' rat, were placed in the enclosure. This animal effectively has a choice between two novel diets, one surrounded by the demonstrators' excretory cues, and the other not. After 24 hours the first observer was removed, its consumption of the two diets monitored, and a second naive observer placed in the enclosure, and so on, until a chain of eight such animals had each been in the enclosure. In the control group, there were no demonstrators, and each enclosure was cleaned before a new subject was placed in the enclosure. Subjects in this group had a choice between two novel diets, with no opportunity for their diet to be influenced by previous occupants of the enclosure. A prior experiment had established that the olfactory cues that surround food sites decay within 72 hours, thus the persistence of dietary traditions in the transmission groups beyond the third animal in the chain would represent evidence for the social transmission of dietary preferences by excretory marking. For further details of the procedure and experimental apparatus see Laland and Plotkin (1991, 1993).

The results of the transmission experiment are illustrated in Figure 10.2a. Control animals exhibited a preference for the cinnamon-flavoured diet, the cocoa diet constituting 25% of their consumption. Subjects in the cinnamon demonstration group consumed significantly more of the cinnamon diet and significantly less of the cocoa diet than both the controls and the subjects in the cocoa demonstration group. This almost exclusive consumption of cinnamon diet is a markedly different pattern of feeding behaviour from that exhibited by the controls. The elevated consumption of the cinnamon diet by this group can only be attributed to the influence of the residual deposits left in the enclosure, constituting strong evidence for social transmission of dietary information. In contrast, the subjects in the cocoa demonstration group showed little evidence for social transmission. Although at the first transmission step they consumed

(a)

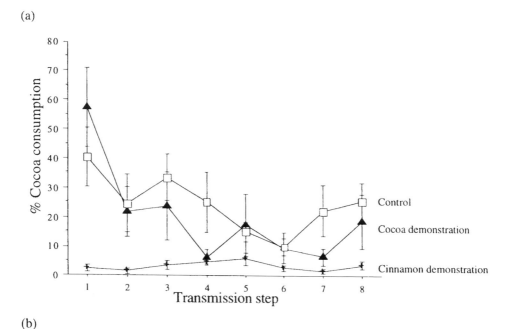

(b)

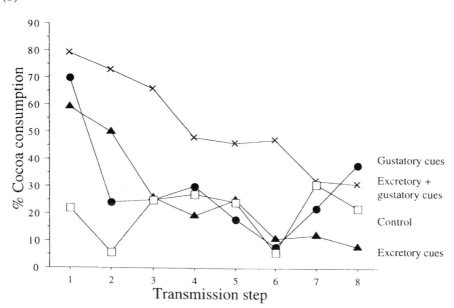

Figure 10.2. Social transmission of dietary preferences. (*a*) Residual faecal and urinary deposits left by previous inhabitants of the enclosure lead rats in the cinnamon demonstration group to maintain a preference for the cinnamon-flavoured diet along a chain of animals, consuming significantly more of this diet than control animals. However, there was no equivalent transmission of a preference for a cocoa-flavoured diet (Laland and Plotkin 1991). (*b*) Excretory deposits and gustatory cues from conspecifics combine to reinforce the stability of a dietary preference (Laland and Plotkin 1993).

more of the cocoa-flavoured diet and less of the cinnamon-flavoured diet than did the controls, this preference decayed rapidly along the chain. This illustrates quite clearly that an ('A → B') demonstration of social learning will not always generate ('A → B → C...') social transmission. One interpretation of the difference in stability between the experimental groups is that social transmission may be more stable when it reinforces a prior preference (i.e. for a more palatable diet) than when it conflicts with one. The differences cannot be attributed to differences in the amount of excretory marking by demonstrators in the two groups. The results suggest that the equilibrium mean proportion of the diet represented by food items for which there is socially transmitted information may be elevated relative to animals' consumption of such food items in the absence of social information. For animals in the cinnamon demonstration group, diet composition depends on historical factors, and cannot be predicted from palatability, profitability, or patterns of reinforcement.

Studies of social learning in Norway rats have uncovered a number of processes that facilitate the communication of food preferences among individuals, including following conspecifics to food sites and picking up cues on the breath of conspecifics that have just eaten (Galef and Clarke 1971, Galef and Wigmore 1983). If these processes operate in natural populations, it is unlikely that each will act in isolation. It is thus conceivable that, in a more natural situation, different communicatory mechanisms may interact to reinforce the social transmission of food preferences.

Laland and Plotkin (1993) went on to investigate whether the unstable transmission of a cocoa diet preference could be bolstered by the introduction of a second process for the communication of diet preferences. The additional process introduced was the communication of food preferences by gustatory cues studied by Galef and co-workers. Galef (1996) reports on a series of investigations of the social learning of food preferences among rats via cues on the breath of conspecifics. The principal finding is that a naive rat allowed to interact with a demonstrator conspecific that has recently eaten a novel diet will pick up gustatory cues on the breath of the demonstrator. When subsequently given a choice between the novel diet that the demonstrator consumed and another novel diet, the subject tends to prefer the former. If this process were to act in concert with excretory cues, the social transmission of a preference for the cocoa-flavoured diets among rats could be more stable.

In this experiment there were four groups. Two of these were equivalent to the cocoa demonstration group, here labelled the excretory-marking group and control group of the previous experiment. Subjects in a second transmission group, labelled the gustatory-cues group, were given the opportunity to acquire dietary information by picking up gustatory cues on the breath of a previous resident of the enclosure, but not to pick up cues from excretory markings. Finally, rats in the third transmission group, labelled excretory-marking-and-gustatory-cues group, could utilise both gustatory cues of the breath of a previous resident and excretory information. The first animals in the gustatory-cues group were given the opportunity to acquire dietary information by interacting for 30 minutes with a demonstrator that had previously eaten the cocoa-flavoured diet, and were then placed in a clean (i.e. unmarked) enclosure. Thereafter, each new subject interacted for 30 minutes with the previous resident of the enclosure, before being placed in a clean enclosure. Animals in the excretory-marking-and-gustatory-cues group had the same procedure as those in the gustatory-cues group, except that the enclosure was not cleaned between transmission steps.

The results are presented in Figure 10.2b. Animals in both the excretory-marking group and the gustatory-cues group showed enhanced cocoa diet consumption on the first transmission step, but by the third link in the chain their consumption was no different from the controls. Thus, in these two groups there was no evidence for transmission of dietary preferences. In contrast, rats in the excretory-marking-and-gustatory-cues group consumed proportionately more of the cocoa diet than the controls for the first six transmission steps in the chain. Although the proportion of cocoa diet consumed by subjects in this group decays over time, there is evidence for the transmission of dietary information. As the original demonstrators' excretory cues lost their informational content within 72 hours, subjects in the earlier transmission steps must be the source of the information guiding the dietary choices of rats at steps 4 to 6. This experiment suggests that social learning processes can sum up to reinforce transmission stability. Introduction of a second social learning process was sufficient to delay the decline in levels of diet consumption. This finding suggests that where a socially transmitted trait is affected by more than one mechanism, it may be stabilised. Moreover, a failure to find social learning and transmission in the laboratory, where only one mechanism is investigated, does not demonstrate that transmission is unlikely under more natural conditions, where a host of mechanisms may operate together.

Galef and Allen (1995) have extended these findings with transmission chain studies of dietary information in rats that employed a 'ABCD → BCDE → CDEF → ...' transmission chain design. Here a population of four demonstrators was gradually replaced, one a day, by naive subjects that had the opportunity to acquire information from conspecifics by gustatory cues. Galef and Allen found that this design allowed for the stable transmission of two arbitrary dietary preferences for 14 days, with significant differences remaining in the feeding behaviour of the two transmission groups. Moreover, Galef and Allen established that the stability of transmission was substantially enhanced by restricting feeding to just three hours. They point out that free-living rats are most active at dawn and dusk, with many individuals feeding for only a few hours each day, implying that under natural conditions the stability of transmission may be similarly enhanced.

Recently, the findings from transmission chain studies in rats have been reinforced by an investigation of the transmission of foraging information in guppies, *Poecilia reticulata*, again using transmission chains (Laland and Williams 1997). In this study, untrained guppies swam with trained conspecifics to feed, and in the process learned a route to a food source. Fish in small founder populations were trained to take one of two alternate routes to a food source, and then founder members were gradually replaced by untrained conspecifics. Three days after all founder members had been removed, populations of untrained fish still maintained strong preferences for the routes of their founders. The experimental results suggest that the tendency to shoal may facilitate a simple form of social learning, which allows guppies to learn about their local environments. This raises the possibility that social processes that underlie aggregation in other species may also facilitate social learning. The study also found that guppy social learning is frequency-dependent, with the likelihood of social learning increasing with the number of trained demonstrators. Similar frequency-dependent learning has been found in rats (Beck and Galef 1989) and pigeons (Lefebvre and Giraldeau 1994). Finally, the results imply that selectively neutral behavioural alternatives may be maintained as traditions in aggregated animal populations by very simple social mechanisms. As guppies are not renouned for their intelligence or the complexity of their social behaviour, this study illustrates that insights into the general processes that underlie social learning in animals may be investigated using less established species.

Conclusions

A series of transmission chain experiments with Norway rats has established the following properties of social transmission:

1 Foraging and dietary information can be socially transmitted along chains of animals.
2 Information can be gained as well as lost in transmission.
3 Innovation does not necessarily require 'clever' individuals, but can accrue through the accumulated activities of many animals.

4   There may be an equilibrium amount of socially transmitted information that can be stably transmitted through a population.

5   The opportunity to reinforce socially learned foraging information through individual experience may allow animals to maintain an elevated equilibrium level of foraging efficiency, and prevent decay in performance.

6   Social transmission may be more stable when it reinforces a prior preference (i.e. for a more palatable diet) than when it conflicts with one.

7   An experimental demonstration of social learning employing the traditional laboratory design (A → B) is not sufficient grounds for the conclusion that social transmission of sufficient penetration to result in the diffusion of a behaviour through a population will inevitably result.

8   The equilibrium mean proportion of the diet represented by food items for which there is socially transmitted information cannot always be predicted from animals' consumption of such food items in the absence of social information. Diet composition may depend on historical factors, and cannot always be predicted from palatability, profitability, or patterns of reinforcement.

9   Two or more social learning processes may sum up to reinforce the stability of social transmission.

10 The stability of the transmission of foraging information may be influenced by the number of individuals in a social group, the rate of change of group composition, and amount of time available to forage.

Note

1   In this experiment there were four groups. However, here I focus on the performance of subjects in just three of the groups (labelled 'standard transmission', 'innovator', and 'control'), since subjects in the 'additional-individual-learning' group generated equivocal results. Subsequent experiments were able to clarify the effects of additional individual learning on the transmission process (Laland and Plotkin 1992).

Acknowledgements

KNL is supported by a Royal Society University Research Fellowship.

References

Barnett, S. A. (1975). *The Rat: A Study in Behaviour.* London: Methuen.

Beck M. and Galef B. G., Jr (1989). Social influences on the selection of a protein-sufficient diet by Norway rats (*Rattus norvegicus*). *J. Comp. Psychol.,* **103**, 132–9.

Calhoun, J. B. (1962). *The Ecology and Sociology of the Norway Rat*. Bethesda, Maryland: U.S. Dept. Health, Educ., Welfare.

Curio, E., Ernst, U. and Vieth, W. (1978). The adaptive significance of avian mobbing II. Cultural transmission of enemy recognition in blackbirds: effectiveness and some constraints. *Z. Tierpsychol.*, **48**, 184–202.

Galef, B. G., Jr (1996). Social enhancement of food preferences in Norway rats: a brief review. In *Social Learning in Animals. The Roots of Culture*, ed. C. M. Heyes and B. G. Galef, Jr., pp. 49–64. New York: Academic Press.

Galef, B. G., Jr and Allen, C. (1995). A model system for studying animal traditions. *Anim. Behav.*, **50**, 705–17.

Galef, B. G., Jr and Beck, M. (1985). Aversive and attractive marking of toxic and safe foods by Norway rats. *Behav. Neur. Biol.*, **43**, 298–310.

Galef, B. G., Jr and Clarke, M. M. (1971). Social factors in the poison avoidance of feeding behaviour of wild and domestic rat pups. *J. Comp. Physiol. Psychol.*, **75**, 341–57.

Galef, B. G., Jr and Heiber, L. (1976). Role of residual olfactory cues in the determination of feeding site selection and exploration patterns of domestic rats. *J. Comp. Physiol. Psychol.*, **90**, 727–39.

Galef, B. G., Jr and Wigmore, S. W. (1983). Transfer of information concerning distant foods: a laboratory investigation of the 'information-

centre' hypothesis. *Anim. Behav.*, **31**, 748–58.

Giraldeau L. A. and Lefebvre, L. (1986). Exchangeable producer and scrounger roles in a captive flock of feral pigeons. *Anim. Behav.*, **34**, 797–803.

Giraldeau L. A. and Lefebvre, L. (1987). Scrounging prevents cultural transmission of food-finding behaviour in pigeons. *Anim. Behav.*, **35**, 387–94.

Jacobs, R. C. and Cambell, D. T. (1961). The perception of an arbitrary tradition through several generations of a laboratory microculture. *J. Ab. Soc. Psychol.*, **62**, 649–58.

Krebs, J. R. and Davies, N. B. (1991). *Behavioural Ecology: An Evolutionary Approach*, 3rd edn. Oxford: Blackwell.

Laland, K. N. and Plotkin, H. C. (1990). Social learning and social transmission of digging for buried food in Norway rats. *Anim. Learning Behav.*, **18**, 246–51.

Laland, K. N. and Plotkin, H. C. (1991). Excretory deposits surrounding food sites facilitate social learning of food preferences. *Anim. Behav.*, **41**, 997–1005.

Laland, K. N. and Plotkin, H. C. (1992). Further experimental analysis of the social learning and transmission of foraging information amongst Norway rats. *Behav. Process.*, **27**, 53–64.

Laland, K. N. and Plotkin, H. C. (1993). Social transmission in Norway rats via excretory marking of food sites. *Anim. Learning Behav.*, **21**, 35–41.

Laland, K. N. and Williams, K. (1997). Shoaling generates so-

cial learning of foraging information in guppies. *Anim. Behav.*, **53**, 1161–9.

Langen, T. A. (1996). Social learning of a novel foraging skill by White-throated Magpie-jays (*Calocitta formosa*): a field experiment. *Ethology*, **102**, 157–66.

Lefebvre, L. (1986). Cultural diffusion of a novel food-finding behaviour in urban pigeons: an experimental field test. *Ethology*, **71**, 295–304.

Lefebvre, L. and Giraldeau, L. A. (1994). Cultural transmission in pigeons is affected by the number of tutors and bystanders present. *Anim. Behav.*, **47**, 331–7.

Lefebvre, L. and Palameta, B. (1988). Mechanisms, ecology, and population diffusion of socially learned, food finding behavior in feral pigeons. In *Social Learning: Psychological and Biological Perspectives*, ed. T. R. Zentall and B. G. Galef, Jr, pp. 141–64. Hillsdale, NJ: Earlbaum.

Slater, P. J. B. (1985). *An Introduction to Ethology*. Cambridge: Cambridge University Press.

Steiniger, F. von (1950). Uber Duftmarkierung der Wanderratte. *Z. Ang. Zool.*, **38**, 357–61.

Telle, H. J. (1966). Beitrag zur Kenntnis der Verhaltensweise von Ratten, vergleichand dargestellt bei *Rattus norvegicus* und *Rattus rattus*. *Z. Ang. Zool.*, **53**, 129–96.

Wilkinson, G. S. (1992). Information transfer at evening bat colonies. *Anim. Behav.*, **44**, 501–18.

11

Social influences on foraging in bats

Gerald S. Wilkinson and Janette Wenrick Boughman

Introduction

Theory developed over the past decade (e.g. Boyd and Richerson 1985, 1988, Laland *et al.* 1993, 1996, Barta and Szep 1995) predicts that the extent to which socially mediated biases are to be expected in the transmission of behaviours among animals should depend on environmental predictability, fitness effects associated with adopting alternative behavioural variants, and social organisation. If the environment fluctuates rapidly, social transmission will fail to track those changes. On the other hand, if the environment changes slowly or predictably, social learning can be favoured if it reduces costs associated with individual learning. The rate at which novel behaviours will spread through a population by social learning is also influenced by the types of interactions permitted by the social organisation. Transmission among unrelated animals in the same generation (horizontal) or among unrelated animals across generations (oblique) permits more rapid change than transmission from parents to offspring (vertical). Consequently, social learning may be especially important when unrelated animals have frequent opportunities to interact while making decisions that affect their survival and reproduction.

With over 900 described species, the order Chiroptera almost certainly contains more species predisposed to social learning than any other order of mammals for at least four reasons. 1. Most bat species exhibit communal roosting during all or part of the year. With over 80% of bat species occurring in tropical regions, communal roosting is often associated with mating or rearing young, rather than hibernating. Because lactating bats often make multiple feeding trips in a night (Wilkinson 1992, Wilkinson and Boughman 1998), communal nursery roosts seem especially likely to provide numerous opportunities for acquiring information about the location of food or other resources. 2. Many bat species also move seasonally between traditional roosts. In some temperate species, mating and hibernation occurs at winter roosts. In some migratory species, such as the lesser long-nosed bat, *Leptonycteris curasoae* (Figure 11.1 Wilkinson and Fleming 1996), individuals may utilise several roost sites separated by as much as 1000 km. Naive young bats almost certainly follow older bats (Wilkinson 1992) to alternative roost sites, as would

Figure 11.1. Lesser long-nosed bats.

be expected given the potential costs associated with mistakes. 3. For their size, bats are exceptionally long-lived and often exhibit female philopatry. For example, banding studies have revealed that 10 g little brown bats, *Myotis lucifugus* (Keen and Hitchcock 1980), and 20 g greater horseshoe bats, *Rhinolophus ferrumequinum* (Jones *et al.* 1995), can survive 30 years or more. Our own studies have indicated that 35 g female common vampire bats, *Desmodus rotundus*, and 80 g female greater spear-nosed bats, *Phyllostomus hastatus*, can live 16 years in the wild (G.S. Wilkinson, unpublished results). Given that most species give birth to only one young, individuals in social groups vary in foraging experience. 4. All microchiropteran species use echolocation for orientation and many use it to capture prey. Because calling rate increases during prey attacks, echolocating bats broadcast their foraging activity to other individuals in their vicinity.

Despite tremendous ecological diversity and apparent opportunities for social learning, few studies have directly considered the degree to which social learning occurs in bats. Therefore, here we describe how three learning problems – choosing a diet, finding resources and communicating while foraging – may involve social learning in some species of bats and discuss evidence for socially mediated biases where available. We then use a recent model for animal social learning (Laland *et al.* 1996) to determine if variation in feeding patch longevity and foraging success predicts differences in following behaviour exhibited by evening bats, *Nycticeius humeralis*, greater spear-nosed bats and common vampire bats. Qualitative agreement between the model predictions and frequency of following across species support using bats as model systems for studying social learning and information transfer at communal roosts.

Opportunities for social learning in bats

Diet preference
Many species of bats form dense roosting clusters in caves, hollow trees or foliage. Despite the darkness of these sites, at least four different situations provide nonvisual cues about dietary preferences to roostmates. First, individuals may be able to smell or taste food which has adhered to the fur of neighbours. For example, in the process of visiting flowers many nectar feeding bats become covered in pollen, which is subsequently ingested while grooming at a communal roost. Some species, e.g. lesser long-nosed bats (Howell 1979) and lesser spear-nosed bats, *P. discolor* (G.S. Wilkinson and J.W. Boughman, unpublished results), groom roostmates and could ingest pollen from their fur. Horizontal transmission of flower preferences will probably depend on female social group stability, as this factor appears to determine allogrooming frequency across species (Wilkinson 1987).

Second, roostmates may acquire dietary information from faeces. Bat pollinated flowers and bat dispersed fruits often produce conspicuous, pungent odours which attract bats (Pijl 1961, Helverson 1993). Because fruit bats frequently defaecate within minutes after feeding (Werner and Gardner 1978), faecal odour can provide a prompt dietary cue. Socially mediated preferences caused by exposure to faecal odours seems plausible given the olfactory sensitivity of many fruit eating bats (Bhatnagar and Kallen 1974, Schmidt 1975).

Third, species which capture or collect large prey items, either fruit or animal, often bring some food items back to a day or night roost for consumption where these items could be smelled, felt or tasted by roostmates. For example, the fishing bat (*Noctilio leporinus*) consumes fish (Bloedel 1955), pallid bats, *Antrozous pallidus*, eat scorpions (Bell 1982), large slit-faced bats, *Nycteris grandis*, eat grasshoppers, fish and frogs (Fenton *et al.* 1981), fringe-lipped bats, *Trachops cirrhosus*, eat frogs (Tuttle and Ryan 1981), short-tailed fruit bats, *Carollia perspicillata*, eat *Piper* fruit (Fleming 1988), false vampire bats, *Vampyrum spectrum*, eat birds with noticeable odours (Vehrencamp *et al.* 1977), and greater spear-nosed bats consume large insects and a variety of fruit items (Wilkinson and Boughman 1998) in roosts. We have observed a female greater spear-nosed bat steal a large grasshopper from the mouth of a roost-mate confirming that dietary information can be shared when food is consumed in the roost.

Finally, dietary preferences may be acquired if parents provision young. For example, nocturnal observations in a hollow tree revealed that an adult male false vampire bat returned with and subsequently shared a ground dove (*Columbina talpacoti*) with a juvenile (G.S. Wilkinson, unpublished results). Similarly, food sharing by regurgitation occurs frequently between females and young in common vampire bats (Wilkinson 1984), and may help juveniles learn the taste and smell of blood consumed by the mother. Common vampire bats have been reported to exhibit preferences for particular breeds of cattle and horses (Turner 1975).

Resource location

Microchiropteran bats find food in one of three ways: searching independently, following other bats from the roost to a feeding site, or approaching bats that are actively hunting or feeding. The latter two situations represent examples of social learning in which one bat attends to echolocation calls or other cues provided by another bat. Social learning can benefit a follower by reducing the energetic costs associated with searching and increasing the probability of finding food. However, following may cost a searcher to the extent that recruitment of another forager reduces individual consumption rates (Vickery *et al.* 1991). Following is expected, therefore, when rich food patches are short-lived and difficult to find (Waltz 1982, Barta and Szep 1995) – ecological situations likely to be common for many species.

Following

Although direct evidence for following is scarce, many species of bats leave communal roosts in groups, as expected if following occurs. For example, many pteropid species depart communal roosts at dusk and fly in groups to feeding areas (Nelson 1965, Thomas and Fenton 1978). Some also make seasonal migrations to track flowering and fruiting plants (Thomas 1983, McWilliam 1986, Richards 1995). Following may occur by visual observation in megachiropteran and some microchiropteran bats (Howell 1979) much as it does in some birds (Brown 1986, Rabenold 1987, Heinrich 1988).

In contrast, aerial insectivores could use echolocation calls for following. Such behaviour would cause clustered departures, as have been observed in many species (Wilkinson 1995). For example, pipistrelles (*Pipistrellus pipistrellus*) depart communal roosts nonrandomly (Speakman *et al.* 1992, 1995) and have been observed following at feeding sites (Racey and Swift 1985). Radio-telemetry studies of evening bats (Wilkinson 1992), common vampire bats

(Wilkinson 1985), lesser spear-nosed bats (G.S. Wilkinson, unpublished results) and greater spear-nosed bats (G.S. Wilkinson and J.W. Boughman, unpublished) have revealed that pairs of individuals sometimes depart together and subsequently feed in the same areas. Evening bats that followed on second foraging trips increased their foraging success, compared with bats that departed independently (Wilkinson 1992). Following can, therefore, facilitate food finding in some species.

With the exception of common vampire bats, in which females will feed simultaneously with their female offspring from the same bite (Wilkinson 1985), little evidence indicates that young bats follow their mothers to feeding sites (Wilkinson 1995). Nevertheless, young bats almost certainly follow adults in other situations. For example, two roost exclusion experiments demonstrated that newly volant evening bats followed adults to new roost sites in hollow trees (Wilkinson 1992). Bats that migrate to feed, hibernate or give birth at traditional sites provide additional examples in which vertical social learning probably occurs. Adults may tolerate the presence of young in these situations because resource competition is weak.

Following also occurs among bats flying in groups. A number of species have been observed flying in tandem within a few feet of each other while feeding on fish, fruit, flowers and insects (Wilkinson 1995). Experimental studies with lesser spear-nosed bats have revealed that naive bats allowed to forage with experienced bats take less time to find hidden food (Wilkinson 1987). On some trials, naive bats flew directly behind knowledgeable bats to the accessible feeding station. We have performed a similar experiment with greater spear-nosed bats, distributing hidden food either evenly or patchily and releasing bats alone or with a 1 min interval between two individuals. After log transformation, analysis of variance revealed a difference between singletons and pairs in the time elapsed to find food ($F_{1,24} = 8.6$; $P < 0.017$). Bats in pairs showed improved food finding only in the patchy environment. Singletons took longer to find food (134.3 ± 1.5 s) than either the first (21.3 ± 1.7 s) or the second bat (29.4 ± 1.8 s). In addition, bats tended to visit the same feeding stations as bats in previous trials, independent of location ($P < 0.007$, binomial test), suggesting that olfactory cues may also have influenced food discovery.

Eavesdropping

Aerial insectivores emit feeding buzzes, i.e. decrease duration and increase repetition rate of their echolocation calls, as they approach prey (Simmons and Stein 1980). Consequently, information about foraging activity is broadcast to other bats and could be used to locate foraging sites and assess prey abundance.

Several studies have confirmed that some species approach playbacks of feeding buzzes (Barclay 1982, Balcombe and Fenton 1988). Such eavesdropping requires that the bats are sufficiently close to hear each other's biosonar. Detection distances under most environmental conditions are likely to be less than 100 m for many temperate insectivorous bats, and sometimes less than 10 m for bats with low amplitude, high frequency biosonar (Griffin 1971, Wilkinson 1995). Eavesdropping is, therefore, likely to be restricted to aerial insectivores that use relatively loud, low frequency echolocation calls.

Advertisement

Some bats could also learn where food is located from conspecifics which use audible vocalisations and behaviour to defend a feeding territory. Many aerial insectivores use nonecholocation calls when chasing intruders away from foraging sites, e.g. Daubenton's bats, *Myotis daubentonii* (Wallins 1961), lesser sac-winged bats, *Saccopteryx leptura* (Bradbury and Emmons 1974), northern serotines, *Eptesicus nilssonii* (Rydell 1986), red bats, *Lasiurus borealis* (Hickey and Fenton 1990), and pipistrelles (Miller and Degn 1981). Calling rate is inversely related to prey abundance in pipistrelles; field playback experiments demonstrate that these calls repel conspecifics (Barlow and Jones 1997). Instead of calling and chasing, several large carnivorous megadermatids, including heart-nosed bats, *Cardioderma cor* (Vaughan 1976), yellow-winged bats, *Lavia frons* (Vaughan and Vaughan 1986) and ghost bats, *Macroderma gigas* (Guppy *et al.* 1985, Tidemann *et al.* 1985), give loud audible calls while perched, apparently to advertise territory ownership.

   In contrast, a few species emit audible calls which attract conspecifics. Field playbacks demonstrate that loud, broad-band screech calls attract greater spear-nosed bats at cave entrances and foraging sites and elicit additional calling (Wilkinson and Boughman 1998). Observations of LED-tagged bats indicate that these calls function to coordinate foraging activity among females from the same social group. Individuals foraging with social partners appear to benefit from improved defense of flowering or fruiting trees as well as from learning food location from others (Wilkinson and Boughman 1998).

Communication

As the previous sections indicate, vocalisations may often mediate how bats learn feeding site location from conspecifics. But, social learning can also influence call production directly. Bats might learn the context in which social calls should be emitted as a consequence of reinforcing social interactions. For example, male hammer-headed bats, *Hypsignathus monstrosus*, increase call repetition rate when females approach (Bradbury 1977). Social reinforcement could occur in this example because only males with high calling rates mate. Alternatively, under several social situations acoustic modification of vocalisations can be expected.

Social modification of echolocation calls

Because echolocation calls are effectively designed for orientation and prey capture (Simmons and Stein 1980), extensive modification of signal form in response to conspecifics seems unlikely. Nevertheless, if eavesdropping or following are costly to a searcher, active foragers would be expected to increase frequency, decrease amplitude or decrease echolocation calling rate to reduce the probability of detection while hunting.

Several studies have reported cases of bats modifying the frequency of their echolocation calls when foraging near conspecifics, e.g. hipposiderids (Pye 1972) and emballonurids (Barclay 1983). Such modification reduces similarity among individuals, which may help avoid interference in echo detection and processing by partitioning acoustic space (Habersetzer 1981). But, when researchers have noted a change in echolocation call frequency, it has increased, not decreased, when conspecifics were encountered, e.g. mouse-tailed bats, *Rhinopoma hardwickei* (Habersetzer 1981), pipistrelles (Miller and Degn 1981), and northern serotines (Rydell 1993). Furthermore, Obrist (1995) found in four species that calls are shorter and given less frequently when conspecifics are present, as expected if foragers modify their calls to reduce inter-individual detection distance.

In addition, geographic, colony or sex differences in echolocation calls (Neuweiler *et al.* 1986, Heller and Helverson 1989) could result from social modification of calls. However, such differences could also reflect population genetic variation in morphology or differences in foraging behaviour (Rydell 1993). Evidence that echolocation calls can be modified to increase resemblance among colony members has only been obtained for greater horseshoe bats (Jones and Ransome 1993). Mothers were more similar to pups than expected if call similarity was due solely to heritable variation. Furthermore, calls of adult females changed in frequency over their lives and pup calls mirrored these changes.

Social modification of communication calls

The acoustic structure of communication calls should be influenced by the social environment if this enhances call function. For example, calls which function to form foraging groups or advertise feeding sites should have characteristics which resist attenuation and maximise locating ability. If group composition matters, social calls should also convey caller identity. While calls that indicate individual identity or kinship can be heritable (Scherrer and Wilkinson 1993), calls that indicate social identity, such as membership in a group or pair, are more likely to be modified by social experience.

Recent work on greater spear-nosed bats indicates that screech call acoustic structure not only conforms to signal design expectations, but also responds to changes in the composition of social groups. Greater spear-nosed bat screech calls are loud and low in frequency, as expected for long distance transmission. These calls also cover a wide frequency range, which improves the ability of other bats to localise the caller. Our observations suggest that females use screech calls not only to coordinate foraging activity (Wilkinson and Boughman 1998), but also to indicate and assess group identity.

Greater spear-nosed bat screech calls from different social groups differ in acoustic structure and individuals within groups sound similar (Boughman 1997). Consequently, screech calls contain information to indicate group membership of the caller, but not individual identity. Bats attend to these acoustic differences, as their response to call playbacks depends on the group membership of the bat giving the calls (Boughman and Wilkinson 1998).

Group differences in screech calls cannot result from heritable variation because *P. hastatus* female group mates are unrelated (McCracken and Bradbury 1981). Our data indicate that screech calls are modified through social experience. In a reciprocal transfer experiment that mimics naturally occurring dispersal, acoustic structure of calls changed in response to changes in group composition resulting in increased similarity among bats in the same social group (Boughman 1998). This modification took place among both first year and older, reproductively mature animals.

Few other published studies have demonstrated vocal learning in mammals. There is good evidence, however, that infant lesser spear-nosed bats (Figure 11.2) change characteristics of their isolation calls to match playbacks of maternal directive calls (Esser 1994). Esser (1994) suggests that increased similarity between mother and offspring facilitates individual recognition during retrievals. Whether the genus *Phyllostomus* is exceptional in the capacity for vocal learning or provides an example of a broader phenomenon in bats is unknown at present.

Lesser spear-nosed bats may share more than an ability for vocal learning with greater spear-nosed bats. Both species forage on pollen and nectar in groups (Sazima and Sazima 1977, McCracken and Bradbury 1981, Wilkinson and Boughman 1998), and both respond to greater spear-nosed bat screech call playbacks by approaching the speaker and calling. Lesser spear-nosed bat screech calls sound similar to greater spear-nosed bat screech calls except they are higher in frequency and more trill-like (unpublished data). Additional

Figure 11.2. Lesser spear-nosed bat.

study is needed to determine if lesser spear-nosed bat foraging calls may also change in response to the social environment.

A test of social learning theory

In a recent paper, Laland *et al.* (1996) developed a model to predict the frequency of social learning at an information centre where animals can acquire information about diet choice and resource location. The frequency of social learning is determined by four variables, c, e, ε, and s, where c is the probability of social transmission, e is a measure of environmental predictability, i.e. the probability of patch disappearance, ε is the probability that a naive forager successfully finds better food by itself and s indicates the fitness advantage associated with adopting the knowledgable conspecific's diet or feeding location. Animals are assumed to search for food every day, but social learners can acquire information about better feeding sites without cost from any knowledgeable conspecific prior to foraging.

Numerical analysis of this model reveals three possible outcomes at equilibrium: 1. all individuals learn independently, 2. a polymorphic region where individual and social learners coexist, and 3. a region in which all animals should use both individual and social learning. Qualitatively, individual learning is favoured when the probability of finding better food by foraging alone is high and the environment changes rapidly. In contrast, socially transmitted information should be used when the probability of finding food by searching is low and the environment changes slowly. Social learning increases mean fitness if $e + \varepsilon < 1$ (Laland *et al.* 1996). Consequently, if the model is valid, we would expect $e + \varepsilon$ to correlate inversely with following.

To test this model we compare $e + \varepsilon$ to the frequency of following in three species: evening bats, greater spear-nosed bats and common vampire bats. Although these three species have very different diets and life histories, all three exhibit following behaviour from a communal roost. Female evening bats follow previously successful foragers at nursery colonies and subsequently improve their own foraging success presumably because they are led to sites where small flies and beetles are temporarily abundant (Wilkinson 1992). Female greater spear-nosed bats follow other females from their social group apparently to defend rich flowering or fruiting trees from other bats (Wilkinson and Boughman 1998). Young female vampire bats fly with their mothers to pastures and sometimes feed from the same wound (Wilkinson 1985). Thus, in each case following can improve access to food.

For all three species, we assume that e is the probability that on any given day a feeding site has below average food availability. For evening bats we used samples of prey insect abundance collected in 1 h at dusk using five automated suction traps located at feeding sites about 1 km from a nursery colony (Wilkinson 1992). For each site we estimated the number of consecutive days that prey biomass density was either above or below the mean of all five sites. Patch duration was 2.55 ± 0.57 (SE) days averaged over a six week period. We then calculated e as the inverse of the mean over all five sites each week. We calculated ε as the mean proportion of bats returning after their initial foraging trip which had gained weight and, therefore, fed successfully. We used video-tape records of bats entering and leaving an attic roost while crawling across a balance to obtain bat weights. Finally, we estimated the proportion of bats which departed within 10 s of another bat on their second or later foraging trip as potential social learners because these bats are close enough to hear and follow the echolocation calls of the previous bat (Wilkinson 1992).

Greater spear-nosed bats consume animal prey and feed on fruit, pollen or nector of more than 30 different plants (Gardner 1977, Gorchov *et al.* 1995). Consequently, to estimate e we first examined faecal collections to determine food items in the diet. During December and January over 50% of the diet of greater spear nosed bats includes pollen and nectar of balsa (*Ochroma lagopus*). In May and June 60–80% of the diet consists of *Cecropia peltata* fruit with the remainder comprising large-bodied insects (Wilkinson and Boughman 1998).

We used daily censuses of the number of flowers on 10 balsa trees over a three week period in December–January to determine the average number of consecutive days each tree had more flowers than the mean (4.12 ± 0.92 days, $n = 25$). Similarly, we used counts of the number of fruits on 10 *Cecropia* trees over a two week period to determine that *C. peltata* patch duration was 6.29 ± 0.82 days ($n = 14$). Although these samples are small, similar phenological patterns have been reported for another population of *C. peltata* (Fleming and Williams 1990). During May–June we also estimated large-bodied insect predictability using 10 Plexiglas impact traps suspended in locations where we had observed greater spear-nosed bats forage. Trap samples indicated a patch duration of 1.65 ± 0.20 days ($n = 24$) for insects. Finally, we weighted estimates of patch duration by the proportion of the diet comprising either insects or *C. peltata*, to obtain an average e for the May–June period of 4.67 ± 0.60 days.

We estimated e for greater spear-nosed bats during each season as the proportion of radio-tagged bats that returned briefly to the cave and subsequently departed on a second foraging trip (unpublished data). Following was obtained from observations of LED-tagged females from the same social group that departed the cave roost and flew away together. Captures of bats attracted to playbacks of screech calls at flowering trees confirmed that females from the same roosting group sometimes foraged at the same feeding site (Wilkinson and Boughman 1998).

We assumed that e for a vampire bat was the reciprocal of the average number of days livestock were present in each pasture, i.e. 6 days (Turner 1975). We estimated ε as the proportion of bats less than 2 years of age that were netted while entering roosts within 1 h of dawn and had successfully consumed a blood meal, i.e. 67% (Wilkinson 1984). Finally, we used night vision scope observations on two nights of vampire bats departing from a hollow tree between dusk and dawn to calculate the proportion of departures in which one bat left simultaneously with another bat. Group departures are likely to indicate following in vampire bats because night vision scope observations of 67 bats feeding on horses revealed seven instances of two bats drinking together from the same bite (Wilkinson 1985).

A plot of e on ε for all three species reveals that no points fell within the region in which all individuals should be social learners (Figure 11.3a). While the feeding environment for greater spear-nosed bats and vampire bats is sufficiently predictable, individual feeding success appears to be too high to favour unconditional social learning. However, vampire bats and greater spear-nosed bats in January are predicted to exhibit a combination of social and individual learning. In contrast, environmental predictability is sufficiently low in evening bats that social learning is not predicted. Greater spear-nosed

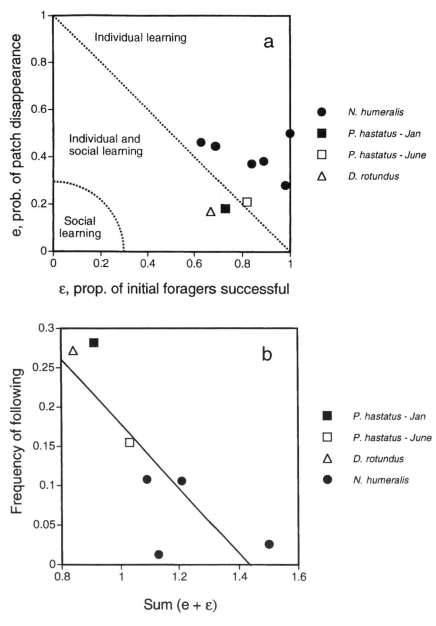

Figure 11.3. (*a*) Comparison of estimates for the probability of patch disappearance, *e*, and proportion of successful foragers, ε, for evening bats, *Nycticeius humeralis*, greater spear-nosed bats, *Phyllstomus hastatus*, and common vampire bats, *Desmodus rotundus*, to equilibrium predictions made by a model for social transmission at an information centre (Laland *et al.* 1996). (*b*) Frequency of following from a communal roost for the same three species and time periods indicated in (*a*) plotted against *e* + ε as a qualitative test of the model.

bats in June are predicted to show less social learning than in January because initial foraging success is higher in June.

Even though social learning among evening bats is not expected, adding environmental predictability to initial foraging success successfully predicted the frequency of following behaviour across species (Figure 11.3b). The Spearman rank correlation between $e + \varepsilon$ and the proportion of bats following was -0.86 ($P = 0.036$). Differences in colony sizes between the species do not contribute to this result. In fact, the evening bat colony contained more individuals (between 40 and 80) than either of the other two species and should, therefore, have exhibited the highest frequencies of following behaviour by chance. At the time of observations, the vampire bat tree roost had 27 individuals while only 20 adult female greater spear-nosed bats, on average, were tagged with the same colour LED.

Although our results provide qualitative support for the model in that $e + \varepsilon$ correlates with following frequency, our data do not conform to predictions in that the evening bat points lie in the region where indivdual learning is expected. We suspect either our estimates of e or ε for evening bats are in error or some assumption of the model is incorrect. If evening bats followed each other on initial foraging trips ε may represent an overestimate because we used foraging success from the first foraging trip to estimate ε. If following occurred on first foraging trips as often as we estimated it did on subsequent trips, then the e on ε points for evening bats would lie within the polymorphic region where individual and social learners are expected.

One obvious assumption of the model that can be questioned is whether either individual or social learning occurs without cost. To put it another way, do both forms of learning provide equally profitable mechanisms for locating food? Animals that follow a knowledgeable conspecific can avoid paying the energetic costs associated with searching. Consequently, the fitness advantage of social learning should be enhanced. This seems likely to increase the parameter space in which social learning will be favoured. Development of models with explicit learning costs is needed to confirm this possibility.

A second possibility is that this model may underestimate the effectiveness of social learning. Because ε can exceed $1 - e$, the model allows the possibility that social learners are less likely to find preferred food than individual learners. However, in reality individuals who fail to find food by utilising social information will revert to searching, and may still find preferred food. Changing the model to accommodate this possibility seems likely to increase the parameter space allowing polymorphism between individual and social learners and decrease the parameter space where only individual learning is favoured.

Conclusions

While few studies have yet provided compelling evidence for social learning in bats, qualitative agreement between model predictions and following frequency in an insectivore, an omnivore and a sanguivore suggests that many bats may experience environments in which social transmission of diet or resource location would be favoured by natural selection. Indeed, unless knowledgeable individuals actively avoid interaction, the communal social organisation of many bat species seems to predispose social learning. Because social transmission of foraging site locations can reduce variation in foraging success among individuals (Wenzel and Pickering 1991), social learning could increase survival. Thus, we conjecture that social learning may have played an important role in permitting the extraordinary longevity, relative to body size, displayed by many bats.

In addition to providing model systems for studying social transmission of diet and resource location, bats may also eventually provide some of the best nonhuman, mammalian examples of vocal learning. Although very few species have been studied in detail, vocal repertoires of bats tend to be large (Gould 1977, Fenton 1985, Kanwal *et al.* 1994), with many vocalisations serving unknown communication functions. Further study is clearly needed to determine if the ability of lesser and greater spear-nosed bats to modify their vocalisations and match a conspecific is unique to the genus or more widespread in the order. In either event, more detailed study designed to characterise the process by which these bats perform call matching may provide important insights into how vocal learning among humans has evolved.

Acknowledgements

We thank H. Alcorn, S. Boughman, P. Miller, S. Perkins, T. Porter, K. Smith and J. Wolf for assistance in collecting data and the Department of Zoological Research at the Smithsonian's National Zoological Park for housing *Phyllostomus hastatus*. We greatly appreciate helpful comments by K. Laland and are grateful for research support from the Chicago Community Trust, the National Science Foundation, the Smithsonian Institution, and the University of Maryland Graduate School.

References

Balcombe, J. P. and Fenton, M. B. (1988). The communication role of echolocation calls in vespertilionid bats. In *Animal Sonar*, ed. P. E. Nachtigall and P. W. B. Moore, pp. 625–8. New York: Plenum Press.

Barclay, R. M. R. (1982). Interindividual use of echolocation calls: eavesdropping by bats. *Behav. Ecol. Sociobiol.*, **10**, 271–5.

Barclay, R. M. R. (1983). Echolocation calls of emballonurid bats from Panama. *J. Comp. Physiol.*, **151**, 515–20.

Barlow, K. E. and Jones, G. (1997). Function of pipistrelle social calls: field data and a playback experiment. *Anim. Behav.*, **53**, 991–9.

Barta, Z. and Szep, T. (1995). Frequency-dependent selection on information-transfer strategies at breeding colonies: a simulation study. *Behav. Ecol.*, **6**, 308–10.

Bell, G. P. (1982). Behavioral and ecological aspects of gleaning by a desert insectivorous bat, *Antrozous pallidus* (Chiroptera: Vespertilionidae). *Behav. Ecol. Sociobiol.*, **10**, 217–23.

Bhatnagar, K. P. and Kallen, F. C. (1974). Cribriform plate of ethmoid, olfactory bulb and olfactory acuity in forty species of bats. *J. Morphol.*, **142**, 71–124.

Bloedel, P. (1955). Hunting methods of fish-eating bats, particularly *Noctilio leporinus*. *J. Mammal.*, **36**, 390–9.

Boughman, J. W. (1997). Greater spear-nosed bats give group distinctive calls. *Behav. Ecol. Sociobiol.*, **40**, 61–70.

Boughman, J. W. (1998). Vocal learning by greater spear-nosed bats. *Proc. R. Soc. Lond. B*, **265**, 227–33.

Boughman, J. W. and Wilkinson, G. S. (1998). Greater spear-nosed bats discriminate group mates by vocalizations. *Anim. Behav.*, **55**, 1717–32.

Boyd, R. and Richerson, P. J. (1985). *The Evolution of Culture*. Chicago: University of Chicago Press.

Boyd, R. and Richerson, P. J. (1988). An evolutionary model of social learning: the effects of spatial and temporal variation. In *Social Learning*, ed. T. R. Zentall and B. G. Galef, Jr, pp. 29–48. Hillsdale: Lawrence Erlbaum Associates.

Bradbury, J. W. (1977). Lek mating behavior in the hammer-headed bat. *Z. Tierpsychol.*, **45**, 225–55.

Bradbury, J. W. and Emmons, L. H. (1974). Social organization of some Trinidad bats. I. Emballonuridae. *Z. Tierpsychol.*, **36**, 137–83.

Brown, C. R. (1986). Cliff swallow colonies as information centers. *Science*, **234**, 83–5.

Esser, K.-H. (1994). Audio-vocal learning in a non-human mammal: the lesser spear-nosed bat *Phyllostomus discolor*. *Neuroreport*, **5**, 1718–20.

Fenton, M. B. (1985). *Communication in the Chiroptera*. Bloomington: Indiana University Press.

Fenton, M. B., Thomas, D. W. and Saseen, R. (1981). *Nycteris grandis* (Nycteridae): an African carnivorous bat. *J. Zool. Lond.*, **194**, 461–5.

Fleming, T. H. (1988). *The Short-Tailed Fruit Bat*. Chicago: Chicago University Press.

Fleming, T. H. and Williams, C. F. (1990). Phenology, seed dispersal, and recruitment in *Cecropia peltata* (Cecropiaceae) in Costa Rican tropical dry forest. *J. Trop. Ecol.*, **6**, 163–78.

Gardner, A. L. (1977). Feeding habits. In *Biology of Bats of the New World Family Phyllostomatidae*, Part II, ed. R. J. Baker, J. K. Jones, Jr and D. C. Carter, pp. 293–350. Lubbock: Texas Tech Press.

Gorchov, D. L., Cornejo, F., Ascorra, C. F. and Jaramillo, M. (1995). Dietary overlap between frugivorous birds and bats in the Peruvian Amazon. *Oikos*, **74**, 235–50.

Gould, E. (1977). Echolocation and communication. In *Biology of Bats of the New World Family Phyllostomatidae*, Part II, ed. R. J. Baker, J. K. Jones, Jr and D. C. Carter, pp. 247–80. Lubbock: Texax Tech Press.

Griffin, D. R. (1971). The importance of atmospheric attenuation for the echolocation of bats (Chiroptera). *Anim. Behav.*, **19**, 55–61.

Guppy, A., Coles, R. B. and Pettigrew, J. D. (1985). Echolocation and acoustic communication signals in the Australian ghost bat (*Macroderma gigas*). *Austr. Mammal.*, **6**, 299–308.

Habersetzer, J. (1981). Adaptive echolocation sounds in the bat *Rhinopoma hardwickei*. *J. Comp. Physiol.*, **144**, 559–66.

Heinrich, B. (1988). Winter

foraging at carcasses by three sympatric corvids, with emphasis on recruitment by the raven, *Corvus corax*. *Behav. Ecol. Sociobiol.*, **23**, 141–56.

Heller, K.-G. and Helverson, O. v. (1989). Resource partitioning of sonar frequency bands by rhinolophoid bats. *Oecologia*, **80**, 178–86.

Helverson, O. v. (1993). Adaptations of flowers to the pollination by glossophagine bats. In *Plant–Animal Interactions in Tropical Environments*, ed. E. A. Barthlott, pp. 41–59. Bonn: Museum Alexander Koenig.

Hickey, M. B. C. and Fenton, M. B. (1990). Foraging by red bats (*Lasiurus borealis*): do intraspecific chases mean territoriality? *Can. J. Zool.*, **68**, 2477–82.

Howell, D. J. (1979). Flock foraging in nectar-feeding bats: advantages to the bats and to the host plants. *Am. Nat.*, **114**, 23–49.

Jones, G., Duverge, P. L. and Ransome, R. D. (1995). Conservation biology of an endangered species: field studies of greater horseshoe bats. *Symp. Zool. Soc. Lond.*, **67**, 309–24.

Jones, G. and Ransome, R. D. (1993). Echolocation calls of bats are influenced by maternal effects and change over a lifetime. *Proc. R. Soc. Lond, B*, **252**, 125–8.

Kanwal, J., Suga, N. and Matsumura, Y. (1994). The vocal repertoire of the moustached bat, *Pteronotus parnelli*. *J. Acoust. Soc. Am.*, **96**, 1229–54.

Keen, R. and Hitchcock, H. B. (1980). Survival and longevity of the little brown bat (*Myotis lucifugus*) in southeastern Ontario. *J. Mammal.*, **61**, 1–7.

Laland, K. N., Richerson, P. J. and Boyd, R. (1993). Animal social learning: toward a new theoretical approach. *Persp. Ethol.*, **10**, 249–77.

Laland, K. N., Richerson, P. J. and Boyd, R. (1996). Developing a theory of animal social learning. In *Social Learing in Animals: The Roots of Culture*, ed. C. M. Heyes and B. G. J. Galef, pp. 129–54. New York: Academic Press.

McCracken, G. F. and Bradbury, J. W. (1981). Social organization and kinship in the polygynous bat *Phyllostomus hastatus*. *Behav. Ecol. Sociobiol.*, **8**, 11–34.

McWilliam, A. N. (1986). The feeding ecology of *Pteropus* in north-eastern New South Wales, Australia. *Myotis*, **23**, 201–8.

Miller, L. A. and Degn, H. J. (1981). The acoustic behavior of four species of vespertilionid bats studied in the field. *J. Comp. Physiol.*, **142**, 67–74.

Nelson, J. E. (1965). Movements of Australian flying foxes (Pteropodidae: Megachiroptera). *Austr. J. Zool.*, **13**, 53–73.

Neuweiler, G., Metzner, W., Heilmann, U., Rubsamen, R., Ekrich, M. and Costa, H. H. (1986). Foraging behaviour and echolocation in the rufous horseshoe bat (*Rhinolophus rouxi*) of Sri Lanka. *Behav. Ecol. Sociobiol.*, **20**, 53–67.

Obrist, M. (1995). Flexible bat echolocation: the influence of individual, habitat, and conspecifics on sonar signal design.

Behav. Ecol. Sociobiol., **36**, 207–19.

Pijl, L. v. d. (1961). Ecological aspects of flower evolution. II. Zoophilous flower classes. *Evolution*, **15**, 44–59.

Pye, J. D. (1972). Bimodal distribution of constant frequencies in some Hipposiderid bats (Mammalia: Hipposideridae). *J. Zool., Lond.*, **166**, 323–35.

Rabenold, P. P. (1987). Recruitment to food in black vultures: evidence for following from communal roosts. *Anim. Behav.*, **35**, 1775–85.

Racey, P. A. and Swift, S. M. (1985). Feeding ecology of *Pipistrellus pipistrellus* (Chiroptera: Vespertilionidae) during pregnancy and lactation. I. Foraging behaviour. *J. Anim. Ecol.*, **54**, 205–15.

Richards, G. C. (1995). A review of ecological interactions of fruit bats in Australian ecosystems. *Symp. Zool. Soc. Lond.*, **67**, 79–96.

Rydell, J. (1986). Foraging and diet of the northern bat *Eptesicus nilssoni* in Sweden. *Holarct. Ecol.*, **9**, 272–6.

Rydell, J. (1993). Variation in the sonar of an aerial-hawking bat (*Eptesicus nilssonii*). *Ethology*, **93**, 275–84.

Sazima, I. and Sazima, M. (1977). Solitary and group foraging: two flower-visiting patterns of the lesser spear-nosed bat (*Phyllostomus discolor*). *Biotropica*, **9**, 213–15.

Scherrer, J. A. and Wilkinson, G. S. (1993). Evening bat isolation calls provide evidence for heritable signatures. *Anim. Behav.*, **46**, 847–60.

Schmidt, U. (1975). Vergleichende Riechschwellen bestimmungen bei neotropischen Chiropteran (*Desmodus rotundus, Artibeus jamaicensis, Phyllostomus discolor*). *Z. Säugetierkd.*, **40**, 269–98.

Simmons, J. A. and Stein, R. A. (1980). Acoustic imaging in bat sonar: echolocation signals and the evolution of echolocation. *J. Comp. Physiol.*, **135**, 61–84.

Speakman, J. R., Bullock, D. J., Eales, L. A. and Racey, P. A. (1992). A problem defining temporal pattern in animal behaviour: clustering in the emergence behaviour of bats from maternity roosts. *Anim. Behav.*, **43**, 491–500.

Speakman, J. R., Stone, R. E. and Kerslake, J. E. (1995). Temporal patterns in the emergence behaviour of pipistrelle bats, *Pipistrellus pipistrellus*, from maternity colonies are consistent with an anti-predator response. *Anim. Behav.*, **50**, 1147–56.

Thomas, D. W. (1983). The annual migration of three species of West African fruit bats (Chiroptera: Pteropodidae). *Can. J. Zool.*, **61**, 2266–72.

Thomas, D. W. and Fenton, M. B. (1978). Notes on the dry season roosting and foraging behaviour of *Epomophorus gambianus* and *Rousettus aegyptiacus* (Chiroptera: Pteropodidae). *J. Zool., Lond.*, **186**, 403–6.

Tidemann, C. R., Priddel, D. M., Nelson, J. E. and Pettigrew, J.

D. (1985). Foraging behaviour of the Australian ghost bat, *Macroderma gigas* (Microchiroptera: Megadermatidae). *Austr. J. Zool.*, **33**, 705–13.

Turner, D. C. (1975). *The Vampire Bat: A Field Study in Behavior and Ecology.* Baltimore: Johns Hopkins University Press.

Tuttle, M. D. and Ryan, M. J. (1981). Bat predation and the evolution of frog vocalizations in the neotropics. *Science*, **214**, 677–8.

Vaughan, T. A. (1976). Nocturnal behavior of the African false vampire bat (*Cardioderma cor*). *J. Mammal.*, **57**, 227–48.

Vaughan, T. A. and Vaughan, R. P. (1986). Seasonality and the behavior of the African yellow-winged bat. *J. Mammal.*, **67**, 91–102.

Vehrencamp, S. L., Stiles, F. G. and Bradbury, J. W. (1977). Observations on the foraging behavior and avian prey of the neotropical carnivorous bat, *Vampyrum spectrum. J. Mammal.*, **58**, 469–78.

Vickery, W. L., Giraldeau, L-A., Templeton, J. J., Kramer, D. L. and Chapman, C. A. (1991). Producers, scroungers, and group foraging. *Am. Nat.*, **137**, 847–63.

Wallins, L. (1961). Territorialism on the hunting ground of *Myotis daubentoni. S. Mittelungea*, **9**, 156–9.

Waltz, E. C. (1982). Resource characteristics and the evolution of information centers.

Am. Nat., **119**, 73–90.

Wenzel, J. W. and Pickering, J. (1991). Cooperative foraging, productivity, and the central limit theorem. *Proc. Natl. Acad. Sci. USA*, **88**, 36–8.

Werner, T. K. and Gardner, A. L. (1978). Observations of fruit consumption in some stenodermine bats. *Bat Res. News*, **19**, 112.

Wilkinson, G. S. (1984). Reciprocal food sharing in vampire bats. *Nature*, **309**, 181–4.

Wilkinson, G. S. (1985). The social organization of the common vampire bat. I. Pattern and cause of association. *Behav. Ecol. Sociobiol.*, **17**, 111–21.

Wilkinson, G. S. (1987). Altruism and cooperation in bats. In *Recent Advances in the Study of Bats*, ed. P. A. Racey, M. B. Fenton and J. M. V. Rayner, pp. 299–323. Cambridge: Cambridge University Press.

Wilkinson, G. S. (1992). Information transfer at evening bat colonies. *Anim. Behav.*, **44**, 501–18.

Wilkinson, G. S. (1995). Information transfer in bats. *Symp. Zool. Soc. Lond.*, **67**, 345–60.

Wilkinson, G. S. and Boughman, J. W. (1998). Social calls coordinate foraging in greater spear-nosed bats. *Anim. Behav.*, **55**, 337–50.

Wilkinson, G. S. and Fleming, T. H. (1996). Migration and evolution of lesser long-nosed bats, *Leptonycteris curasoae. Mol. Ecol.*, **6**, 131–9.

12

Social transmission of information in a eusocial rodent, the naked mole-rat (*Heterocephalus glaber*)

Chris G. Faulkes

Introduction

The African mole-rats of the family Bathyergidae are subterranean hystricomorph rodents endemic to sub-Saharan Africa. They occur in a wide range of physically and climatically divergent habitats (from mesic to xeric) and show a diversity of social systems (Jarvis *et al.* 1994, Faulkes *et al.* 1997b, Lacey and Sherman 1997). Of a total of 18 or more species, four are solitary dwelling and plural occupancy of burrows only occurs during the breeding season. At least nine taxa are colonial and, depending on the species, burrows may contain up to approximately 14 individuals. Some, or perhaps all, of these colonial mole-rats are cooperative breeders with colonies composed of a breeding pair and their offspring, who delay dispersal and reproduction until favourable environmental conditions occur (Jarvis and Bennett 1991, Jarvis *et al.* 1994, Faulkes *et al.* 1997b).

Sociality reaches a pinnacle in two other species, the naked mole-rat, *Heterocephalus glaber*, and the Damaraland mole-rat, *Cryptomys damarensis*. Both of these species are cooperative breeders with a high reproductive skew (Sherman *et al.* 1995), as only a small percentage of individuals gain the opportunity to breed. Both species, but the naked mole-rat in particular, have been compared to the social invertebrates like bees, ants and termites. They exhibit the characteristics that define such 'eusocial' insect species (Michener 1969, Wilson 1971), i.e. colonies contain overlapping generations, there is cooperative care of offspring, and a there is clear division of reproductive labour in which only a few individuals produce offspring (Jarvis 1981, Jarvis and Bennett 1993). The reproductively suppressed, subordinate group members cooperate in rearing offspring and protecting and maintaining the colony (Jarvis 1981, 1991, Lacey and Sherman 1991, Faulkes *et al.* 1991a). One large, dominant female (the 'queen'; Jarvis 1981) controls the reproduction of all the other males and females in a colony that may commonly contain around 100 individuals. The queen mates with one, two or sometimes three specific breeding males (Jarvis 1981, 1991, Lacey and Sherman 1991). Even though these small (20–80 g), highly specialised subterranean rodents can live in excess of 20 years in captivity, in the wild most stand no chance of reproducing in their lifetime, because there is only very limited opportunity to attain queen or breeding male status, or to disperse from their natal colony (Jarvis *et al.* 1994, Brett, 1991a, Sherman *et al.* 1992, O'Riain *et al.* 1996).

Naked mole-rats live in a closed society, where their subterranean habitat, eusociality, extreme specialisation to a fossorial lifestyle (e.g. degenerate vision and poikilothermy) and xenophobic aggression to conspecifics from sur-rounding colonies, lead to very limited dispersal and outbreeding (O'Riain *et al.* 1996), and consequently to high levels of inbreeding (Reeve *et al.* 1990, Faulkes *et al.* 1990a, 1997a). Because of the extremely high genetic relatedness within a colony, nonbreeding naked mole-rats contribute to their own fitness indirectly by helping their parents or siblings rear offspring (inclusive fitness, Hamilton 1964).

Within the family Bathyergidae, social group size is correlated with rainfall and the distribution of food. Solitary and less social species tend to be found in mesic areas with higher, predictable rainfall and uniformly distributed food in the form of underground roots and tubers, which form the staple diet of mole-rats. The two eusocial species are found in arid regions with unpredict-able rainfall. Geophytes (bulbs, corms and tubers) are patchy and widely dispersed, and the soil is generally hard, making burrowing energetically expensive (Lovegrove 1991, Jarvis and Bennett 1993, Jarvis *et al.* 1994). These ecological constraints impose a high cost on dispersal and individual repro-duction and, together with kin selection, have probably played a major role in the evolution of cooperative breeding strategies and eusociality in mole-rats (Jarvis *et al.* 1994, Faulkes *et al.* 1997b).

The subterranean niche

Naked mole-rats inhabit the arid regions of Kenya, Ethiopia and Somalia and live totally underground in extensive burrow systems that can comprise 3–4 km of tunnels. Most of the burrow is composed of foraging tunnels that are dug in search of underground roots and tubers (Brett 1991a,b). The central part of this labyrinth contains communal nest and toilet chambers, important focal points which may act as 'information centres' where individuals can interact socially on a regular basis.

Sensory cues derived from the social environment of an individual have important influences on behaviour and subsequent learning. The subterranean environment clearly imposes constraints on the mode of communication among individuals. In common with other subterranean animals, the visual system of the naked mole-rat is degenerate compared with terrestrial diurnal mammals (Eloff 1958), and social information is transmitted via other sensory modalities, by the use of tactile, olfactory and auditory cues, although even airborne sound may be restricted in small tunnels (Heffner and Heffner 1993).

Undoubtedly touch, odour and sound are all important routes of communication in naked mole-rats. Although naked mole-rats lack fur, their bodies are covered in vibrissae (Thigpen 1940), which could potentially give a highly developed tactile sense. The importance of information conveyed via the whiskers is well documented in rats and mice, where each individual facial whisker has its own barrel field, a discrete area in the cortex of the brain dedicated to processing incoming stimuli (Woolsey *et al.* 1975). These barrel fields have also been reported in a subterranean mammal, the starnosed mole (*Condylura crista*, Catania and Kaas 1995). Sensory vibrissae may function to enable naked mole-rats to orientate when travelling down the narrow tunnels of their burrow, and to attenuate other tactile cues.

Mole-rats spend a large amount of time in close body contact, huddling in the communal nest chamber (Lacey *et al.* 1991). One particular tactile cue, 'shoving', is a characteristic behaviour that occurs when two animals press their muzzles together, and one aggressively pushes the other backwards, sometimes a distance of up to one metre (Lacey *et al.* 1991). Shoving behaviour is elicited mainly by the breeding queen and may have a number of important roles, discussed later.

In terms of olfaction, one would expect mole-rats to have a well developed sense of smell, given their poor visual sense. Although little research has focused on the anatomy and physiology of the olfactory sense in naked mole-rats, in keeping with other rodents they do possess a vomeronasal organ (D.H. Abbott, pers. commun.). This accessory olfactory system is specifically involved in the chemoreception of pheromonal signals (Harrison 1987). Learned responses to odour cues are important in a number of aspects of naked mole-rat social behaviour.

The underground niche also has an influence on the auditory environment. In keeping with other subterranean species, the hearing capabilities of the naked mole-rat are degenerate compared with terrestrial rodents, and restricted in their range. Naked mole-rats have a limited ability to detect sound, have limited high frequency hearing, and have difficulty in localising sound, compared with surface dwelling rodents (Heffner and Heffner 1993). Despite this, naked mole-rats have the largest known vocal repertoire of any rodent; 19 vocalisations are known, most of which are associated with particular contexts or behaviours (Pepper *et al.* 1991, Judd and Sherman 1996).

Social transfer of information – finding food

Since the first report of eusociality in naked mole-rats (Jarvis 1981), research into their ecology, behaviour, physiology and genetics, and comparative studies with other bathyergids, have progressed steadily (Sherman *et al.* 1992, Jarvis *et al.* 1994, Faulkes and Abbott 1997). Moreover, given the lifestyle and characteristics of naked mole-rats there is a number of questions that may be asked with reference to the occurrence of social transfer of information.

A good example of social learning recently shown in captive groups of naked mole-rats relates to foraging behaviour (Judd and Sherman 1996). In a way that bears an interesting similarity to eusocial bees dancing to communicate the whereabouts of food, it has been shown that naked mole-rats recruit colony mates to food sources, apparently by laying down an odour trail. In captivity, naked mole-rats are kept in artificial burrow systems mimicking those in the wild (albeit on a much smaller scale), consisting of a network of interconnecting Perspex tunnels, with Perspex nest and toilet chambers, and other chambers where food is introduced. Judd and Sherman (1996) withheld food in captive colonies of naked mole-rats for around 16 hours. They then found that individual foragers, when finding a particular food source for the first time after food deprivation, usually gave a characteristic 'chirp-like' vocalisation on returning to the nest chamber (on 74% of occasions). These specific chirp vocalisations not only alert colony members that a food source is present, but may also be individual-specific, thus identifying the forager. Subsequently, other colony members are able to find a food source discovered by a colony mate by following its pathway through the tunnels of the artificial burrow system, irrespective of whether the direction of tunnelling was the same or had been changed by modifying the layout of the tunnels in the captive colony (i.e. by swapping the Perspex tunnels around). Also, these pathway preferences were lost if the Perspex tunnels were cleaned, implying that recruits follow an odour trail laid down by the forager, probably by skin contact between the forager and the tunnel surface. The source of odour may come from urine smeared over the head and shoulders by the feet, a characteristic behaviour exhibited by mole-rats following urination in the communal toilet chamber (Lacey *et al.* 1991).

The chirp vocalisation and associated odour cue means that an individual who has just successfully foraged is recognised and only this particular trail is followed amongst the myriad of other odours within the colony. It has already been noted that in the habitat of the naked mole-rat, food (underground roots and tubers) is patchily distributed and widely dispersed (Brett 1991b), and the condition of the soil makes burrowing for food time consuming, risky and energetically expensive (Lovegrove 1991). Socially learnt foraging behaviour could therefore certainly have an adaptive function in the wild, aiding navigation through complex burrows with the minimum expenditure of energy, increasing the efficiency of foraging and thereby benefiting the colony and ultimately the inclusive fitness accrued by nonbreeding colony members. Similar foraging trails have also been described for terrestrial rodents (e.g. Norway rats, Galef and Buckley 1996, reviewed by Judd and Sherman 1996).

This example of social learning shows the importance of both auditory and olfactory cues in a subterranean mammal, and the functioning of the communal nest chamber as a centre where exchange of information occurs. Apart from this clear-cut example, there is a number of cases in which predominantly circumstantial evidence from naked mole-rat studies opens up questions with respect to functional implications of the social transfer of information among natural patterns of behaviour.

Social transfer of information within a behavioural division of labour

In addition to the clear cut partitioning of reproduction within colonies of naked mole-rats, there is a behavioural division of labour among nonbreeding animals that appears to be related to body size and age although, unlike social insects, discrete morphological castes are not apparent (Faulkes *et al.* 1991a, Jarvis *et al.* 1991, Lacey and Sherman 1991). Small nonbreeding animals generally carry out more of the maintenance activities of the colony; these include bringing food back to the nest chamber from foraging sites, cleaning and maintaining tunnels, and fetching material for 'bedding' and nest building. Larger individuals display higher frequencies of defense related activities such as nest guarding, patrolling the burrow, and threatening snakes and unfamiliar conspecifics (Faulkes *et al.* 1991a, Jarvis *et al.* 1991, Lacey and Sherman 1991).

In the absence of distinct behavioural castes containing individuals with irreversible morphological specialisations, the behavioural division of labour in naked mole-rat colonies has been shown to be dynamic. Among captive colonies, the relationship between behaviour, and age and body mass, shows some variation, leading to confusion as to whether behavioural role depends strictly on age (age polyethism) or on body mass (body mass polyethism). Differences arise because while age and body mass are correlated in naked mole-rats, there are significant differences in growth rates between different litters, with a trend towards a lower asymptotic body mass with increasing litter order (O'Riain 1996). Interestingly, while body size may remain constant in some individuals for prolonged periods of time, growth may restart and body mass change in response to changes in the social environment of the colony, particularly following death or removal of breeding animals (O'Riain 1996, Lacey and Sherman 1997, Clarke and Faulkes 1997). The flexibility in growth patterns and lack of morphological specialisation among nonbreeders means that naked mole-rats can show considerable behavioural plasticity in response to changing social and environmental conditions. This ability to optimise the behavioural and morphological profile of the colony may well be adaptive as tasks can be matched to body size, and colony efficiency and therefore inclusive fitness of individuals will be maximised (O'Riain 1996).

Given this dynamic state of affairs, what part does social learning play in determining the behavioural role of a particular individual in a naked mole-rat colony? At present, this is a question we cannot fully answer. We know that behaviour is correlated with body size, and to some extent age. Also, dominance is significantly correlated with body mass and the attainment of reproductive status (Clarke and Faulkes 1997) and it seems likely that dominance, growth, behavioural role and reproductive success may all be mechanistically linked. Despite an innate component to some of the specific behaviours observed in naked mole-rats, social interactions do play an important part in modulating the behavioural dynamics of colonies. In captivity, new colonies can be formed by separating a pair of nonbreeding animals of as little as five months of age, who become reproductively active and commence breeding on removal from the suppressing influences of their parent colony (Faulkes and Abbott 1991b). Colonies formed in this way often show rapid recruitment (litters average around 12 animals, interbirth intervals average 79 days; Jarvis 1991), and individuals display the full repertoire of behaviours (O'Riain 1996). However, this observation does not preclude the possibility that, apart from an innate component, the successive litters of offspring also learn their behavioural repertoire, initially from the parent breeding pair, then from the progressively more experienced adults. One example of this has been described by O'Riain (1996), where he found that the effficiency and the frequency with which juveniles performed tasks improved with increasing age, and that juveniles often solicited digging or gnawing sites that were frequently worked by more experienced adults. To accomplish this, the juveniles physically encroached on the work space of the adult whilst emitting a characteristic 'loud chirp' vocalisation often associated with mild conflict (Pepper *et al.* 1991; Figure 12.1), and subsequently displaced the latter to take over digging or gnawing. This has a functional significance, as the new work force is able to focus attention on tasks prioritised by more experienced workers. Together

Figure 12.1. Competition for work in a captive colony of naked mole-rats, a behaviour often observed in juveniles who encroach on the work space of adults. During these encounters a characteristic 'loud chirp' vocalisation is emitted and often mild conflict occurs, in this case 'tugging' behaviour (Lacey *et al.* 1991). Thus, juvenile recruits to the colony work force tend to focus their efforts in areas prioritised by more experienced individuals, indicating a learned component to task selection (O'Riain 1996).

with the food recruitment chirp, these are two examples of a vocalisation being associated with a learning process in naked mole-rats, and emphasises the importance of this sensory modality in the subterranean niche.

Social learning and reproductive suppression

Although distinct behavioural castes are absent in naked mole-rats, the reproductive division of labour is very clear, and in almost all colonies, both in captivity and in the wild, reproduction is restricted to a single breeding queen and one to three specific breeding males. During oestrus the queen may solicit mating from one or all the breeding males (in colonies where more than one is present), and in captivity at least, multi-male paternity of litters is possible (Faulkes *et al.* 1997a). The monopoly of reproduction is remarkable given that wild colonies often contain around 100 individuals, and it has been estimated that over 99% of naked mole-rats will never breed (Jarvis *et al.* 1994). It is reasonable to consider that socially aquired information plays a role in maintaining this asymmetry in reproduction. Both pheromones (an odour signal emitted by one member of a species that may elicit a behavioural and/or physiological response in a recipient of the same species) and stress responses arising from behavioural harassment of subordinate nonbreeding group members by dominants have been shown to operate in suppression of reproduction in other mammals. For example, in common marmoset monkeys, *Callithrix jacchus*, reproductive suppression in socially subordinate females appears to arise from psychological conditioning involving social learning. Associative learning of olfactory (pheromonal) and visual cues from dominant females leads to an anovulatory state in the nonbreeding females (Abbott *et al.* 1997). In naked mole-rats, a similar mechanism may operate. We have been unable to find any evidence for a primer pheromone effect in reproductive suppression, whereby the queen suppresses nonbreeders with specific urinary chemosignals, to which the latter are exposed in the communal toilet (Faulkes and Abbott 1993, Smith *et al.* 1997). Our research so far suggests that reproductive suppression in the nonbreeding animals of both sexes is imposed by social cues that involve direct contact between the dominant queen and the nonbreeders, possibly involving an element of conditioning and social learning (Faulkes and Abbott 1993). These social cues are translated into a disruption of the normal patterns of secretion of luteinising hormone (LH), and possibly also follicle stimulating hormone (FSH), from the anterior pituitary gland. In turn this results in inadequate hormonal stimulation of the gonads by these pituitary gonadotrophins, and ultimately in a state of infertility (for review see Faulkes and Abbott 1997). In females, impaired gonadotrophin secretion ultimately results in a failure of ovarian cycles and ovulation, and the reproductive tract and ovaries of suppressed nonbreeders remain in a pre-pubescent state (Faulkes *et al.* 1990b,c). Similarly, among male naked mole-rats, altered secretion of pituitary LH gives rise to clear physiological differences between breeders and nonbreeders (Faulkes *et al.* 1991b,c, 1994).

The exact nature of the proximate factors, and the mechanism that translates behavioural cues from the queen into the neuroendocrine and physiological inhibition of normal reproductive function in nonbreeding naked mole-rats, remain to be elucidated. Dominance and agonistic interactions are undoubtedly important, as upon their removal or death, both breeding queens and breeding males are succeeded by high ranking nonbreeders (Jarvis 1991, Lacey and Sherman 1991, Clarke and Faulkes 1997, Clarke and Faulkes, unpublished). Overt aggression in naked mole-rat colonies is rare except during periods of the succession of the queen, and it is at these times that the social hierarchy is rearranged and the new queen establishes her dominance over the other colony members (Clarke and Faulkes 1997). These may be critical periods when aggression leads to suppressed reproduction and condi-

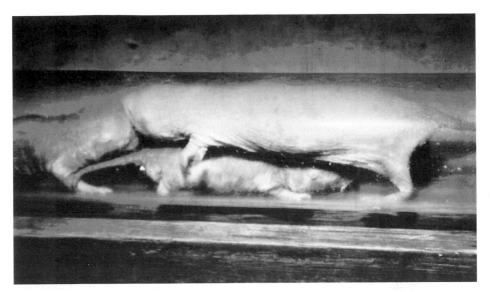

Figure 12.2. A dominant breeding queen naked mole-rat passes over a subordinate nonbreeding animal in a tunnel of a captive colony, and 'shoves' another subordinate nonbreeder. Both these behaviours may be important cues in learned association of dominance and subsequent suppression of reproduction in nonbreeding animals.

tioning, so that subsequently more subtle cues from the queen are sufficient to maintain suppression. There is now good evidence that a mechanism such as this occurs in cooperatively breeding common marmosets (Abbott *et al.* 1997), and if this is also the case in naked mole-rats, then social learning and conditioning play a central role in the most extreme example of socially induced reproductive suppression in mammals.

One possible behaviour observed in naked mole-rat colonies that is implicated in this conditioning by social learning and maintenance of suppression is 'shoving', a behaviour almost entirely restricted to the breeding animals (Figure 12.2). The queen initiates shoving with the highest frequency, and she directs this behaviour at larger individuals of both sexes (Reeve and Sherman 1991, Clarke and Faulkes 1997).

Currently, there are two hypotheses regarding the function of shoving behaviour. The threat-reduction hypothesis proposes that shoving reduces the threat of challenges for the queen's reproductive status (Reeve 1992) and is a key behaviour in the establishment and imposition of dominance (Clarke and Faulkes 1997). The activity-incitation hypothesis proposes that shoving serves to incite activity in 'lazy workers' (Reeve and Sherman 1991, Reeve 1992). These two hypotheses are by no means mutually exclusive, and behavioural data support both. From an evolutionary point of view, maintaining social order, inhibiting reproduction in subordinates of both sexes, and encouraging work-related activity would all be expected ultimately to increase the reproductive success of the queen. Toleration of this would be adaptive in the nonbreeders because the high costs of dispersal preclude individual reproduction, but they accrue inclusive fitness benefits when the queen reproduces. The two hypotheses for the function of shoving are also consistent with the observation that in captive colonies containing animals of mixed kinship, the queen shoves larger, less related individuals more frequently (Reeve and Sherman 1991). These individuals would be expected to have more to gain by escaping suppression and attaining reproductive status, and less to gain by helping/working, because as nonbreeders their inclusive fitness benefits are lessened by their reduced degree of relatedness with the queen.

Shoving also has several associated vocalisations which may serve to reinforce the behaviour and possibly to identify the individuals involved in the interaction. During mild conflict between colony members, including shoving, a 'loud chirp' may be emitted, although interestingly this has not been observed in breeding queens and therefore would normally be restricted to individuals that are the recipients of shoving (Pepper *et al.* 1991). An 'upsweep trill' may also be emitted following aggressive encounters between the queen and nonbreeding conspecifics. Colony members respond to these upsweep trills by markedly increasing their activity and rates of vocalising, mostly with 'soft chirps', and often also by pressing noses with the breeding female (Pepper *et al.* 1991). This observation of nose pressing is similar to the behaviour observed during passing in tunnels. When individuals meet face to face they also press noses and sniff, implying that individual recognition may be occurring. Following this, one animal will generally climb over the top of the other (occasionally side to side passes occur) in a consistent relationship, where the socially dominant animal passes over the top and the subordinate beneath (Clarke and Faulkes 1997; Figure 12.2). It seems plausible that shoving (and also the rarer forms of agonistic interactions), as well as more subtle cues such as tunnel passing, together with individual recognition by both odours and vocalisation, all serve to enforce dominance relationships by social learning. Dominance status, suppression of reproduction and possibly also growth, body size and behavioural role, may all be interlinked and dependent on these socially learnt relationships between individuals.

Social learning, xenophobia and inbreeding in naked mole-rats

Naked mole-rats are highly xenophobic to foreign conspecifics, even though they may be genetically similar and share common maternal ancestors; adjacent colonies in the wild may have arisen by the splitting and budding of an existing colony (Brett 1991a, Reeve *et al.* 1990, Lacey and Sherman 1997, Faulkes *et al.* 1990a, 1997b). Naked mole-rats appear to have mechanisms based on learnt recognition of individual and colony odours that enable them to recognise unfamiliar conspecifics (O'Riain and Jarvis 1997). This ability to discriminate on the basis of odour, together with identification by individual-specific vocalisations and vocalisations specific to encounters with foreign conspecifics, mentioned previously (Pepper *et al.* 1991, Judd and Sherman 1996), emphasises the importance of social aquisition of information in the recognition of individuals in colonies.

Unfamiliar individuals from foreign colonies will be attacked and deaths often result. The learnt recognition of familiar colony members appears to be acute and may need frequent reinforcement by continual exposure of individuals to each other. In captivity it is often difficult to reintroduce a mole-rat that has been removed from its colony for periods of as little as 12 hours, as the absentee is treated as foreign and attacked (O'Riain and Jarvis 1997). Despite these xenophobic encounters, certain individuals, almost always males, are able to infiltrate foreign colonies and disperse, resulting in occasional out-breeding events (O'Riain *et al.* 1996). Putative dispersers are morphologically, behaviourally and physiologically distinct, with larger fat reserves than colony mates of similar skeletal proportions, a tendency to solicit mating with non-colony members, and levels of reproductive hormones similar to breeding males. Disperser males appear to circumvent the normal behavioural reaction from unfamiliar conspecifics to being foreign by not acting aggressively when meeting an unfamiliar individual, and thus are able to integrate into the colonies into which they migrate (O'Riain *et al.* 1996, F.M. Clarke, unpublished).

Conclusions

Colonies of eusocial naked mole-rats, complex social systems with overlapping generations, provide opportunities for social learning, and there are specific examples now emerging from research. Some of these socially learnt behaviours are remarkably similar to those seen in eusocial insects, for example communicating the location of food and the recognition of nestmates. What is becoming apparent is that intra- and inter-colony recognition of individuals is critically important in underpinning the process of social learning in mole-rat colonies, and is central in maintaining and modulating the social hierarchy and the reproductive and behavioural roles of individuals. While the possibility of individual-specific vocalisations exists, odour signals are known to be involved in recognition in mole-rats (O'Riain and Jarvis 1997). The role of the major histocompatibility complex of genes (MHC) and its breakdown products as a genetic determinant of individual specific odours has received much attention (e.g. in mice; Potts *et al.* 1991). Whether such a mechanism operates in naked mole-rats is at present unknown. O'Riain and Jarvis (1997) failed to identify a genetic component to colony recognition and proposed that familiarity is the criterion for the xenophobic behaviour in naked mole-rats. Furthermore, colonies of naked mole-rats are known to lack genetic variation at normally highly polymorphic loci, including minisatellite DNA and MHC genes (Reeve *et al.* 1990, Faulkes *et al.* 1990a, 1997b), potentially reducing the effectiveness of these loci in producing individual specificity. Perhaps one of the most intriguing questions remaining to be tackled in our studies of naked mole-rats is how and to what extent individuals are recognised and identified in colonies that may sometimes contain hundreds of animals? One potential mechanism could be to learn more generally what is familiar (the group odour), rather than to identify each individual within a colony (specific recognition), and react to the unfamiliar. This would certainly explain the xenophobic nature of naked mole-rats, but not the intra-colony interactions that imply learnt recognition of individuals on a large scale. While research is fairly advanced in the naked mole-rat, less is known about the extent of social learning in other species in the family. Given the range of social strategies exhibited, future comparative studies within the African mole-rat family could provide a unique opportunity to investigate the development of social learning in both an evolutionary and ecological context.

Acknowledgements

I would like to thank Frank Clarke, Mike Bruford, Nick Oguge and Paul Sherman for comments on the manuscript; Mike Llovet, Mandy Gordon, and Jake Rozowski for care of our captive colonies of naked mole rats; Hilary Box for inviting and inspiring me to write this paper and the Institute of Zoology and the Royal Society for financial support.

References

Abbott, D. H., Saltzman, W., Schultz-Darken, N. J. and Smith, T. E. (1997). Specific neuroendocrine mechanisms not involving generalized stress mediate social regulation of fe-male reproduction in cooperatively breeding mar-moset monkeys. *Ann. NY Acad.*

Sci., **807**, 219–38.

Brett, R. A. (1991a). The population structure of naked mole-rat colonies. In *The Biology of the Naked Mole-Rat*, ed. P. W. Sherman, J. U. M. Jarvis and R. D. Alexander, pp. 97–136. New York: Princeton University Press.

Brett, R. A. (1991b). The ecology of naked mole-rat colonies: burrowing, food and limiting factors. In *The Biology of the Naked Mole-Rat*, ed. P. W. Sherman, J. U. M. Jarvis and R. D. Alexander, pp. 137–84. New York: Princeton University Press.

Catania, K. C. and Kaas, J. H. (1995). Organization of the somatosensory cortex of the star-nosed mole. *J. Comp. Neurol.*, **351**, 549–67.

Clarke, F. M. and Faulkes, C. G. (1997). Hormonal and behavioural correlates of dominance and queen succession in captive colonies of the eusocial naked mole-rat, *Heterocephalus glaber. Proc. R. Soc. Lond. B*, **264**, 993–1000.

Eloff, G. (1958). The functional and structural degeneration of the eye of the South African rodent moles, *Cryptomy bigalkei* and *Bathyergus maritimus. S. Afr. J. Sci.*, **54**, 293–302.

Faulkes, C. G. and Abbott, D. H. (1991). Social control of reproduction in both breeding and non-breeding male naked mole-rats, *Heterocephalus glaber. J. Reprod. Fertil.*, **93**, 427–35.

Faulkes, C. G. and Abbott, D. H. (1993). Evidence that primer pheromones do not cause so-cial suppression of reproduction in male and female naked mole-rats, *Heterocephalus glaber. J. Reprod. Fertil.*, **99** 225–30.

Faulkes, C. G. and Abbott, D. H. (1997). Proximate mechanisms regulating a reproductive dictatorship: a single dominant female controls male and female reproduction in colonies of naked mole-rats. In *Cooperative Breeding in Mammals*, ed. N. G. Solomon and J. A. French, pp. 302–34. New York: Cambridge University Press.

Faulkes, C. G., Abbott, D. H. and Mellor, A. (1990a). Investigation of genetic diversity in wild colonies of naked mole-rats by DNA fingerprinting. *J. Zool. Lond.*, **221**, 87–97.

Faulkes, C. G., Abbott, D. H. and Jarvis, J. U. M. (1990b). Social suppression of ovarian cyclicity in captive and wild colonies of naked mole-rats, *Heterocephalus glaber. J. Reprod. Fertil.*, **88**, 559–68.

Faulkes, C. G., Abbott, D. H., Jarvis, J. U. M. and Sherriff, F. (1990c). LH responses of female naked mole-rats, *Heterocephalus glaber*, to single and multiple doses of exogenous GnRH. *J. Reprod. Fertil.*, **89**, 317–23.

Faulkes, C. G., Abbott, D. H., Liddell, C. E., George, L. M. and Jarvis, J. U. M. (1991a). Hormonal and behavioural aspects of reproductive suppression in female naked mole-rats *Heterocephalus glaber.* In *The Biology of the Naked Mole-Rat*, ed. P. W. Sherman, J. U. M. Jarvis and R. D. Alexander, pp. 426–45. New York: Princeton University Press.

Faulkes, C. G., Abbott, D. H. and Jarvis, J. U. M. (1991b). Social suppression of reproduction in male naked mole-rats, *Heterocephalus glaber. J. Reprod. Fertil.*, **91**, 593–604.

Faulkes, C. G., Trowell, S. N., Jarvis, J. U. M. and Bennett, N. C. (1994). Investigation of numbers of spermatozoa and sperm motility in reproductively active and socially suppressed males of two eusocial African mole-rats, the naked mole-rat (*Heterocephalus glaber*), and the Damaraland mole-rats (*Cryptomys damarensis*). *J. Reprod. Fertil.*, **100**, 411–16.

Faulkes, C. G., Abbott, D. H., O'Brien, H. P., Lau, L. Roy, M., Wayne, R. K. and Bruford, M. W. (1997a). Micro- and macro-geographic genetic structure of colonies of naked mole-rats, *Heterocephalus glaber. Molec. Ecol.*, **6**, 615–28.

Faulkes, C. G., Bennett, N. C., Bruford, M. W., O'Brien, H. P., Aguilar, G. H. and Jarvis, J. U. M (1997b). Ecological constraints drive social evolution in the African mole-rats. *Proc. R. Soc. Lond. B*, **264**, 1619–27.

Galef, B. G. and Buckley, L. L. (1996). Use of foraging trails by Norway rats. *Anim. Behav.*, **51**, 765–71.

Hamilton, W. D. (1964). The genetical evolution of social behaviour. I, II. *J. Theor. Biol.*, **7**, 1–52.

Harrison, D. (1987). Preliminary thoughts on the incidnce, structure and function of the mammalian vomeronasal organ. *Acta Otolaryngol. Stockholm*, **103**, 489–95.

Heffner, R. S. and Heffner, H. E. (1993). Degenerate hearing and sound localization in naked mole-rats (*Heterocephalus glaber*), with an overview of central auditory structures. *J. Comp. Neurol.*, **331**, 418–33.

Jarvis, J. U. M. (1981). Eu-sociality in a mammal – cooperative breeding in naked mole-rat Heterocephalus glaber colonies. *Science, NY*, **212**, 571–3.

Jarvis, J. U. M. (1991). Reproduction of naked mole-rats. In *The Biology of the Naked Mole-Rat*, ed. P. W. Sherman, J. U. M. Jarvis and R. D. Alexander, pp. 384–425. New York: Princeton University Press.

Jarvis, J. U. M. and Bennett, N. C. (1991). Ecology and behaviour of the family Bathyergidae. In *The Biology of the Naked Mole-Rat*, ed. P. W. Sherman, J. U. M. Jarvis and R. D. Alexander, pp. 66–96. New York: Princeton University Press.

Jarvis, J. U. M. and Bennett, N. C. (1993). Eusociality has evolved independently in two genera of bathyergid mole-rats – but occurs in no other subterranean mammal. *Behav. Ecol. Sociobiol.*, **33**, 253–60.

Jarvis, J. U. M., O'Riain, M. J. and McDaid, E. J. (1991). Growth and factors affecting body size in naked mole-rats. In *The Biology of the Naked Mole-Rat*, ed. P. W. Sherman, J. U. M. Jarvis and R. D. Alexander, pp. 358–83. New York: Princeton University Press.

Jarvis, J. U. M., O'Riain, M. J., Bennett, N. C. and Sherman, P. W. (1994). Mammalian eusociality: a family affair. *Trends Ecol. Evol.*, **9**, 47–51.

Judd T. M. and Sherman, P. W. (1996). Naked mole-rats recruit colony mates to food sources. *Anim. Behav.*, **52**, 957–69.

Lacey, E. A. and Sherman, P. W. (1991). Social organization of naked mole-rat colonies: evidence for a division of labor. In *The Biology of the Naked Mole-Rat*, ed. P. W. Sherman, J. U. M. Jarvis and R. D. Alexander, pp. 275–336. New York: Princeton University Press.

Lacey, E. A. and Sherman, P. W. (1997). Cooperative breeding in naked mole-rats: implications for vertebrate and invertebrate sociality. In *Cooperative Breeding in Mammals*, ed. N. G. Solomon and J. A. French, pp. 267–301. New York: Cambridge University Press.

Lacey, E. A., Alexander, R. D., Braude, S. H., Sherman, P. W. and Jarvis, J. U. M. (1991). An ethogram for the naked mole-rat: nonvocal behaviours. In *The Biology of the Naked Mole-Rat*, ed. P. W. Sherman, J. U. M. Jarvis and R. D. Alexander, pp. 209–42. New York: Princeton University Press.

Lovegrove, B. G. (1991). The evolution of eusociality in mole-rats (Bathyergidae): a question of risks, numbers and costs. *Behav. Ecol. Sociobiol.*, **28**, 37–45.

Michener, C. D. (1969). Comparative social behaviour of bees. *Annu. Rev. Entomol.*, **14**, 299–342.

O'Riain, M. J. (1996). Pup ontogeny and factors influencing behavioural and morphological variation in naked mole-rats Heterocephalus glaber (Rodentia, Bathyergidae). PhD thesis, University of Cape Town, South Africa.

O'Riain, M. J. and Jarvis, J. U. M. (1997). Colony member recognition and xenophobia in the naked mole-rat. *Anim. Behav.*, **53**, 487–98.

O'Riain, M. J., Jarvis, J. U. M. and Faulkes, C. G. (1996). A dispersive morph in the naked-rats. *Nature*, **380**, 619–21.

Pepper J. W., Braude, S. H., Lacey, E. A. and Sherman, P. W. (1991). Vocalizations of the naked mole-rat. In *The Biology of the Naked Mole-Rat*, ed. P. W. Sherman, J. U. M. Jarvis and R. D. Alexander, pp. 243–74. New York: Princeton University Press.

Potts, W. K., Manning, C. J. and Wakeland, E. K. (1991). Mating patterns in seminatural populations of mice influenced by MHC genotype. *Nature*, **352**, 619–21.

Reeve, H. K. (1992). Queen activation of lazy workers in colonies of the eusocial naked mole-rat. *Nature*, **358**, 147–9.

Reeve, H. K., Westneat, D. F., Noon, W. A., Sherman, P. W. and Aquadro, C. F. (1990). DNA 'fingerprinting' reveals high levels of inbreeding in colonies of the eusocial naked mole-rat. *Proc. Natl. Acad. Sci. USA*, **87**, 2496–500.

Sherman, P. W., Jarvis, J. U. M. and Braude, S. H. (1992).

Naked mole-rats. *Sci. Am.*, **267**, 42–8.

Sherman, P. W., Lacey, E. A., Reeve, H. K. and Keller, L. (1995). The eusociality continuum. *Behav. Ecol.*, **6**, 102–8.

Smith, T. E., Faulkes, C. G. and Abbott, D. H. (1997). A behavioural mechanism involving direct contact with the queen plays a major role in the suppression of reproduction in subordinate female naked mole-rats (*Heterocephalus glaber*). *Hormones Behav.*, **31**, 277–88.

Thigpen, L. W. (1940). Histology of the skin of a normally hairless rodent. *J. Mammal*, **21**, 449–56.

Wilson, E. O. (1971). *The Insect Societies.* Cambridge, MA: Harvard University Press.

Woolsey, T. A., Welker, C. and Schwartz, R. H. (1975). Comparative anatomical studies of the SmI face cortex with special reference to the occurrence of 'barrels' in layer IV. *J. Comp. Neurol.*, **164**, 79–94.

Part 4

Social learning among species of
terrestrial carnivores

Editors' comments

HOB and KRG

These chapters discuss social and physical influences upon socially mediated learning among a diversity of terrestrial carnivores. We include examples from the bears, the cats and the canids. Among the bears, for example, and with the exception of females and their cubs, adults spend most of the year alone. They are frequently described as being among the most 'solitary' mammals. They also have large brains and may live long lives. Females produce few offspring at a time and invest a good deal in their care. The cubs spend a long period (as long as three years in some species) of their early lives in close association with their mothers. There are opportunities for information to be transmitted vertically between mother and offspring, about resources, including food and shelter, and hazards – particularly about potential predators. However, it is also relevant to note that, apart from long and close associations with their mother, young bears may also spend a subsequent period in coalitions with their siblings. This provides both protection and the possibility to share information about their environments. We need to know much more about these situations. Barrie Gilbert (chapter 13) provides a useful discussion about opportunities and advantages of social learning among bears. For instance, many bears eat a wide range of plant and animal material. Learning about different foods includes avoiding toxic plants such as poisonous fungi, dealing with animals such as porcupines and obtaining energetically rich foods such as salmon. Survival and fitness are closely associated with resources that are accumulated from a wide variety of foods that are widely distributed in time and space. Information that is acquired about these foods gives significant advantages of fitness to both the adult females and to their young. The high energy requirements needed in the autumn for the costly demands of long dormancy in order to survive the harsh sub-arctic winters presents a critical example. Hence, females lead their offspring to seasonally available sources of food that must be remembered by the young when they live independently. Gilbert's research site on the Alaskan peninsula provides excellent opportunities to observe closely brown bears at open sites in which most bears are well habituated to humans. Interestingly, for example, bears, and commonly females with their cubs, swim across open sea for distances of more than ten miles to feed on lipid rich eggs of sea bird colonies.

The importance of the mother in providing opportunities for her offspring to learn functionally relevant behavioural strategies is discussed further by Kitchener in his chapter on cats (chapter 14). Unlike many of the mammalian groups discussed in this volume, however, there has been a long standing interest in the acquisition of information from conspecifics in cats – at least in some very few species. Moreover, experiments with domestic cats began early in this century. They were not always relevant to natural behaviour but they did show a propensity for the social transfer of information. Further, there has been a variety of experiments in which the salience of the mother as a demonstrator has been emphasised. In fact, experimental studies of socially mediated learning in natural contexts, as in dealing with prey, have provided excellent models of social learning under controlled conditions. The importance of mothers and the motivating role of siblings have been discussed. Moreover, landmark studies of cheetahs in more natural conditions have provided important insights and criteria for ways in which we describe aspects of socially mediated behaviour. Kitchener refers to all these cases, but an important concern is also to discuss natural opportunities for social learning in wider contexts, both taxonomically and functionally. However, despite such excellent paradigms of study that are available, field workers have not shown substantive interest in studying the development of behavioural strategies from a social learning perspective. Species of the cats are predominantly nocturnal and/or crepuscular and have a dispersed, 'solitary' social organisation. Understandably then, studies of cats in nature have mainly involved species such as lions, cheetahs and leopards living in open habitats. Kitchener focuses mainly upon the influence of socially mediated learning in the development of predatory behaviour – because this is where the bulk of the information exists – both experimentally and, less systematically, among wild cats.

Young cats must refine innate prey killing propensities so that they are efficient in the skills of prey-catching by the time that they become independent. There are variations in the length of time that this takes that are associated with larger species that take larger prey. Further, cats are specialised hunters and the young must also acquire information from their mothers about the relative availability of prey species in particular areas, their energetic gain, and how to deal with them. Importantly, in larger species the permanent dentition takes longer to develop, and young cats are dependent upon their mothers for longer; this provides opportunities to learn about dealing with larger and potentially more injurious prey.

In addition to information in the contexts of recognition and hunting of prey, Kitchener considers the potential influence of socially mediated learning in a variety of other contexts and for which much more detailed information is required. Examples include cooperative hunting, the inheritance of territories, the recognition and avoidance of predators, scent marking, communication and cases in which males occasionally leave food at, or close to, the dens in both wild and captive animals. The discussions raise functionally important directions for future research.

Among the canids the social context of behavioural development critically involves the influence of adult males. The most common mating system, for example, is that of long term monogamy. This is a group in which the reproductive success of males is influenced by the quality of their paternal care. Canids are unusual among mammals in that both parents provision the cubs, as when prey is regurgitated or brought back to the pups intact. In some cases parents are also assisted by the young of a previous litter. Moreover, paternal investment may involve more than provisioning of the young. For instance, male bat-eared foxes typically guard as well as initiate the young into foraging strategies during the nursing period – when females are foraging independently.

In his chapter on social learning in canids (chapter 15), Jan Nel emphasises a number of opportunities for social learning to enhance behavioural strategies in response to local environmental conditions that are advantageous to both parents and their offspring. Importantly, there is significant flexibility in both social systems and in diet. There is also frequent exposure of the pups to foraging, feeding, home range, territories and social behaviour. As with so many mammalian groups, however, there is little detailed information for a majority of species. This is also understandable in that many species of canids are nocturnal and of small/medium size. They often hunt in very dense vegetation. Information that is relevant to social learning mainly refers to learning about food. Hence, regurgitation and provisioning with intact prey provide opportunities for induction into preferences for prey in particular areas. Again, in some species individuals hunt or forage together as social units before the pups disperse. In other species of large social canids, many pups stay in the pack. In wolves, for instance, large and potentially dangerous prey are hunted as a pack (in which individuals live for prolonged periods); this provides opportunities to learn how to kill and defend prey. Nel considers various conditions under which opportunities for social learning might occur.

There are some interesting hypotheses for future development. He also gives a series of detailed observations. In bat-eared foxes males initiate their pups into foraging at around the time that the pups are weaned. They lead them further and further from the den. When they capture their invertebrate prey they use a specific call to attract the pups to them. They then release the prey for the young animals to capture for themselves, and repeat the release when the cubs are not successful. Further, cubs stay with their parents for several months after they are weaned and forage in close association with them. Hence, in all, there are opportunities for the young to learn about species of prey as well as the behavioural strategies used to deal with them.

The extents to which such opportunities to learn are critical in the development of behavioural strategies of such animals requires a good deal of detailed substantiation, as Nel emphasises. Such cases, however, provide good information for the development of studies. Nel also refers to experimental studies that do not consider natural situations, but that do provide conditions for relatively close examinations of socially mediated learning in practical situations – as in the use of dogs in the detection of narcotics, and the control of black-backed jackals in preying on domestic livestock. Once again, as with contributions in other parts of this volume, practical applications of interests in social learning studies are a welcome addition to the body of information that is being built up.

13

Opportunities for social learning in bears

Barrie K. Gilbert

Introduction

The Ursidae – the bears – is a family of carnivores that is geographically widespread. Ursids live in a diversity of habitats from the Andean cloud forests to polar ice flows, and eat a wide variety of foods.

The extant bears consist of seven species (see O'Brien 1993): sun bear (*Ursus malayanus*), sloth bear (*Melursus ursinus*), American black bear (*Ursus americanus*), Asiatic black bear (*Ursus thibetanus*), brown bear (*Ursus arctos*), polar bear (*Ursus maritimus*), and spectacled bear (*Tremarctos ornatus*). Six of these are of recent contemporaneous origin. The spectacled bear is a vestige of an earlier lineage.

The smallest and most arboreal of bears is the sun bear (*Ursus malayanus*), a long-clawed omnivore that inhabits lowland rainforests throughout Southeast Asia from Burma across Laos and Cambodia south to the Islands of Sumatra and Borneo. This tropical habitat may account for its short, sleek coat. Sun bears are omnivorous, eating termites, small mammals, birds, wild bees and softer vegetable matter.

The sloth bear, slightly larger, is found mainly in the forests of India and Sri Lanka. Females weigh 55 to 95 kilograms (120 to 210 pounds), males 50% more. One of the most specialised feeders among the bears, the sloth bear diet is over 90% insects. Morphological adaptations for sucking up ants and termites include absence of incisor teeth, hairless, protrudable lips and a raised elongated palate (Joshi *et al.* 1997).

Another forest bear, the American black bear (*Ursus americanus*) is highly variable in size, depending on the food base. With a rich, thick coat varying from white, bluish, browns to black, this adaptable species occurs from sea level rainforests to alpine peaks in the Rocky Mountains. Healthy populations are also scattered throughout eastern Canada and the USA. Males can range up to 300 kilograms (660 pounds), females perhaps half that. Adults and young use their strong curved claws to climb trees as well as in ripping logs apart. Thought to be largely herbivorous, recent research shows them to be active predators on ungulates and fish.

The Asiatic black bear (*Ursus thibetanus*) is a medium-sized black-coloured bear with large ears and a large white blaze on the chest, sometimes in the shape of a large V. Fond of forests, the Asiatic black bear is widely distributed across southern Asia in mountainous regions of northeastern China and adjacent Russia through Pakistan, northern India and west to Afganistan. Like the sun bear this black bear feeds on a wide range of invertebrates, small vertebrates and fruits, sometimes preying on domestic livestock. Little is known about their natural history. Although they persist over a wide area they may no longer occupy their typical range. Little is known about the social system of these species in the wild.

The brown bear (*Ursus arctos*), or grizzly as it is sometimes called in interior North America, is the largest and most widespread of the terrestrial bears. It occupies a wide range of habitats, but has been extirpated from former ranges in settled areas of the Northern Hemisphere. Small local populations occur in mountainous terrain in Europe and in the northern United States but are widespread in western Canada, Alaska and across northern Asia and parts of Japan. Their diet is markedly diverse, ranging from roots, fruits and other plant parts to insects, fish and any mammal they can capture. Carrion is also common in their diet, especially winter-killed large mammals. The appeal of these bears and their role in the history of human settlement for example, has led to a good deal of information about their behaviour and social system.

Polar bears (*Ursus maritimus*), recognisable immediately by their whitish coat, are the largest land carnivore, having evolved from brown bears about 1.0 million years ago (Waits *et al.* 1998). Males occasionally weigh up to 800 kilograms (1760 lb). A specialised carnivore of the polar oceans, this bear feeds almost entirely on marine mammals. Polar bears have the lowest reproductive rates of any carnivore (Bunnell and Tait 1980, Derocher and Stirling 1996).

The South American spectacled bear (*Tremarctos ornatus*) is more distantly related to the other bears and is placed in a separate subfamily. Small and dark, its eye-rings are characteristic of this formerly wide-ranging bear. Now it is found mainly in forested mountains, from Venezuela down the Cordillera to Bolivia. Although the spectacled bear adapted to wet and dry forests, steppes and coastal scrub deserts prior to persecution by people, it is now largely restricted to high elevations remote from settlements. Its omnivorous diet may have drawn it to crops and livestock, as it eats a wide variety of foods as does its slightly smaller northern cousin, the American black bear. It is vocal with its young in captivity, but its social system in the wild is relatively unknown.

The social nature of solitary bears

As a family of carnivores, Ursids are generally ranked among the more solitary mammals. Thus, they may appear to be a paradoxical and unlikely choice for a search for social learning. However, as I hope to demonstrate, there is a good proportion of the life of bears that is highly social and provides social learning opportunities, especially the prolonged association of bear cubs with their mothers and siblings. If there is evidence of social learning in bears, then we would conclude that the genetic imperative of advanced herd or pack social structure is not the *sine qua non* for cultural transmission of learned behaviour.

Information on the social organisation in the wild of the sun bear, sloth bear, Asiatic black bear, and spectacled bear is meagre or nonexistent. However, studies of the brown bear, American black bear and polar bear have provided growing insights about their social structure. These bears spend most of the year alone, except for females with cubs.

Perhaps the greatest opportunities for social learning in bears occur within this mother–cub unit. The young of these intelligent, large-brained animals have a long social period with their mother and siblings, approaching three or even three and a half years in some cases. This prolonged mother–cub association provides a critical context in which cubs can acquire, via vertical transmission, skills and information pertaining to foraging opportunities, predators and other bears (see Figure 13.1). After leaving their mothers, coalitions of bear siblings may remain together for some time, providing mutual protection and potential information sharing (personal observations). These coalitions between subadults are common and provide an opportunity for horizontal information transmission of successful strategies. In these features of long maternal–young association and of coalitions between related subadults, bears resemble a host of other carnivores, especially felids (North American lynx, European lynx, bobcat, cougar, snow leopard, jaguar and tiger; Caro 1994).

One example of the potential for information transfer in the mother–cub unit is provided by the carnivorous polar bears. Cubs remain with their mothers up to three and a half years (Ramsay and Stirling 1988). This prolonged period of dependence requires significant maternal investment, but also provides considerable opportunity for social learning in a relatively protected setting. The learning and practice of hunting techniques is critical to the survival of polar bear cubs, because, upon achieving independence, the young will face the serious challenges of capturing seals and other prey that are difficult to catch on the sea ice (Ramsay and Stirling 1988). The newly independent bear that has previously shared maternal prey must now kill its own food, making this the hardest period of its life. These considerations

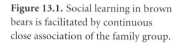

Figure 13.1. Social learning in brown bears is facilitated by continuous close association of the family group.

indicate that polar bears would be an excellent species in which to investigate the role of social learning in survival.

Associations of adults among bears are far less common and intense than associations of mothers and cubs, but they do exist and do provide some opportunities for social information transfer. Of the bear species that have been studied intensively, the home ranges of females generally do not overlap with each other; male home ranges are much larger, encompassing four to six female home ranges. During the breeding season males may accompany females from several days (black bears) up to two weeks (brown bears) for mating. Competition among males may be intense since females only come into oestrous every three years. With approximately three times as many adult males as females prepared to breed, aggression with loud vocalisation is typical during the mating season.

Adult bears foraging at seasonal salmon aggregations at Brooks River in Katmai National Park, Alaska encounter each other. Adult bears approach one another cautiously and with apparent suspicion in the river. In one of these instances I saw one bear relax, throw its head back in an exaggerated, circular motion, and then the pair nipped faces and wrestled gently. I can only speculate that encounters like this followed return to a traditional feeding stream. After a low-intensity tussle, not escalating as a competition over an estrous female does, the pair parted ways and resumed fishing.

Marking of trees by scratching, biting and scent rubbing has been observed in brown bears, black bears and sloth bears in the wild. Marks are concentrated along bear trails (Green *et al.* 1998) and correlate in time with the breeding season and/or with the aggregation of bears at salmon feeding sites. While the social functions of marking still await systematic investigation, observations of marking by the large males during mating competition suggests that marking functions to communicate social dominance in areas around females and other scarce resources. No studies of these species suggest that these marks relate to defense of a feeding territory in the sense of defending and marking the borders of a specific area. Territorial behaviour is variable among black bear populations and appears to be stimulated by limitations in food resources (Powell *et al.* 1997).

The Alaskan brown bear

The remainder of this chapter focuses on brown bears, their nutrient needs and their foraging strategies. It demonstrates that their foraging habits place demands on their abilities to learn and remember the locations of seasonal food sources, including sites on the river where the current is slow and salmon accumulate invisible from the surface (Gilbert, personal observation). They must also master methods of catching fish underwater and learn trails and other pathways to food sources. This information is best acquired during the period of mother–cub association.

Nutrient demands of winter dormancy
Brown bears, who are among the world's largest carnivores, face a serious threat of winter starvation in cold northern climates. They meet this challenge with periods of dormancy of up to six months. Bear dormancy, however, is an expensive adaptation that saves relatively minimal energy because, during this period, bears maintain their body temperature at only a few degrees below normal. They meet their energy demands by a predormancy period of hyperphagia or excessive appetite. Hyperphagic bears can be considered energy maximisers in the sense of Ebersole (1980) and of Kruuk and McDonald (1985) in that they convert abundant food resources into fat to increase survival in winter. Prior to dormancy these carnivores consume an extremely large volume of food. For example, in a 1.5 hour period, I observed a 500 kg (1100 lb) male catch and ingest most of 23 sockeye salmon (*Onchorynchus nerka*). The caloric content of each salmon, when ready to enter fresh water, averages 5400 kcal. If this bear consumed even one-half of these fish, it consumed about 41 000 kcal per hour of feeding time. Grizzly bears in Denali

National Park are reported to feed 14 hours per day (Stelmock and Dean 1986), and some bears may consume over 200 000 berries per day (Pearson 1975). Bears accumulating stored energy as body fat during the autumn hyperphagia show a distinct preference for lipids, fats and oils, but not for protein (Pritchard and Robbins 1990, Gilbert and Lanner 1995). This may relate to the rate and efficiency of digestion and incorporation of triglycerides into body fat.

Brown bears in coastal areas meet their hunger and nutrient needs by searching out the most abundant and accessible concentrations of all five species of spawning salmon, an exceptionally rich source of fat. Some populations feed almost totally on salmon (Hilderbrand *et al.* 1996). The significance of the rich salmon resources to bear population density is manifest in the substantially greater densities of bears in coastal populations compared with inland mountain populations (Miller *et al.* 1997). Coastal populations of bears, who feed on salmon, are approximately 25 times as dense as interior populations.

The nutrient demands of bear dormancy suggest that those bears best able to quickly find, process and ingest lipid rich foods will have the greatest chance of surviving the dormancy period. The remainder of this chapter describes situations where social contacts facilitate various aspects of the learning required for feeding.

Brown bears in Katmai National Park: opportunities for social learning in the exploitation of nutrient rich patches

Among brown bears of Alaska's Pacific coast, aggregations occur each summer on streams where migrating salmon are vulnerable to predation. In Katmai National Park on the Alaskan peninsula, my students and I have studied a population of bears along the Brooks River. These bears are not hunted and are well-habituated to tourists. Midway on the 2 km stretch of river is Brooks Falls, where bears catch fish as they leap the 2 m falls or swim below in dense aggregations. The openness of the sites and extreme habituation of most bears to humans provides an opportunity to gain insights into the potential for social learning.

Bears along the Brooks River have acquired highly individualistic behavioural traits including diving completely underwater to pursue and capture fish. A study of fishing behaviour of brown bears at nearby McNeil River discovered that individual bears used 9 to 28 techniques to capture salmon at a waterfall (Luque and Stokes 1976). Hence, foraging success may depend, in part, on the ability of bears to master the effective techniques for specific fishing conditions.

Some individual bears maximise their seasonal intake of salmon by learning to exploit seasonal concentrations of fish at three to four different sites in succession during the summer. They begin feeding on salmon fresh from the sea at Brooks Falls. Salmon at this stage in their life cycle have the highest body fat content of all spawning fish. Mature male bears have especially high capture rates, in part because of their ability to dominate the most productive fishing sites, and, in part, because of the proficient fishing techniques that they have learned and perfected (Olson 1993).

A second, more dispersed phase of bear feeding begins shortly after fish have progressed upriver to spawning streams. Bears follow them to upper watersheds some 50 km distant, travelling on ancient, worn trails. By learning these patch sites bears are able to exploit further these same populations of salmon, but, in this case, in shallower water on spawning sites. Again, the highest rate of energy return would accrue to those individual bears that learn where the highest concentrations of fish are spawning and how to exploit them efficiently.

Late in the season one final foraging opportunity occurs downstream when massive concentrations of post-spawning salmon die. Bears intercept these post-spawning fish. The dead salmon are carried by the currents and collect on the stream bottom in dense piles. Here, motivated by intense hyperphagia, bears gorge themselves prior to dormancy. A large male may consume 5–6 lb of salmon in as little as 3 to 7 minutes (Olson et al. 1977).

We may infer from their tracking behaviours that bears learn the precise sites and times when these foods are available. For example, one adult female returned to the fall aggregation on Brooks River, Katmai National Park, on the same day three years in succession (Olson 1993).

We may also infer that the demands of proficient energy acquisition confer highest fitness on those individuals who learn and remember a complex sequence of patch choices that provide close to the maximum rate of energy storage. In particular, maximum intake rates depend on remembering when and where runs of anadromous fish are available and at which sites, such as shallow areas and waterfalls, fish are vulnerable to capture. Learning is very important in attaining high rates of fish capture and the capture rates among bears are highly variable (Luque and Stokes 1976, Olson 1993).

In addition to salmon, brown bears feed on a wide range of other foods that place major demands on their memories and learning capacities. Other spatiotemporal challenges include digging clams in the marine intertidal zone and swimming up to 10 miles across the open sea to seabird colonies to feed on eggs. Bears must also learn to discriminate energetically rich foods and to avoid toxic items (e.g. poisonous fungi) or risky prey, such as porcupines. In the

Rocky Mountains, bear families seasonally migrate to alpine scree slopes to dig and gorge on moths that are hidden in considerable numbers in the rock crevices, and the grizzly bears of Yellowstone National Park are noted for obtaining white-bark pine seeds from squirrel middens, a fat source that is their primary source of calories for the year.

Given these foraging challenges, one would expect that those brown bears able to gain information from other bears about seasonally available food, distributed widely over the landscape, would have a significant fitness advantage. Given the relatively solitary lifestyles of adult bears, the most likely context for the social transmission of foraging behaviours is the mother–cub unit and, it is highly likely that females improve the fitness of their offspring by leading them to annually variable, rich resource patches. If continued over several generations, such maternal behaviours would eventually lead to the formation of travel and foraging traditions.

Foraging traditions among bears: two examples
The role of learning and social transmission of behaviour from mothers to cubs can be illuminated by examining two bear traditions: swimming to remote islands and feeding on refuse. These two examples of the spread of innovative feeding traditions have characteristics of cultural transmission.

Swimming to remote islands
In the Shelikoff Straits of Alaska lie a series of islands where bear visitation and predation on burrowing seabirds has extirpated the populations (Bailey and Faust 1984). Although some of these islands are more than 10 miles offshore, over 40 bears were observed swimming between them. Females with cubs were especially common. Diggings and other signs of bears were encountered on nearly all islands. Bears, however, were only observed along the coast in areas without human settlements – areas abandoned by native people when the volcanic eruption of Mt Katmai in 1912 buried several villages under ash. Subsequent re-settlement was prevented by the designation of that part of the coastline as a national park. Bailey and Faust (1984) felt that the lack of human habitation probably accounted for the widespread appearance of bears on islands in the region. Further down the coast, where settlement extends into prehistory, the island swimming trait does not occur or at least does not persist, presumably because swimming bears are vulnerable to subsistence killing.

Island swimming appears to be a prime example of the ability of bears to learn about locations of rich, novel food sources. While one cannot exclude the possibility that each adult bear independently discovered the islands (perhaps by smell, or vision if the islands were high above the sea surface), it seems more likely that these traditions were learned by the young while accompanying their mothers.

Feeding on refuse

In North America an extensive literature exists on the behaviour of national park bears, especially black bears, at refuse tips or sanitary landfills. These animals quickly learn to exploit visitors' food, a habit that spreads rapidly through the bear populations. In Yellowstone National Park a population of bears that fed on garbage and campground food had population parameters (birth rates, inter-breeding interval, age of maturity) above those of bears that did not (Craighead *et al.* 1995). However, this bear tradition eventually became intolerable to managers, because of its threat to life and property. Hence, many bears were removed or killed. The result was compromised reproductive fitness among those females that were habituated to people and ate their refuse. A study of the fate of these matrilineal lines showed zero fitness in all females that developed the trait (Meagher and Fowler 1989). All offspring appeared to learn the same aggressive feeding style and were eventually removed or destroyed by managers.

Social learning in bears: an hypothesis

Here the question of social learning is addressed from the perspective of a field ethologist using paradigms from behavioural ecology or biological anthropology (Smith and Winterhalder 1992). The conceptual framework is similar to that of E. O. Wilson in explaining the cultural transmission of learned behaviour in large-brained mammals (Wilson 1975). This hypothesis is based on a synthesis from observations of colleagues and my own work over the last 25 years with bears in North America.

Northern bears are foraging generalists that utilise a very wide variety of plant and animal species both living and dead. These feeding behaviours would seem to originate in the energy demands of a large mammal whose excessive autumn feeding and long, costly dormancy is an adaptation to the rigors of the sub-Arctic winter. The survival and fitness of bears is thus assumed to be closely linked to the accumulation of sufficient energy reserves to carry the animal over 6–7 months in dormancy. This presents significant challenges, because bears must acquire their food from a wide variety of patchy sources that may be temporally and spatially unpredictable.

This hypothesis implicates social learning in bear species that express exceptional breadth of food selection and that rank their foods according to caloric value. In this context, those family groups that are able to learn to exploit a series of rich foods, remember their location and transfer that knowledge to their offspring (social learning) are presumed to have a significant fitness advantage. Predictions for characteristics of these species compared with others in the Ursidae are:

- long juvenile period compared with mean life span;
- highest frequency of manipulative games in their play repertoire;
- highest levels of attachment/observational learning with the mother;
- high degree of individual specialisation in foraging tactics within a local population (i.e. high phenotypic plasticity in foraging);
- well-developed memory ability for spatio-temporal distribution of rich food patches (cognitive map);
- high levels of motor mimicry of parental behavioural traits, especially those traits not seen in other members of the population (i.e. idiosyncratic social learning).

Acknowledgements

I am most grateful to the editors for much helpful discussion and comment, and especially to Kathleen Gibson for improving the content and organisation of this chapter.

References

Bailey, E. P. and Faust, N. H. (1984). Distribution and abundance of marine birds breeding between Amber and Kamishak Bays, Alaska with notes on interactions with bears. *Western Birds*, **15**, 161–74.

Bunnell, F. L. and Tait, D. E. N. (1980). Bears in models and reality – implications to management. *Int. Conf. Bear Res. Manage.*, **4**, 15–23.

Caro, T. M. (1994). *Cheetahs of the Serengeti Plains: Group Living in an Asocial Species.*

Chicago: University of Chicago Press.

Craighead, J. J., Sumner J. S. and Mitchell, J. A. (1995). *The Grizzly Bears of Yellowstone: Their Ecology in the Yellowstone Ecosystem, 1959–1992.* Washington, DC: Island Press.

Derocher, A. E. and Stirling, I. (1996). Aspects of survival in juvenile polar bears. *Can. J. Zool.*, **74**, 1245–52.

Ebersole, S. P. (1980). Food density and territory size: an alternative model and test on the

reef fish *Eupomacentrus leucostictus. Am. Nat.*, **115**, 492–509.

Fagen, J. M. and Fagen, R. (1994). Interactions between wildlife viewers and habituated brown bears, 1987–1992. *Nat. Areas J.*, **14**, 159–64.

Gilbert, B. K. and Lanner, R. N. (1995). Energy, diet selection and restoration of brown bear populations. In *Proc. Ninth International Conference of Bear Research and Management*, pp. 231–40. Grenoble: Natural History Museum.

Green, G. I., Mattson, D. J. and Swalley, R. A. (1998). Use of rub trees by Yellowstone Grizzly Bears. *Ursus*, **11** (in press).

Hilderbrand, G. V., Farley, C. T. Robbins, C. T., Hanley, T. A., Titus, K. and Servheen, C. (1996). Use of isotopes to determine diets of living and extinct bears. *Can J. Zool.*, **74**, 2080–8.

Joshi, A. R., Garshelis, D. L. and Smith, J. L. D. (1997). Seasonal and habitat-related diets of sloth bears in Nepal. *J. Mammal.*, **78**(2), 584–97.

Kruuk, H. and Macdonald, D. (1985). Group territories of carnivores: empires and enclaves. In *Behavioural Ecology, British Ecological Society Symposium*, ed. R. M. Sibly and R. H. Smith, **25**, pp. 521–36. London: Blackwell Scientific Publications.

Luque, M. H. and Stokes, A. W. (1976). Fishing behaviour of Alaska brown bear. In *Int. Conf. Bear Res. and Manage.* **3**, pp. 71–8. IUCN Publ. New Ser. 40, Morges, Switzerland.

Meagher, M. and Fowler, S. (1989). The consequences of protecting problem bears. In *Bear–People Conflicts: Proceedings of a Symposium on Management Strategies*, ed. M. Bromely, pp. 141–4. Yellowknife, NWT, Canada: Northwest Territories Dept. Renew. Res.

Miller, S. D., White, G. C., Sellers, R. A., Reynolds, H. V., Schoen, J. W., Titus, K., Barnes, V. G., Jr, Smith, R. B., Nelson, R. R., Ballard, W. B. and Schwartz, C. C. (1997). Brown and black bear density estimation in Alaska using radiotelemetry and replicated mark-resight techniques. *Wildlife Monogr.*, **133**, 1–55.

O'Brien, S. J. (1993). The molecular evolution of bears. In *Bears: Majestic Creatures of the Wild*, ed. I. Stirling, pp. 26–35. Emmaus, PA: Rodale Press.

Olson, T. L. (1993). Resource partitioning among brown bears in Katmai National Park. MSc thesis, Utah State University, Logan.

Olson, T. L., Gilbert, B. K. and Squibb, R. C. (1997). The effects of increasing human activity on brown bear use of an Alaskan river. *Biol. Conserv.*, **82**, 95–7.

Pearson, A. M. (1975). The northern interior grizzly bear (*Ursus arctos*) *Canadian Wildlife Service Report*, Series No. 34, Ottawa, 86 pp.

Powell, R. A., Zimmerman, J. W. and Seaman, D. E. (1997). *Ecology and Behaviour of North American Black Bears.* London: Chapman and Hall.

Pritchard, G. T. and Robbins, C. T. (1990). Digestive and metabolic efficiencies of grizzly and black bears. *Can. J. Zool.*, **68**, 1645–51.

Ramsay, M. A. and Stirling, I. (1988). Reproductive biology of female polar bears (*Ursus maritimus*). *J. Zool. (Lond.)*, **214**, 601–34.

Smith, E. A. and Winterhalder, B. (1992). *Evolutionary Ecology and Human Behavior.* New York: Aldine de Gruyter.

Stelmock, J. J. and Dean, F. C. (1986). Brown bear activity and habitat use, Denali National Park – 1980. *Int. Conf. Bear Res. Manage.*, **6**, 155–67.

Waits, L. P., Talbot, S. L., Ward, R. H. and Shields, G. F. (1998). Mitochondrial DNA phylogeography of the North American brown bear and implications for conservation. *Conserv. Biol.*, **12**(2), 408–17.

Wilson, E. O. (1975). *Sociobiology: The New Synthesis.* Cambridge, MA: Harvard University Press.

Wozencraft, C. W. and Hoffmann, R. S. (1993). How bears came to be. In *Bears: Majestic Creatures of the Wild*, ed. I Stirling, pp. 14–22. Emmaus, PA: Rodale Press.

14

Watch with mother: a review of social learning in the Felidae

Andrew C. Kitchener

Introduction

There are about 37 species or so of the Felidae (Kitchener 1991). They range in size from the 2 kg rusty-spotted cat, *Prionailurus rubiginosus*, to the 200+ kg tiger, *Panthera tigris*, but most species weigh less than 20 kg (81%). Felids are also predominantly nocturnal and/or crepuscular (89%), live in mainly closed habitats (76%), and have a dispersed ('solitary') social system (95%) (Kitchener 1991, Seidensticker and Lumpkin 1991). Only the domestic cat, *Felis catus*, the lion, *P. leo*, and the cheetah, *Acinonyx jubatus*, can be regarded as showing a high degree of sociability (Schaller 1972, Bradshaw 1992, Caro 1994), although the tiger and the lynxes, *Lynx* spp., do show similar social tendencies under certain conditions (Thapar 1986, Kitchener 1991). Therefore, for the vast majority of felid species, there are considerable difficulties to overcome in carrying out studies of social learning in nature, and with some landmark exceptions (e.g. Caro 1994) field workers on wild felids have not looked at the development of behaviours from a social learning perspective – it has not often been considered to be important to the cats that are mostly solitary. However, it is relevant to consider that social learning plays an important role in the development of flexible behavioural strategies, in order for cats to cope with a variety of local ecological and social conditions, in both solitary and sociable species.

There are three main social groups in most cat species; these are (1) mother and young, (2) male and female in oestrus, and (3) dispersing siblings (Kitchener 1991). It is the most enduring unit, i.e. mother and young (up to three years in the lion, Schaller 1972), that offers most opportunities for social learning. For the mother it is important that she is able to ensure that her young develop into competent, territorial predators as quickly as possible, in order that her current condition and her future reproductive potential are not compromised. It is also essential that the young are able to learn a variety of appropriate behavioural strategies for dealing with local ecological conditions. The mother cat must also monitor the behavioural and physical development of her young, in order to ensure they are provided with appropriate opportunities for learning particular behavioural strategies at appropriate times so that her efforts will be not wasted and her young do not risk injury.

Larger species have additional problems to overcome in their behavioural development, regardless of their degree of sociability. The longer development period required for the permanent dentition to develop (Leyhausen 1979) increases the period of dependence on the mother and, consequently, the opportunities for social learning (Schaller 1972, Caro 1994). This is even more important when considering the greater difficulties and risks involved in dealing with larger and more dangerous prey. Therefore, large felid species take advantage of an extended period of development to optimise prey size and minimise injuries/mortalities from the capture of large prey through social learning.

Given the difficulties involved in studying wild felids, it is perhaps unsurprising that questions about the social acquisition of behaviour have not yet been addressed. To date most research has centred on the sociable species, either in laboratory conditions (domestic cat) or in the field in open habitats (e.g. lion, cheetah, leopard *Panthera paidus*) (Kruuk and Turner 1967, Schaller 1972, Turner and Bateson 1988, Bradshaw 1992, Caro 1994), with the notable exception of Leyhausen (1979) and Mellen (1993), who studied the behaviours of a variety of cat species mostly in captivity. However, the similarity in the lifestyles and life-history strategies of all felids means that experimental studies on the development of predatory behaviour in domestic cats provide useful models for studying and interpreting the behavioural development of wild felids. Moreover, the sociable wild cats offer many opportunities to study social learning because they tend to live in open habitats and readily habituate to the presence of observers.

In this chapter I focus mainly on the role of social learning in the development of the predatory behaviours and prey recognition in domestic and wild felids, using a variety of experimental studies on domestic cats and data from field studies which have not been previously interpreted from a social learning perspective. The discussion is then extended to other behaviours, where fewer data are available and more speculation is involved. These include predator avoidance and recognition, scent marking and territorial inheritance, mate recognition and sexual behaviour, and, potentially, grooming and play (Leyhausen 1979, Kitchener 1991, Mellen 1988). Throughout this review the role of social learning in providing behavioural strategies for dealing with locally varying social and ecological conditions is emphasised against a background of relevant life-history constraints (e.g. deciduous tooth replacement). Finally, field studies provide good data for sociable species, where there are not

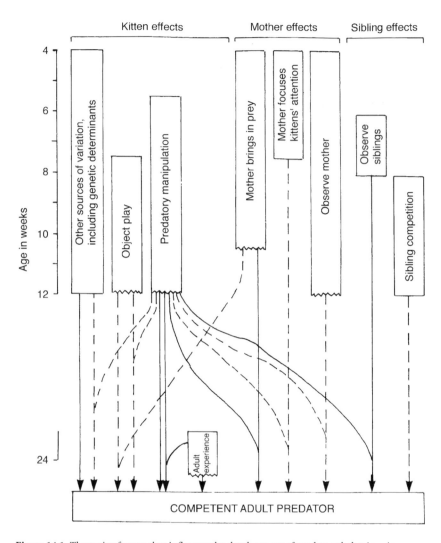

Figure 14.1. The major factors that influence the development of predatory behaviour in domestic cats (after Martin and Bateson 1988). Solid lines indicate experimental observations, broken lines indicate hypothesised influences. In wild felids the mother effects are likely to be the most important influences (see text).

only excellent opportunities for social learning from mothers, but also from other conspecifics, that provide flexible behavioural strategies for cooperative hunting, male provisioning of food and intra-group communication.

Predatory skills and pay recognition

Interestingly, very early experimental studies on domestic cats (Hobhause 1901, Berry 1908, Teyrovsky 1924, Herbert and Harsh 1944, Adler 1955, Chesler 1969, John *et al.* 1968) demonstrated the potential for social learning of behaviour that had no direct bearing on the natural behaviours of domestic cats, but they did demonstrate the potential for acquiring information from other cats and, in particular, emphasised the role of the mother as a demonstrator (Ewer 1969).

In fact, it has been shown that there are many influences on the development of predatory skills in domestic cats. Competent predators may develop via many different routes, summarised in Figure 14.1 (Caro 1979, Martin and Bateson 1988). A cat may learn its predatory skills either as a kitten (at varying ages) or as an adult, with age of learning having little effect on its ultimate efficiency and competence. However, it could be argued that social, particularly maternal, influences in wild species are likely to be particularly important influences on the development of prey-killing behaviours. Hence, although domestic cats in experimental situations may learn to kill without any previous experience of prey, in the wild they would risk starvation by being initially too timid in approaching prey, or by not recognising it. Again, they could suffer injury by tackling inappropriate prey. For example, three captive-bred adult male cheetahs were released into two different private nature reserves in the eastern Transvaal Lowveld of South Africa (Pettifer 1981). Their only previous experience of killing prey was that of a single captive Barbary sheep, *Ammotragus lervia*, although they were fed on dead impala, *Aepyceros melampus*, which was the natural prey in the release areas (Pettifer 1981). However, once released they turned their attention to the calves of giraffes, *Giraffa camelopardalis*, which they hunted cooperatively in pairs; one animal would hook the calf's rump with its dewclaw, while the other would jump onto its shoulders, bringing it to the ground. In this way the cheetahs achieved a hunting success of 41.7% ($n = 12$) for giraffes compared with 9.3% ($n = 97$) for impala, which would have been expected to be their natural prey (Pettifer 1981). Therefore, these cheetah did not recognise their natural prey and were incompetent at hunting it. Although giraffe calves were hunted very successfully, the cheetahs risked increased injury from hunting large prey (individuals were injured trying to hunt an African buffalo, *Syncerus caffer*, a zebra, *Equus quagga*, and a wildebeest, *Connochaetes taurinus*). The cheetahs spent an average of 32 hours on each kill (cf. 2 hours maximum in the Serengeti, Schaller 1972), so that they could also have suffered severe losses of prey due to competitive scavenging from lions and hyaenas, *Crocuta crocuta*, had these been abundant in the area.

In addition the slow-breeding giraffe population may have been vulnerable to extinction due to the selective hunting of juveniles, thereby denying the cheetahs of a future source of prey. It would appear that mother cheetahs provide essential opportunities for their cubs to know which prey to hunt. For example, Caro (1994) noted that when following their mothers, cheetah cubs would often become distracted by inappropriate prey (including giraffe calves in 10% of inappropriate 'hunts' at 8.5 months or older), but as they grew older their attention became focused on the correct prey, owing to the intervention of their mothers.

For many cat species, there is a typical pattern in the development of predatory skills, in which the mother plays an important role (Ewer 1969, Leyhausen 1979, Caro 1980b,c, Bradshaw 1992, Caro and Hauser 1992). First, the mother brings dead prey to her young and eats it in front of them. She then brings live prey for them to play with and kill, and only intervenes to stop the prey escaping or to alert her kittens to the presence of the prey if they have lost interest. Finally, she allows her kittens to follow her on hunts, where they are allowed to kill prey themselves (e.g. wildcat, *Felis silvestris*, Kolb 1977, lynxes, Haglund 1966, Parker *et al.* 1983, tiger, Schaller 1967, leopard, Turnbull-Kemp 1967, cheetah, Caro 1994, lion, Schaller 1972). Lions and other big cats differ from smaller species in that prey are not generally brought to the young, because the prey are too large to be carried long distances (see below) (Leyhausen 1979, Caro 1994), and so young big cats may only observe the killing of prey for the first time when accompanying their mothers (Leyhausen 1979).

Therefore, there are two aspects to the development of predatory skills. First, there is the refinement of innate prey-killing behaviours so that young cats develop efficient prey-catching skills when they become independent (Leyhausen 1979). This appears to take much longer for larger species of cats feeding on larger, more difficult prey (see below). For example, Caro (1994) has observed that even though cheetah cubs are able to kill a variety of prey at 15 months of age, they are not necessarily proficient killers (only 7% of 185 kills without mothers were successful); they are still almost entirely dependent on their mothers for their full nutritional needs, and may not become fully independent until they are 21 months old.

Second, cats learn from their mothers what prey species are available, and hence, through direct experience of them, how to deal with the killing and eating of them (see above). Cats are specialised hunters. Kruuk (1986) showed that felids hunt, on average, four key prey species compared with six or more for canids. Therefore, it is important that young cats learn from their mothers which prey species are the most available with the highest energetic returns in that particular area. It has been suggested that rather than literally teaching kittens about prey capture or prey recognition, the mother may be offering her young the opportunity to learn for themselves (Ewer 1969). This has been called 'opportunity learning' by Caro and Hauser (1992).

A developmental constraint?: replacement of deciduous dentition

In many cases cats are able to kill prey, apparently competently, well before they become independent. For example, lion and tiger cubs can kill at 5–6 months of age, but may be two or three years old before they are independent (Schaller 1972, Leyhausen 1979). In part this long developmental period is related to the replacement of the deciduous dentition. Although the permanent dentition develops so that cats are never without canines or carnassials (dP^3/dP_4 changes to P^4/M_1), there is a period when the canines are not functional, owing to resorption of the roots of the deciduous canines as the permanent ones develop. Therefore, young cats are dependent on their mothers for food during this period (Leyhausen 1979). Leyhausen (1979) recorded a tiger that attempted to kill a dwarf zebu, *Bos taurus indicus*, but left no mark on the zebu's skin with its canines. Soon after the tiger was killed and found to have two sets of non-functional canines and a dislocated foreleg. Evidently, it had been abandoned by its mother because it was unable to keep up with her. Unfortunately, it is difficult to obtain good data for most cat species on the replacement of their deciduous dentition. In small cat species (e.g. the wildcat, the serval, *Leptailurus serval*, and the caracal, *Caracal caracal*), the young cats become independent soon after the complete development of the permanent dentition, but in larger species the period of time before independence is much greater (Table 14.1). This suggests that the young of larger species need an extended opportunity to learn how to deal with difficult and potentially injurious large prey. For example, lion cubs do not become fully independent until at least 9 months after their permanent dentition has developed and 18–19 months after they are able to kill their own prey (Schaller 1972, Leyhausen 1979, see also cheetah example above (Pettifer 1981).

Experiments on domestic cats

There were several well-known early experiments on the development of predatory skills and prey recognition in domestic cats, which have been reviewed by Caro (1980a). Berry (1908) and Kuo (1930) apparently demonstrated that domestic cats could only learn predatory skills through previous

Table 14.1. *Comparative data on ages of weaning, first kill, independence and replacement of deciduous dentition in felids. All ages are given in days after birth. References are given in parentheses*

Species	A	B	C	D	E	E – D
Wildcat	42–120		105–120	116–148	131–180	15–32
Felis	(1,16)		(20)	(20)	(1,11)	
silvestris						
Serval	180			180	180–240	0–60
Leptailurus	9			(17)	(10)	
serval						
Caracal	70–180			300	274–365	– 24–65
Caracal	(1,3,8,			(2)	(1,9)	
caracal	16,17)					
Puma	28–105		165–180	240	540–720	300–480
Puma concolor	(1,16)		(5,12)	(12)	(1,12)	
Cheetah	84–255	75–360	240	270	465–540	195–270
Acinonyx	(1,2,3,16)	(2,8)	(1,2)	(1,2)	(1,3)	
jubatus						
Lynx	60–180		120–129	240–300	300–426	60–126
Lynx lynx	(1,15,16)		(3,5,15)	(19)	(1,15,18)	
Leopard	84–130	330	210	< 365	348–600	– 13–235
Panthera	(1,2,3,16)	(2)	(5)	(7)	(1,2,3,	
pardus					7,16)	
Tiger	90–180	150–180	249–365	330–450	540–1080	90–750
Panthera	(1,14)	(4)	(5,4,14)	(4,14)	(1,3,14)	
Lion	165–450	150–180	300	330–450	630–1080	180–750
Panthera leo	(1,3,16)	(4)	(5)	(4,6)	(1,2,3)	

Key:

A, Age at weaning (may vary considerably due to different definitions); B, Age when first kills; C, Age when permanent dentition begins to appear; D, Age when permanent dentition is complete or permanent canines fully erupted; E, Age at independence; E–D, Difference in age between when permanent dentition is first complete and age of independence.

References:

1, Caro (1994); 2, Skinner and Smithers (1991); 3, Ewer (1973); 4, Leyhausen (1979); 5, Hemmer (1976); 6, Schaller (1972); 7, Bailey (1993); 8, Estes (1991); 9, Haltenorth and Diller (1980); 10, Geertsema (1985); 11, Easterbee (1991); 12, Currier (1983); 13, Caro (1980b); 14, Mazak (1981); 15, Tumlison (1987); 16, Hayssen *et al.* (1993); 17, Rosevear (1974); 18, Nowell and Jackson (1996); 19, Hemmer (1993); 20, Stahl and Leger (1992).

learning experiences involving their mother. For example, Kuo (1930) showed that 85% of kittens could become rat killers just by watching their mothers kill a single rat, *Rattus norvegicus*. However, Berry's and Kuo's methodologies were criticised by Yerkes and Bloomfield (1910), Leyhausen (1979) and Caro (1980a).

More importantly Kuo's (1930, 1938) series of experiments demonstrated that mothers had a strong influence on prey recognition and preference. Kittens reared in isolation with mice or rats did not recognise them as prey and so would not kill them, although they may have killed other strains of rats. If reared with other kittens and rats, the kittens preferred to socialise with other kittens and were ambivalent to the rats, but did not kill them. Kittens reared as vegetarians would kill as many prey as normal kittens, but ate the prey less often.

Other researchers have apparently confirmed this. For example, Wyrwicka (1978) trained female cats to eat either mashed potato or banana. Their kittens began to imitate their mother's food preference as early as 35 days old, but for most kittens this began at 49–56 days and they retained this food preference over meat pellets after weaning at 9–27 weeks old. In a set of similar experiments Wyrwicka and Long (1980) showed that in the presence of their mother, kittens would start eating a new unusual food of either tuna or cereal much sooner (mean 0.2 days after seeing her feed) than if they were presented with the food on their own (mean 4.8 days).

These experiments show that domestic cats are flexible in their choice of prey, so that they can respond to what natural prey is available. They also show that the mother's prey choice is the key determinant of prey recognition by kittens, although siblings may play a role to a lesser extent.

Subsequently, Caro (1979, 1980a,b,c, see also Caro and Hauser 1992) rigorously investigated the development of prey capture skills in domestic cats, including the influence of the mother and siblings (Caro 1980b,c), and the interrelationships between the development of play and prey capture skills (Caro 1981).

His experiments demonstrated that female domestic cats play an important role in the early development of predatory skills in their kittens by providing opportunities to learn. Interestingly, siblings were also found to play a role in social behaviour development, including the interrelationships between the development of play and prey capture skills (Caro 1981).

In Caro's experiments live prey were introduced to kittens from the ages of 4–12 weeks old, either in the presence or absence of their mother to provide a balanced experimental design. His key results, particularly with respect to the influence of the mother, are summarised below.

Mother cats played a significant role in attracting 4–5 week old young to prey by distinct vocalisations, or by watching prey, or by carrying the prey directly to the kittens. Mothers would modify their behaviour as the kittens' prey capture skills developed. At first they would demonstrate how to kill prey and later they would rarely kill prey, allowing the kittens to learn through direct experience of live prey, although they might still attract their kittens' attention to prey. Mothers would appear also to monitor the rate of development of their young, by varying the time at which they introduced prey to kittens aged between 4 and 6 weeks.

In the presence of their mothers, kittens learnt to kill more prey sooner and interact with it longer compared with kittens without their mothers. These kittens would also kill five times more prey at age 12 weeks and used the nape bite on mice significantly more times than kittens without their mothers. Kittens tended to follow the prey preference of their mothers and approached these prey more confidently than other prey. This shows the important role that mothers have in promoting the development of prey capture skills and prey choice.

Finally, siblings have an important motivating role in the development of prey capture skills. Up to 6–8 weeks of age kittens were more likely to watch prey together, suggesting that siblings played an important role in focusing attention on and motivating interactions with prey (see also Leyhausen 1979). As they became older, kittens became competitors and would not respond to released prey together. Siblings ignored each other if one had captured the released prey.

Given the similarity of prey capture behaviours in all felids, these studies provide a useful model for studying the role of social learning in the development of prey capture and recognition in wild species.

Intraspecific variation in diets and behaviour

Although it is relatively easy to investigate the development of behaviour in captive domestic cats in experimental situations, this is much harder for wild species. Firstly, wild felids are often unsuitable as laboratory captives where high degrees of observability and manipulation are required. Secondly, they are often rare and difficult to obtain, so that sample sizes may be very small. For example, Leyhausen (1979) employed a comparative approach on small numbers of a wide range of species in laboratory conditions, but not from a social learning perspective. However, for most of our information regarding the potential of social learning in the development of predatory skills we are dependent on inferences based on the similarity in behaviour between all felids and long-term studies of cats that live in open habitats (see below).

A notable exception to this is recent work carried out by Caro (1994), who employed similar rigour, where possible, in the study of behavioural development of wild cheetahs as he did for domestic cats. When their cubs were about 1.5 months old, mother cheetahs brought prey back to them and killed it. When the cubs were 2.5–3.5 months of age, their mothers tended to let live or injured prey go in the presence of their cubs, but they would kill it after about 10 minutes. This increased to about 30% of prey being released when the cubs were 4.5–6.5 months old, and when they would begin to suffocate their own prey. As the cubs grew older, they tended to disembowel the prey or tear it apart while a family member held on to the throat, so that by ten months of age, less prey were released. Although from 2 months cubs accompanied their mothers on 72% of hunts, it was not until they were more than 12 months old that they began to initiate hunts. Most prey released to cubs was either hares or neonate Thomson's gazelles, *Gazella thomsonii*, which were easier for mothers' to catch and cubs to kill without any risk of injury. For example, 30.9% of released prey were neonate gazelles compared with 4% adults, which were usually debilitated by the mother in some way. The costs of this opportunity learning were high for the mothers, because prey often escaped from the cubs; cubs killed only 46.2% ($n = 13$) of hares, *Lepus capensis*, released to them compared with a hunting success of 84.8% ($n = 33$) for their mothers. To help reduce the chances of escape, hares were often injured to prevent them from escaping. Mother cheetahs were also much more likely to kill prey if they were hungry (as indicated by belly size), rather than let their cubs learn predatory skills. Therefore, in many respects the role of the mother in the development of predatory skills of cheetah cubs is strikingly similar to that found in the domestic cat (Caro 1980a,b,c, Caro 1994).

This is a very important study among mammalian field studies of social learning. Many studies have shown that individuals may be influenced by the behaviour of other individuals, but this study shows that the mother modifies her behaviour with reference to that of her offspring, i.e. to the rate of development of their predatory skills. Moreover, it emphasises the interaction between behavioural development in the young and the fluctuating energetic demands of the mother. In other words, this maximises opportunities for the cubs to learn, while minimising the risks to the mother's energetic survival.

There are many other examples of mother cats bringing live prey to their young, or in the case of big cats, bringing their cubs to the prey. In the savannas of East Africa serval mothers call their kittens from up to 50 metres away to come and receive the prey they have caught (Estes 1991). Schaller (1967) recorded a female tiger knocking down a buffalo, *Bubalus bubalis*, so that her cubs could attack it. Schaller (1972) also recorded a Thomson's gazelle fawn being brought alive to its cubs by a lioness. Leopard cubs receive prey from their mothers from 101 days old, accompany their mothers on hunts from 9.5 months of age and do not kill their first impala until 12.5 months old (Skinner and Smithers 1990). Turnbull-Kemp (1967) recorded a leopard cub accompanying its mother on a hunt, which on a vocal command stayed back while it watched its mother kill prey. Similarly, lion cubs follow their mothers from 5–7 months of age, join in hunts at 11 months, kill gazelles at 15–16 months and larger prey from about two years. There seems to be a striking similarity in the role of the mother in the development of predatory skills in all cats studied so far. Exceptions may include the black-footed cat, *Felis nigripes,* which Leyhausen (1979) recorded bringing a live mouse to its kittens, but the mother did not kill it and only prevented the mouse from escaping. In this case the kittens never observed their mother killing prey, before learning to kill it for themselves by trial and error. However, it would seem likely that opportunity learning is most important for small felid species that feed on large difficult prey or large felid species that feed on large prey, where the young need to practice manipulation of potentially dangerous prey without risking injury to themselves.

One obvious inference that can be drawn is that for felids with wide geographical distributions, there may be a need to specialise in different prey in different areas. For example, in the Serengeti, leopards prey mostly on Thomson's gazelles (Kruuk and Turner 1967, Schaller 1972), but since this species is restricted to East Africa, the leopard must prey on a wide range of other species elsewhere in its range from North East Asia to South Africa. Therefore, in the Kalahari of southern Africa the commonest prey was porcupines, *Hystrix africaeaustralis,* in the Parc National de Tai in the Ivory Coast it was duikers, *Cephalophus* spp., and in the Himalayas it was wild goats, *Capra aegagrus* (see Kitchener 1991). Regional variations in the diet of the cheetah are given in Wrogemann (1975). Therefore, the mother plays a vital role in introducing the locally available prey species to her young.

Most cats kill their prey with a neck or nape bite, which allows the canine teeth to dislocate the cervical vertebrae (Ewer 1973, Leyhausen 1979). Domestic cats are able to become proficient killers of mice after just one attempt at this (Caro 1980, Bradshaw 1992). Large felids kill large prey with a throat bite, which causes suffocation (Ewer 1973, Leyhausen 1979). From the extensive captive studies by Leyhausen (1979), there is evidence that the neck bite is innately derived, although fine tuning of this killing technique benefits from direct experience of prey. However, the throat bite is probably learned. Most larger cat species are encouraged to follow their mother on hunts and so would see the throat bite in action. It is important that cats have alternative strategies for killing prey, because it is sometimes not possible to assess accurately the size and, hence, the difficulty involved in killing a particular prey. For example, Fox and Chundawat (1988) observed a snow leopard, *Uncia uncia*, killing a domestic goat, *Capra hircus*. At first the snow leopard grabbed the goat by the neck for a typical killing bite at the nape, but realising that this prey was more difficult, it switched to the suffocating throat bite. Seidensticker and McDougal (1993) analysed the killing techniques of tigers in Nepal and showed that whereas adult tigers tended to kill prey using a nape bite, which crushed the cervical vertebrae, young animals killed prey of similar size with a throat bite. Sunquist (1981) observed that adult tigers tended to use a throat bite when the prey was more than 50% of their own body weight. Perhaps subadult tigers use the throat bite on average prey, because of their inability to judge accurately the size of prey and their relative inexperience in killing. Therefore, the learned throat bite allows subadult tigers to kill their prey safely and with more certainty.

Local traditions may develop in the method of killing of prey, which are suggestive of social learning. For example, in the Kalahari Gemsbok National Park in Botswana, lions are often killed by gemsbok, *Oryx gazella*, who stab them with their long sharp horns. In one area Eloff (1973) observed that lions avoided risking injury or death in this way by biting the lower back of these antelopes, thereby breaking the sacro-lumbar joints.

Another example of a local tradition may be 'man-eating'. In the past it was widely believed that tigers only became 'man-eaters' because they were old or injured, and so were forced to switch to slow-moving, easy-to-catch humans. However, many 'man-eaters' are known to be young, healthy animals (McDougal 1987). It seems that in these cases, as the natural prey base is eroded by humans, so female tigers are forced to switch to domestic livestock and humans as prey, thereby passing their prey preference onto their cubs (McDougal 1987).

Servals have apparently developed specific hunting strategies for different fossorial prey (Ewer 1973), which are probably passed on to kittens through social learning. To hunt mole-rats of the genus *Tachyoryctes* which dig shallow burrows, servals use their large ears to listen for their movements within a burrow. Once pin-pointed, the mole-rat is dug up. However, mole-rats of the genus *Cryptomys* are dealt with differently. They dig deep burrows, so that servals damage the entrance to the burrow system and wait for the resident mole-rat to come to the surface to repair the damage. Hence, social learning is implicated not only in what species of prey are available locally, but also in the methods the cats employ in capturing and killing difficult prey.

Cooperative hunting may also represent a strategy that the young learn

Table 14.2. *The effect of group size on hunting success and interkill distance for Canada lynx,* Lynx canadensis, *preying on snowshoe hares,* Lepus americanus

Group size	Hunting success (%)	Interkill distance (km)
1	14	7.6
2	17	4.9
3	38	2.9
4	55	0.5

From Parker *et al.* (1983).

from their mothers. Canada lynx, *Lynx canadensis,* and Eurasian lynx, *L. lynx,* mothers and cubs normally travel through boreal forest in single file, but when hunting together they fan out in a line to catch any hares, *Lepus* spp., that are flushed from cover by them (Haglund 1966, Parker *et al.* 1983). This brings obvious benefits to the lynx, in that hunting success is increased and interkill distance is decreased (Table 14.2). It is clear that this hunting technique is learnt, and that social learning from the mother would seem to be a likely route. However, it is unknown how this is achieved and individual trial and error may be a primary influence. Adult Eurasian lynx have also been observed to hunt like lions (see below), where one animal waits in ambush while another drives a hare towards it (Haglund 1966). Again social learning may be involved, but this requires further investigation.

Social learning may also be involved in what cats do with their prey after killing it. For example, three captive-bred cheetahs were given dead impalas to feed on prior to their release as part of a re-introduction project (Pettifer 1981; see above), but it was not until they received the fourth carcase that they had learnt an effective way into the carcases. Some species stay at the killing site and consume the prey, others drag the prey into cover, while others cache it under piles of dead leaves, in the snow or in trees (see Kitchener 1991 for examples). For example, in many parts of Africa leopards cache their prey in trees well away from scavengers (despite aerial vultures!), but in Sri Lanka leopards do not need to do this and cache their prey on the ground (Eisenberg and Lockhart 1972, Bailey 1993).

Most felids scavenge meat regularly from dead animals, which again could be learnt from the mother (Kitchener 1991). However, cheetahs rarely scavenge, except when adult males steal food from a female and her cubs (see Caro 1994), probably because they are unable to defend carcasses from other scavengers, including vultures (Schaller 1972, Caro 1994). It is possible that cubs could learn from their mothers to avoid dead meat as a potential source of food.

Other behaviours

Given the possibilities regarding the social acquisition of information for hunting strategies and recognition of prey, it is important to consider other behaviours that may be influenced by social learning. However, experiments and fieldwork are not so well developed and we can only suggest where social transmission of information may play an important role in behavioural development. Most speculation involves the possible role of social learning mediated by mothers and siblings in the development of grooming patterns (see Darwin in Whiten and Ham 1992), and play (Barrett and Bateson 1978, Mendl 1988, Bateson and Young 1979, Martin and Caro 1985, Deag *et al.* 1988), particularly in South American species, *Leopardus* spp., with a usual litter size of one (see Fagen 1981 for observations on the development of play in a hand-reared margay, *L. wiedii*, kitten). There is more information regarding the possible role of social learning in the recognition and avoidance of predators, scent marking, territorial inheritance, and mate recognition and sexual behaviour, but much more research is required.

Recognition and avoidance of predators

It seems likely that young cats learn from their mothers about which species are predators and how to avoid them. For the older cubs of most species, the mothers hiss at and slap their young to force them to return to the nest or hide in cover (Leyhausen 1979), so that they learn directly what species should be avoided. However, the black-footed cat has an alarm call which causes the kittens to scatter into cover and freeze; the kittens only return to their mother when she gives the 'all-clear' vocalisation (Leyhausen 1979). Cheetah cubs also scatter when their mother rushes at a predator (Caro 1994). Cheetah cubs do not recognise predators at 2 months old and display fear towards a wide variety of species at 4 months of age, before eventually learning from her which species to avoid at 5 months (Caro 1994).

Hence, in some species, at least, young cats become conditioned to respond to predators by association with their mother's behaviour, or in the case of the black-footed cat, by her alarm call.

Scent marking

Cats use a variety of methods for scent marking, including faeces, urine and glandular secretions. These are often combined with a visual mark, to draw the attention of passing cats (Kitchener 1991). For example, most cat species, including, for example, the tiger, spray urine on to tree trunks, leaving a clear visual mark at a height convenient for others to see and sniff.

Domestic cats learn much about scent marking by experimentation and apparently observing their mothers (Neville 1992). In two groups of semi-feral domestic cats, Feldman (1994) found that cats of all ages and both sexes scratched the bark of trees. Scratching is thought to combine the functions of a manicure with scent marking by the interdigital glands. This scent mark is made visually conspicuous by scratching (Kitchener 1991). Since the cats selected only certain trees, particularly those with soft bark (Feldman 1994), it is possible that kittens learnt this behaviour, as well as which trees to use, at least partly from conspecifics.

Depending on the species and the area in which it lives, cats may vary intra- and interspecifically in the pattern of scent marking (e.g. at the boundaries of the home ranges of tigers, Smith *et al.* 1989, and at intersections of trails within the ranges of Spanish lynx, *Lynx pardinus* Robinson and Delibes 1988), and the type of scent mark used (Kitchener 1991). For example, Smith *et al.* (1989) showed that tigers selected particular tree species for spraying urine onto and they mostly selected the underside of trees which leaned over by 8–24° from the vertical, so that the marks were protected from the rain. Although this could be learnt by trial an error, it is plausible that cubs learn from their mothers to select the underside of leaning trees for scent marks.

Territorial inheritance

Young cats often accompany their mothers on hunting trips in her home range (Ewer 1973, Leyhausen 1979). This allows kittens to learn not only about their mother's home range, but also in general about the components that make up a territory or home range, including hunting sites, resting sites, access to water and cubbing dens (Kitchener 1991). Indeed, female young often take over part of their mother's territory as they grow older, either temporarily or perma- nently, while male kittens disperse widely, probably due to intermale intoler- ance. This kind of territorial inheritance has been observed for feral domestic cats, tigers and cheetahs (of up to 2.5 years in age) (Izawa and Ono 1986, Smith *et al.* 1989, Caro 1994). The advantages to mother and daughter are clear; the

Table 14.3. *The effect of different rearing regimes on mean duration of mating and reproductive success of female domestic cats*

	Hand-reared alone	Hand-reared with siblings	Mother-reared
Number of kittens	7	7	7
Breeding success (%)	28.6	42.8	85.7
Mean duration of mounts (minutes)	9.62	5.52	3.63

From Mellen (1988).

young cat learns from her mother by observation of what are the important features of a home range which she may eventually inherit to promote her survival in the long term, and the mother promotes the survival of her genes by providing her daughter with a familiar territory (even if only temporarily, particularly in the vulnerable period just after independence). Both mother and daughter may benefit by combined territorial defence from other females by having an ally close at hand, thereby reducing the area each animal defends directly.

Sexual behaviour and mate recognition

In captivity, hand-reared wild felids are often hyper-aggressive to their human keepers and reproduce poorly compared with mother-reared animals (Mellen 1988). However, some private institutions deliberately hand rear young wild felids, so that they are habituated to people, and achieve successful breeding as a result. Mellen (1988) investigated experimentally different rearing regimes on the subsequent sexual behaviour of domestic cats. Cats hand-reared in isolation were always hyper-aggressive to humans and other cats and did not like to be handled. When placed with an experienced male, they would often show appropriate oestrous behaviour, but would attack as the male approached. These cats mated infrequently compared with cats hand reared with siblings and mother-reared cats (Table 14.3). However, they often mated for longer and often did not achieve intromission because their lordosis was incorrect and consequently they produced fewer litters (Table 14.3) (Mellen

1988). This suggests that mate recognition, lordosis and sociability are influenced socially and may involve a component of social learning. For example, the highly sexually dimorphic clouded leopard, *Neofelis nebulosa*, is difficult to breed in captivity, because the very large males often kill and injure the small females (Richardson 1992, G.Law, pers. commun.). This may be because most captive animals have been hand reared, whether they were born in captivity or obtained from the wild (G. Law, pers. commun.). These animals seem unable to recognise conspecifics and react inappropriately, resulting in an inappropriate response from the female who is then treated as a prey animal by the male. I have observed a wild-caught female showing oestrous behaviour to a male, but when the animal's human keeper left for a few minutes, she stopped rolling and watched for his return. On the keeper returning she demonstrated typical oestrous behaviour again. A subsequent introduction to the male resulted in a severe injury to the female. This suggests that mate recognition is socially learnt and that hand-reared cats are liable to become attached to humans if reared in isolation (Leyhausen 1979).

Sociable cats

Even though the domestic cat, lion and cheetah are truly sociable, they show the complete spectrum of sociality from living in groups to living on their own depending on habitat and availability of food (Leyhausen 1988, Kitchener 1991). Therefore, there are certain behaviours which vary within species depending on the social situation, and which may be influenced by social learning. These are discussed below.

Cooperative hunting

In female fission–fusion groups (prides) the females hunt large prey together. There has been some debate as to whether cooperative hunts arise accidentally or whether they require pride members to coordinate their movements and hunting strategy with respect to each other (Kruuk and Turner 1967, Schaller 1972). Recently, Stander (1992; see also Stander and Albon 1993) studied the cooperative hunts of lionesses in the Etosha National Park in Namibia. In the open habitat the individual positions of lionesses were observed in 486 cooperative hunts. It was found that there were two basic kinds of hunters in a group hunt; those in the outer positions (the wings) surrounded the prey and chased it, while those in the centre (the centres) ambushed it (Figure 14.2). Stander (1992) found that each animal had a preferred position in the hunting array and that hunting success was highest if animals occupied their preferred

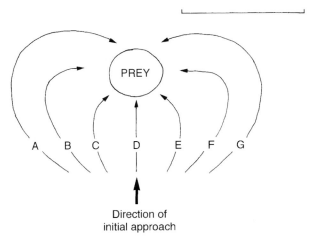

Figure 14.2. A schematic diagram to show the most typical stalking routes of members of lion prides that hunt cooperatively in the Etosha National Park, Namibia (Stander, 1992). 'Centres' are C–E; the rest are 'wings'. Scale bar represents 100–200 metres.

hunting position. Some animals influenced the position of others depending on their presence or absence, so that it seems likely that lions learn their cooperative hunting strategy from other pride members and adopt a hunting position which is most appropriate to their learned skills.

Lions keep a look-out for the behaviour of vultures. If they see them circling above the ground and descending to a carcase, they will investigate in order to take advantage of an easy source of food (Schaller 1972). It is possible that lion cubs learn from other pride members how to watch out for congregating vultures.

Male provisioning of food

Although most cats are polygynous so that males would not be expected to share in the provisioning of food to the female and young (Kitchener 1991), there are accounts of males leaving food at or near the dens of both captive and wild-living animals (e.g. tigers, lynxes, fishing cat, *Prionailurus viverrinus* (Table 14.4), and wildcats (Leyhausen 1979, Thapar 1986, Tomkies 1991)). Why some males should do this is unknown, but future studies could investigate the particular social and ecological conditions that affect male reproductive strategies in these ways to see if there are functional advantages under these conditions.

However, one example may indicate that a male's reproductive strategy is influenced by his potential lifetime reproductive success. Male lions may differ in their tolerance of cubs at a kill. In the Serengeti where the average tenure of a

Table 14.4. *The time spent by a male fishing cat with his kittens in captivity during more than 13 hours of observation*

	Kittens alone	F with kittens	M with kittens	M and F with kittens	Total time
Time (minutes)	126	159	71	459	815
% total time	15.5	19.5	8.7	56.3	

From Leyhausen (1979).

male with a pride is about two years, the males tolerate cubs feeding at the same kill (Schaller 1972). However, in the Ruwenzori National Park, where males may hold on to a pride for up to 10 years, the males do not tolerate cubs at a kill (van Orsdol 1986). It has been suggested that this difference may have arisen because each cub represents a significant proportion of the total reproductive output of a Serengeti male compared with a Ruwenzori male (van Orsdol 1986). However, it is difficult to know whether this difference is adaptive, or whether there are different cultural traditions in different areas.

Communication

Lions share a common home range which they defend against other prides (Schaller 1972). Unlike most other felids, lions do not scent mark with faeces, but use only urine sprays and foot scrapes. Lions and wolves, *Canis lupus*, are both sociable species that advertise their ownership of a group territory vocally through roaring and howling, respectively (Schaller 1972, Harrington and Mech 1979). This difference in communication of home range ownership may be socially learnt from conspecifics. With up to 20 or more pride members living in an open habitat with few prominent features (Schaller 1972), it would be difficult to use faeces as an effective scent marking system. Most big cats advertise their home ranges and/or attract mates to some extent with loud, long range calls (Schaller 1967, Hoogesteijn and Mondolfi 1992, Bailey 1993), so that it does not seem unusual that this normal behaviour should extend to all members of a social group. In lions roaring is thought to function as a territorial advertisement, to allow pride members to locate each other, to intimidate rivals and to strengthen social bonds in the pride (Estes 1991). It would be interesting to investigate whether lions living alone or in smaller social groups than in the Serengeti communicate use faeces and rely much less on roaring (Leyhausen 1988), but no data are currently available.

Conclusion

As a group felids have been little studied from a social learning perspective because they are mostly small, solitary and live in closed habitats. However, the need for flexible behavioural strategies for coping with local ecological conditions suggests that social learning from the mother plays an important role in the behavioural development of the young.

Despite the difficulties of studying social learning in the field, it is possible to extrapolate from studies of domestic cats in laboratory conditions to wild felids because much felid behaviour is similar. For example, there have been several experimental investigations into the development of predatory skills and prey preferences in domestic cats, which indicate that these are influenced to a great extent by watching the mother or being offered opportunities to practise hunting by her. Although domestic cats can learn to hunt through a number of different routes, even as adults, in the wild the mother cannot risk the survival of her young to serendipity. It is important to ensure the development of predatory skills and prey recognition through opportunity learning.

Larger species of felid have additional problems in dealing with large prey without risking injury. The learning process in these species is extended beyond the constraint imposed by the later development of the permanent dentition, whereas in smaller species dispersal occurs soon after this. This developmental constraint, therefore, provides the opportunity for learning how to deal with large, potentially injurious, but highly rewarding prey.

Similar social learning can be inferred for wild felids from interspecific similarities and differences in behaviour and development, and dietary variations throughout wide geographical ranges. Social learning from mothers may also play an important role in the development of grooming, play, predator recognition and avoidance, scent marking, territorial inheritance, mate recognition and sexual behaviour. However, these require further investigation.

The sociable cats offer opportunities for social learning within groups from all group members. The possible social influences on the development of behaviour include cooperative hunting, male provisioning of food and methods of communication.

Therefore the felids offer good, but very challenging, opportunities for the investigation of social learning, whether in captivity or in the wild. However, we know currently very little about social learning in felids with the exception of the role of the mother in the development of predatory skills and prey recognition, but even here this is mostly restricted to the domestic cat. In particular, the sociable species offer excellent opportunities for studying intra-specific variation in behaviours owing to a range of social and environmental variables affecting social group size and structure.

Acknowledgements

I am very grateful to Hilary Box for giving me the opportunity to look at cats in a completely different way and for inviting me to give this paper. I am most grateful to John Deag who helped me find my way into the domestic cat literature and to Hilary Box, Peter Kertesz, Graham Law, John Skinner and Rob Young who made many helpful suggestions.

References

Adler, H. E. (1955). Some factors of observational learning in cats. *J. Genet. Pyschol.*, **86**, 159–77.

Bailey, T. N. (1993). *The African Leopard.* New York: Columbia University Press.

Barrett, P. and Bateson, P. (1978). The development of play in cats. *Behaviour*, **55**, 106–20.

Bateson, P. and Young, M. (1979). The influence of male kittens on the object play of their female siblings. *Behav. Neur. Biol.*, **27**, 374–8.

Berry, C. S. (1908). An experimental study of imitation in cats. *J. Comp. Neurol. Psychol.*, **18**, 1–25.

Bradshaw, J. W. S. (1992). *The Behaviour of the Domestic Cat.* Wallingford: CAB International.

Caro, T. M. (1979). Relations between kitten behaviour and adult predation. *Z. Tierpsychol.*, **51**, 158–68.

Caro, T. M. (1980a). The effects of experience on the predatory patterns of cats. *Behav. Neur. Biol.*, **29**, 1–28.

Caro, T. M. (1980b). Effects of the mother, object play, and adult experience on predation in

cats. *Behav. Heur. Biol.* **29**, 29–51.

Caro, T. M. (1980c). Predatory behaviour in domestic cat mothers. *Behaviour*, **74**, 128–48.

Caro, T. M. (1981). Predatory behaviour and social play in kittens. *Behaviour*, **76**, 1–24.

Caro, T. M. (1994). *Cheetahs of the Serengeti Plains.* Chicago: Chicago University Press.

Caro, T. M. and Hauser, M. D. (1992). Is there teaching in nonhuman animals? *Q. Rev. Biol.*, **67**, 151–74.

Chesler, P. (1969). Maternal influence in learning by observation in kittens. *Science*, **166**, 901–3.

Currier, M. J. P. (1983). *Felis concolor. Mammal. Sp.*, **200**, 1–7.

Deag, J. M., Manning, A. and Lawrence, C. E. (1988). Factors influencing the mother–kitten relationship. In *The Domestic Cat*, ed. D. C. Turner and P. Bateson, pp.23–40. Cambridge: Cambridge University Press.

Easterbee, N. (1991). Wildcat, *Felis silvestris.* In *The Handbook of British Mammals*, 3rd edn., ed. G. B. Corbet and S. Harris, pp. 431–7. Oxford: Blackwell

Scientific.

Eisenberg, J. F. and Lockhart, M. (1972). An ecological reconnaissance of Wilpattu National Park, Ceylon. *Smithson. Contrib. Zool.*, **101**, 1–118.

Eloff, F. C. (1973). Ecology and behaviour of the Kalahari lion. In *The World's Cats*, Vol. 1, ed. R. L. Eaton, pp. 90–126. Washington, DC: Carnivore Research Institute.

Estes, R. D. (1991). *The Behaviour Guide to African Mammals.* Berkley: University of California Press.

Ewer, R. F. (1969). The 'instinct to teach'. *Nature*, **222**, 698.

Ewer, R. F. (1973). *The Carnivores.* New York: Cornell University Press.

Fagen, R. (1981). *Animal Play Behaviour.* Oxford: Oxford University Press.

Feldman, H. N. (1994). Methods of scent marking in the domestic cat. *Can J. Zool.*, **72**, 1093–9.

Fox, J. L. and Chundawat, R. S. (1988). Observations of snow leopard stalking, killing and feeding behaviour. *Mammalia*, **52**, 137–40.

Geertsema, A. A. (1985). Aspects of the ecology of the serval *Lep-*

tailurus serval in the Ngorongoro Crater, Tanzania. *Neth. J. Zool.*, **35**, 627–10.

Haglund, B. (1966). Winter habits of the lynx (*Lynx lynx* L.) and wolverine (*Gulo gulo* L.) as revealed by tracking in the snow. *Viltrevy*, **4**, 81–310.

Haltenorth, T. and Diller, H. (1980). *A Field Guide to the Mammals of Africa Including Madagascar.* London: Collins.

Harrington, F. H. and Mech, L. D. (1979). Wolf howling and its role in territory maintenance. *Behaviour*, **68**, 207–49.

Hayssen, V., van Tienhoven, A. and van Tienhoven, A. (1993). *Asdell's Patterns of Mammalian Reproduction.* Ithaca and London: Cornell University Press.

Hemmer, H. (1976). Gestation period and postnatal development in felids. In *The World's Cats*, Vol. 3, ed. R. L. Eaton, pp. 143–65. Washington, DC: Carnivore Research Institute.

Hemmer, H. (1993). *Felis lynx* – Luchs. In *Handbuch der Säugetiere Europas. Band5/11 Raubsäuger (Teil II).* ed. M. Stubbe and K. Krapp, pp. 1119–67. Wiesbaden: AULA.

Herbert, M. J. and Harsh, C. M. (1944). Observational learning by cats. *J. Comp. Psychol.*, **37**, 81–95.

Hobhouse, L. T. (1901). *Mind in Evolution.* London: Macmillan.

Hoogesteijn, R. and Mondolfi, E. (1992). *The Jaguar.* Caracas: Arimitano Publications.

Izawa, M. and Ono, Y. (1986). Mother–offspring relationship in the feral cat population. *J. Mammal Soc. Japan*, **11**, 27–34.

John, R. E., Chesler, P., Bartlett, F. and Vicker, I. (1968). Observational learning in cats. *Science*, **59**, 1489–90.

Kitchener, A. (1991). *The Natural History of the Wild Cats.* London: Helm.

Kolb, H. H. (1977). Wild cat. In *The Handbook of British Mammals*, 2nd edn., ed. G. B. Corbet and H. N. Southern, pp. 375–82. Oxford: Blackwell Scientific.

Kruuk, H. (1986). Interactions between Felidae and their prey species. In *Cats of the World*, ed. S. D. Miller and D. D. Everett, pp. 353–74. Washington, DC: National Wildlife Federation.

Kruuk, H. and Turner, M. (1967). Comparative notes on predation by lion, leopard, cheetah and wild dog in the Serengeti area, East Africa, *Mammalia*, **31**, 1–27.

Kuo, Z. Y. (1930). The genesis of a cat's response to the rat. *J. Comp. Psychol.*, **11**, 1–35.

Kuo, Z. Y. (1938). Further study of the behaviour of the cat towards the rat. *J. Comp. Psychol.*, **25**, 1–8.

Leyhausen, P. (1979). *Cat Behavior.* New York and London: Garland STPM Press.

Leyhausen, P. (1988). The tame and the wild – another just-so story? In *The Domestic Cat*, ed. D. C. Turner and P. Bateson, pp. 57–66. Cambridge: Cambridge University Press.

McDougal, C. (1987). The man-eating tiger in geographical and historical perspective. In *Tigers of the World*, ed. R. L. Tilson and U. S. Seal, pp. 435–48. New Jersey: Noyes.

Martin, P. and Bateson, P. (1988). Behavioural development in cat. In *The Domestic Cat*, ed. D. C. Turner and P. Bateson, pp. 9–22. Cambridge: Cambridge University Press.

Martin, P. and Caro, T. M. (1985). On the functions of play and its role in behavioural development. *Adv. Study Behav.*, **15**, 59–103.

Mazak, V. (1981). *Panthera tigris. Mammal. Sp.*, **152**, 1–8.

Mellen, J. D. (1988). The effects of hand-raising on sexual behaviour of captive small felids using domestic cats as a model. *Assoc. Am. Zool. Parks Aquaria Annu. Proc.*, pp. 253–9.

Mellen, J. D. (1993). A comparative analysis of scent-marking. Social and reproductive behaviour in 20 species of small cats (*Felis*). *Am. Zool.*, **33**, 151–66.

Mendl, M. (1988). The effects of litter-size variation in the development of play behaviour in the domestic cat: litters of one and two. *Anim. Behav.*, **36**, 20–34.

Neville, P. (1992). Behaviour patterns that conflict with domestication. In *The Behaviour of the Domestic Cat*, ed. J. W. S. Bradshaw, pp. 187–204. Wallingford: CAB International.

Nowell, K. and Jackson, P. (eds.) (1996). *Wildcats. Status Survey and Conservation Action Plan.* Cambridge: IUCN.

van Orsdol, K. G. (1986). Feeding behaviour and food intake of lions in Ruwenzori National Park, Uganda. In *Cats of the World*, ed. S. D. Miller and D. D. Everett, pp. 377–88. Washington, DC: National

Wildlife Federation.

Parker, G. R., Maxwell, J. W. and Morton, L. D. (1983). The ecology of the lynx (*Lynx canadensis*) on Cape Breton Island. *Can. J. Zool.*, **61**, 770–86.

Pettifer, H. L. (1981). The experimental release of captive-bred cheetah (*Acinonyx jubatus*) into the natural environment. In *Worldwide Furbearer Conference Proceedings*, ed. J. A. Chapman and D. Paisley, pp. 1001–24. Falls Church, VA: R.R. Donnelley and Sons.

Richardson, D. (1992). *Big Cats.* London: Whittet.

Robinson, I. H. and Delibes, M. (1988). The distribution of faeces by the Spanish lynx (*Felis pardina*). *J. Zool. Lond.*, **216**, 577–82.

Rosevear, D. R. (1974). *The Carnivores of West Africa.* London: British Museum (Natural History).

Schaller, G. B. (1967). *The Deer and the Tiger.* Chicago: Chicago University Press.

Schaller, G. B. (1972). *The Serengeti Lion.* Chicago: Chicago University Press.

Seidensticker, J. and Lumpkin, S. (eds.) (1991). *Great Cats.* London: Merehurst.

Seidensticker, J. and McDougal, C. (1993). Tiger predatory behaviour, ecology and conservation. *Symp. Zool. Soc. Lond.*, **65**, 105–26.

Skinner, J. D. and Smithers, R. H. N. (1990). *The Mammals of the Southern African Subregion.* Pretoria: University of Pretoria Press.

Smith, J. L. D., McDougal, C. and Miquelle, D. (1989). Scent marking in free-ranging tigers, *Panthera tigris. Anim. Behav.*, **37**, 1–10.

Stahl, P. and Leger, F. (1992). *Le chat sauvage d'Europe.* Encyclopédie des Carnivores de France No. 17. Nort s/Edre: SFEPM.

Stander, P. E. (1992). Co-operative hunting in lions: the role of the individual. *Behav. Ecol. Sociobiol.*, **29**, 445–54.

Stander, P. E. and Albon, S. D. (1993). Hunting success of lions in a semi-arid environment. *Symp. Zool. Soc. Lond.*, **65**, 127–43.

Sunquist, M. E. (1981). The social organization of tigers (*Panthera tigris*) in Royal Chitawan National Park, Nepal. *Smithson. Contrib. Zool.*, **336**, 1–98.

Teyrovsky, V. (1924). Studies on the intelligence of the cat I. Imitation. *Publ. Faculte Sci. Univ. Masaryk.*, **41**, 1–21. (In Czech; cited in Leyhausen, 1979.)

Thapar, V. (1986). *Tiger: Portrait of a Predator.* London: Collins.

Tomkies, M. (1991). *Wildcats.* London: Whittet.

Tumlison, R. (1987). *Lynx lynx. Mammal. Sp.*, **269**, 1–8.

Turnbull-Kemp, P. (1967). *The Leopard.* Cape Town: Howard Timmins. (Cited in Skinner and Smithers, 1990.)

Turner, D. C. and Bateson, P. (eds.) (1988). *The Domestic Cat.* Cambridge: Cambridge University Press.

Whiten, A. and Ham, R. (1992). On the nature and evolution of imitation in the Animal Kingdom: reappraisal of a century of research. *Adv. Study Behav.*, **21**, 239–83.

Wrogemann, N. (1975). *Cheetah under the Sun.* Johannesburg: McGraw-Hill.

Wyrwicka, W. (1978). Imitation of mother's inappropriate food preference in weaning kittens. *Pav. J. Biol. Sci.*, **13**, 55.

Wyrwicka, W. and Long, A. M. (1980). Observations on the initiation of eating of new food by weanling kittens. *Pav. J. Biol. Sci.*, **15**, 115–22.

Yerkes, R. W. and Bloomfield, D. (1910). Do kittens instinctively kill mice? *Psychol. Bull.*, **7**, 253–63. (Cited in Caro, 1980a.)

15

Social learning in canids: an ecological perspective

Jan A. J. Nel

Introduction

The family Canidae is composed of some 34–35 species, assigned to 13–15 genera (Ginsberg and Macdonald 1990, Sheldon 1992). Current opinion regards the domestic dog and dingo as subspecies of the grey wolf – *Canis lupus familiaris* and *C. l. dingo* respectively (Sheldon 1992, Corbett 1995). In size canids range from the *c.* 1–1.5 kg Blandford's and Fennec foxes (*Vulpes cana* and *Fennecus zerda*) to the > 60 kg grey wolf, *Canis l. lupus* (Ginsberg and Macdonald 1990). They occupy a broad spectrum of habitats, which includes arid deserts (e.g. the Fennec fox and Blandford's fox), ice fields (arctic fox, *Alopex lagopus*) and lowland rainforests (the small-eared dog or zorro, *Dusicyon microtis* (Ginsberg and Macdonald 1990, Sheldon 1992). Canids also have a very wide geographical distribution. In the past the grey wolf, *C. l. lupus*, and at present the red fox, *Vulpes vulpes*, enjoy the widest geographical distribution of any nonhuman mammal (Ginsberg and Macdonald 1990).

Food and food acquisition

As a group, canids utilise a wide spectrum of food sources, including fresh meat, carrion, insects and fruits, which they obtain either by foraging singly, as pairs, as family groups or as packs (Macdonald 1984a). Apart from active hunting, scavenging and gleaning are also employed. Some species, e.g. the black-backed jackal, *Canis mesomelas*, are facultative cooperative hunters, which increases their capture success (Wyman 1967, Lamprecht 1978a). A few, such as the wolf, *C. l. lupus*, and the African wild dog, *L. pictus*, are obligatory cooperative hunters, killing prey much larger than themselves (Moehlman 1989).

Communal feeding occurs in many species (Kleiman and Eisenberg 1973), ranging from the insectivorous bat-eared fox, *Otocyon megalotis*, to the flesh-eating African wild dog, *Lycaon pictus*.

Social units, including mating systems

The most common mating system is long-term monogamy, a trait rare amongst mammals (Kleiman 1977). However, in smaller canids (i.e. < 6 kg) there is a tendency towards polygyny, perhaps due to a reduced need for paternal care (Moehlman 1989). Medium-sized canids are more strictly mon-ogamous, while in larger canids, e.g. African Wild dogs and wolves, polyan-drous mating also occasionally occurs (e.g. Van Lawick 1973, Reich 1981, Harrington *et al.* 1982, Moehlman 1986). Social systems vary from single foragers, pairs and family groups to packs; for at least some species the nature of the food supply strongly influences the size of social groupings. The kind and dispersion of food sources utilised by a species strongly influences social organisation not only in carnivores in general (e.g. Kruuk 1975, Mills 1978, 1990) but also in African ungulates (Jarman 1974), bats (Bradbury and Ve-hrencamp 1976) and primates (Clutton-Brock and Harvey 1977). For example, in coyotes, *Canis latrans*, Bekoff and Wells (1980) and Bowen (1981) have found that when carrion (elk carcasses) is freely available, packs form, but when carrion is scarce and rodents plentiful, single individuals or pairs forage.

Corresponding effects of the influence of food on social structure have been reported for golden jackals, *Canis aureus*, in Israel by Macdonald (1979). The otherwise prevalent pair structure is replaced by larger groups at feeding stations in these animals. Macdonald (1981) has also demonstrated another consequence of the food supply: home-range and territory size in red foxes, *Vulpes vulpes*, are dictated by the dispersion of food patches, while the number of foxes in a territory depends on the amount of food available. However, there will always be only one adult dog fox, but one or more vixens. Scent marking of ranges or territories may also correlate with group size, e.g. in golden jackals (Macdonald 1979), packs mark, while solitary individuals do not.

Litter size and helpers

Canids typically produce one litter per year, with mean litter size in most species correlating positively with female weight (Table 15.1). An exception is the maned wolf, *Chrysocyon brachyurus* (*c.* 20–23 kg), where litter size varies between one and five, with a mean of two. This is the only large canid, however, that regularly feeds on plant material and rodents (Dietz 1984, 1985).

Within a species, litter size can also vary depending on the food source and availability. For example, in Arctic foxes, *A. lagopus*, mean litter size varies from 10.1 in northwestern Canada, where lemmings, *Discrostonyx torquatus* and *Lemmus sibiricus*, can be very abundant when cubs are born (MacPherson 1969), to 4.0–5.3 in Iceland, where food resources are patchily distributed but reasonably constant in quantity (Hersteinsson 1984, Ginsberg and Macdonald 1990). A trait unique to canids (Moehlman 1986) is that as female weight increases, average total litter weight increases slightly faster. This means that, contrary to other mammal species, as a canid species gets larger, not only does litter weight increase but individual neonates are smaller, and thus more dependent, and there are more of them (Moehlman 1986). However, using phylogenetic considerations these correlates do not all hold up; although neonate weight and litter size correlate with female size, larger canids do not have relatively smaller neonates, and there are no correlations between neonate weight and litter size (Geffen *et al.* 1996). Canids are also unusual amongst mammals in that provisioning of cubs by both parents and, in several species, also by helpers (offspring from a previous litter) are found; intact prey are carried back to the cubs or regurgitation occurs. Paternal investment can extend beyond provisioning, however; in bat-eared foxes males typically guard the young and initiate them into foraging during the suckling period while the female forages on her own (Nel 1978). Communal care, where more than one litter occupies a den, is also exercised by some species (Gittleman 1985).

Period of dependence: time available for social learning

The period from birth to weaning (end of suckling) varies between species but adequate data for most are lacking, as are data on the effect of available food on the duration of this period, and infant mortality. Even more important when investigating learning in a social context, is the time that elapses from the cessation of suckling up to the dispersal of the cubs. The longer the period of exposure to parental influence while accompanying them on foraging trips, the more opportunity there will be for social learning. The duration of this potential social learning period is very poorly documented but is possibly not correlated with mean species weight: in the fairly similarly sized Cape fox, *Vulpes chama*, and bat-eared fox, this period is *c.* 1 month, and 5–7 months respectively (Bester 1982, Nel 1996). The Cape fox is predominantly carnivorous and forages singly, while the insectivorous bat-eared fox forages as a family group. In general, the young of the larger and more social (often communally hunting) species tend to become independent and disperse at a later age (Bekoff *et al.* 1981). Even so, dispersal age of pups within a species can also vary, and is affected by food supplies; coyote pup dispersal is delayed when elk carrion is plentiful (Camenzind 1978, Bekoff and Wells 1980, Bowen 1981). This also results in changes in group size and social structure due to differences in age of dispersing young (Kleiman and Brady 1978).

Territoriality

The nature of the prey taken can affect the occurrence of territoriality. Actively hunting species (e.g. coyotes, red foxes, some jackals) are all territorial (see e.g. Moehlman 1989), whereas species consuming a large percentage of insects, e.g. bat-eared foxes, or plant material, e.g. the crab-eating fox or zorro, *Cerdocyon*

Table 15.1. *The genera and species of extant canids, and some characteristics affecting sociality[a]*

Genus and species[b]	Weight (kg)	Litter size (range)	Litter size (mean)	Weaning age (weeks)	Dispersal age (months)	Communal care	Diet	Social group	Communal foraging	Cooperative hunting
Alopex lagopus	3.0	6–16	6.3	10	< 6	Y	O	2 + y	Y	Y
Canis adustus	8.3	3–6	3.4	10	11	—	O	2 + y – fam	—	—
C. aureus	9.0	5–6	5.5	10	—	Y	O	fam-pack	Y	Y
C. latrans	10.9	6–18	5.3	7	7–10	Y	C	2 + y-pack	Y	Y
C. l. lupus	16–60	1–11	6.0	10	—	Y	C > O	pack	Y	Y
C. l. dingo	8.3–17	1–10	6	8–9	6–8	Y	C/O	pack	Y	Y
C. mesomelas	7–11	1–8	3.5	10	4–5	Y	O	2 + y – fam	Y	Y
C. rufus	18–41	1–11	6	10	—	Y	C	pack	Y	Y
C. simensis	11–19	2–6	4	10	> 15	Y	C	7–pack	—	—
Cerdocyon thous	5–8	3–6	3.5	13	18–24	N	O	fam-pack	N	Y
Chrysocyon brachyurus	20–23	2–6	2.2	16	—	Y	O	1	N	N
Cuon alpinus	10–13	8–9	7.7	10	—	—	C > O	pack	Y	Y
Dusicyon culpaeus	5–13.5	3–8	5.5	5	—	—	O	—	—	—
D. griseus	4.4	2–6	4.0	—	—	—	O	—	—	—
D. gymnocercus	4–6.5	3–5	3.4	—	—	—	O	2 + y	—	—
D. microtis	9–10	—	—	—	—	—	O	1	—	—
D. sechurae	4–5	—	—	—	—	—	?O	—	—	—
D. vetalus	2.7–4	—	—	—	—	—	—	—	—	—

Species										
Fennecus zerda	1–1.5	1–5	3.5	9–10	—	Y	O	2–sg	—	—
Lycaon pictus	25	2–19	10.1	10	18(♀)	—	C	pack	N	Y
Nyctereutes procyonoides	7.5	4–6	6.6	4–5	—	Y	O	2+y	Y	N
Otocyon megalotis	3–5	1–6	4.1	10	6–7	—	I>O	2+y	Y	N
Speothos venaticus	5–7	1–6	3.8	8	—	Y	C>O	fam/pack	Y	Y
Urocyon cinereoargenteus	3.3	1–10	4.2	9	<12	—	O	2+y	—	N
U. littoralis	1.3–2.8	1–10	4.2	9	<12	—	O	2+y	—	N
Vulpes bengalensis	1.8–3.2	2–4	—	—	—	—	O	1–2	—	—
V. cana	0.9–1.5	1–3	20	9	—	N	I–O	1–2	N	N
V. chama	2.6	3–5	3.5	16–18	5	N	C/O	2	N	N
V. corsac	—	2–11	—	—	—	?Y	O	sg	—	—
V. ferrilata	3–4	2–5	—	—	—	—	C	?2	—	—
V. macrotis	1.9	3–4	4.5	—	09–10	—	C	2–3+y	—	—
V. pallida	2–3.6	3–5	—	—	—	—	O	2+y	—	—
V. rueppelli	1.2–3.6	2–3	2.5	—	—	—	O	2–sg	—	—
V. velox	1.8–3	3–6	—	7	5	—	O	2+y	—	—
V. vulpes	3.9	4–5	4.0	8	6–10	Y	O	2+	N	N

Y, Yes; N, No; y, young; sg, small group; —, not recorded/unknown; fam, family (can include individuals from a previous litter); O, omnivorous; C, carnivorous; I, Insectivorous.

aData from, *inter alia*, Bekoff (1975), Bekoff *et al.* (1981), Biben (1983), Macdonald (1984a,b), Gittleman (1985), Moehlman (1986), Ginsberg and Macdonald (1990), Corbett (1995), Geffen *et al.* (1996), Macdonald (1996), Macdonald and Courtenay (1996) and references therein.
bFollowing Ginsberg and Macdonald (1990). However, the swift fox, *Vulpes macrotis*, is regarded as a distinct species (McGrew 1979).

thous, are less so. Territoriality can also operate only during certain seasons depending on prey taxon, as in the crab-eating fox (Ginsberg and Macdonald 1990).

Categorising canid social structure is complicated not only by the pervasive influence of food source and abundance on group size (see e.g. Bekoff *et al.* 1984), but also by the variation within a species of time of dispersal of young, the extension of families into small groups, and the possibility of cooperative hunters coalescing family, or extended family, groups into large packs. To further complicate matters some canid species, e.g. the red fox (Macdonald. 1984b), can live in groups but forage singly within a territory. This phenomenon also occurs in some other carnivores, e.g. brown hyaenas (*Hyaena brunnea*) (Mills 1990).

Opportunities to acquire information from conspecifics

Considering the extended period of paternal care in many canid species, provisioning of young by both the parents as well as by helpers, foraging as a family for extended periods, and the flexibility shown both in social systems and prey taken, the potential role of social learning in enhancing adaptations to local conditions (see e.g. Galef 1995) would seem to be obvious. But what would seem obvious is certainly not reflected in the literature – the paucity of references on behaviour that can be construed as akin to social learning is surprising. However, even if actual instances of social learning have been overlooked or poorly documented, the possibilities of social learning occurring seems promising.

As many field studies have shown (see e.g. Macdonald 1984a and references therein), carrying either intact prey items back to pups, or provisioning them by regurgitation, is common; even insectivorous species like bat-eared foxes at times provision cubs with geckos (*Pachydactylus* spp.) (Pauw 1997). Social induction of prey preferences (Galef 1993, Galef and Whiskin 1994) could therefore be common, and the start of social learning in most, if not all, canid species, but experimental studies are sorely lacking. Many canid species also undertake group foraging, or hunting, before pups disperse (Ginsberg and Macdonald 1990), while in the large, social canids (wolf, *C. l. lupus*, dhole, *Cuon alpinus*, and African wild dog, *L. pictus*) many pups (usually males, in the two lastnamed) remain in the pack permanently. Constant exposure of these pups to foraging, feeding and social behaviour (interactions with mates, or other families of packs) is regular, as is learning home range or territory boundaries and the position of marking sites. Perhaps it is the very fact that such group activities are so common, and often observed, that has dulled the senses of observers to the possibilities of social learning in these contexts.

A big factor in the extent to which behaviours, or consequences of behaviour, e.g. marking range or territory boundaries, are learnt socially, could be the duration of the period – from weaning to dispersal – that pups spend with their parents. As Table 15.1 shows, considerable variation in this period occurs, even between species of much the same size/weight that occur sympatrically. For example, Cape foxes, *V. chama*, are weaned at 4 months, and disperse at 5 months, and no parental help or guidance in capturing prey has been documented (Bester 1982). The similar-sized bat-eared fox *O. megalotis*, by contrast, exhibits extended family coherence, with pups remaining with their parents for at least a further 5–7 months post-weaning (Nel 1978, 1996). This difference occurs under the same local conditions in the southwestern Kalahari Desert. Considering the flexible social systems found in a number of canid species, and the effect of food supplies on the time of dispersal in some, e.g. coyotes (Bowen 1981), the figures given in Table 15.1, as culled from the literature, must be regarded as only an approximation of the actual duration of this 'dependent' or 'exposure' period. Given the variation in local conditions, especially food availability at different locales, the duration of this period obviously will differ from place to place and year to year. But whatever the flexibility shown, extended exposure to maternal and paternal influence, and to that of other group members, can provide the input or source for learning in a social context.

Group foraging species encounter many opportunities for social learning including acquisition of knowledge about territories and other geographical spaces, environmental objects and appropriate behaviours (Giraldeau 1997). A few canid species, e.g. red foxes (Macdonald 1980), dingoes (Corbett 1995) and bat-eared foxes (Nel 1978, 1996), also exhibit behaviours that appear to involve opportunity teaching as defined by Caro and Hauser (1992) either by the females (red foxes and dingoes) or males (bat-eared foxes). These cases all involve prey capture, and can perhaps be regarded also as an manifestation of social induction to food, but in many other species, e.g. wolves (Mech 1970), the pack (including the pups) are brought to kills. Sometimes references are more specific: '. . . young are brought to the kill at [age] 103 days' and also 'pups assist adults in killing fawn' (the Asiatic wild dog or dhole, *Cuon alpinus*; Johnsingh 1982: 453). Under such circumstances the potential for social learning exists. It is not clear, however, to what extent the social learning process would involve young animals observing adults, one adult observing another, or one pup learning from socially interacting with its peers.

Vincent and Bekoff (1978) have shown that, at least under experimental conditions, social interactions between coyote pups contributed little to the development of prey-killing success in these coyotes. However, the latency to kill was significantly lower in the presence of conspecifics, i.e. the threshold excitation for killing was lower, probably due to social facilitation. This phenomenon could be involved in pair- or group-hunting by coyotes (Bekoff 1977).

Such explicit or inferred occurrence of opportunity teaching therefore could be more widespread than has been documented so far, and indeed should be expected (Caro and Hauser 1992). But even more poorly known is the extent of cultural transmission of behavioural traits to form traditions (Heyes 1993), as in the case of African wild dog packs hunting zebra (Malcolm 1984).

Factors influencing the extent of social learning

Which factors, then, could we expect to influence the extent of social learning amongst the canid species? The most important probably is food – both the nature of the prey and its dispersion, and the effect these have on the size of the foraging group, be it facultative or obligatory group foraging. Social learning can have important benefits to members of social groups, especially if their diet is opportunistic and diversified (Giraldeau *et al.* 1994).

It is perhaps significant that in all canid species that are obligatory cooperative hunters, such as the African wild dog, wolf and dhole, large and potentially dangerous prey are always taken by the pack operating as a unit (even if not all members participate in every hunt). In all these species some cubs, of one or both sexes, do not disperse but remain in the natal pack and are gradually inducted into cooperative hunting. These species, as primarily flesh-eaters, have higher than predicted (on the basis of Kleiber's equation) meta-bolic rates, and capturing large prey cooperatively can, but not necessarily always will, result in a higher per capita food intake. Pack hunters range over very large areas while foraging and phylogenetically there may not have been more 'room' than for one such foraging niche in canids per continent; South America seems to lack such a pack-living canid regularly killing very large prey. Packs would of course also be more successful in defending such large kills against competing species (see e.g. Lamprecht 1978b). The long period of interaction with adult pack members therefore would accentuate the value to pups derived from learning, in a social context, of how to help kill and defend prey.

The dispersion of prey could also be an important factor facilitating the extent of social learning. Preying on insects or fruits, both of which can be more abundant than for the needs of a single individual, would place a selective value on socially learning both the identity, and location of prey sources, insofar as pup survival is concerned. Here the experience of adult family members can be transferred to cubs; in such species, e.g. bat-eared foxes, nuclear family groups can stay intact for several months (Nel 1978).

However, a catholic diet does not necessarily imply an extended period, or large amount, of socially learned foraging and feeding behaviours. In the red fox (Macdonald 1987) some social learning may well take place, but as experiments with hand-reared individuals show, the full repertoire of foraging and feeding behaviours can develop even in individuals isolated from con-specifics at an early age.

Of perhaps lesser importance could be learning range or territory bound-aries, usually scent-marked, the location of water points, and the extent and use of warning signals as well as group signals such as communal howling.

Considering the number of field studies on the ecology or behavioural ecology of canids, of which only some have been referred to above, the dearth in documented cases of social learning is frustrating. Keeping in mind the lifestyle of canids, however, this paucity is perhaps understandable and excus-able – canids, like most carnivores except the social ones, are often nocturnal, and of small to medium size. They frequently hunt singly or in pairs in thick vegetation so that continuous direct observation, except in very open terrain, is difficult if not impossible. In consequence, reports on field studies of canids seldom contain any reference, oblique or direct, of social learning taking place.

However, the following examples are given to indicate the ways in which the social acquisition of information may enable individuals to improve their success in the acquisition of prey.

The red fox, *Vulpes vulpes*

The red fox, *Vulpes vulpes*, is catholic in its diet and uses different methods to secure different prey taxa. Rodents are first located and then captured using a 'mouse jump': the forequarters rise high and the forefeet and nose descend vertically on the prey. By contrast, when these foxes hunt earthworms, beetles and other invertebrates they move very slowly over a potential food patch, pausing and changing direction at intervals, and then rapidly plunge the snout into the grass and grasp the earthworm with the incisors (Macdonald 1980). Macdonald (1980) describes one case where a cub,who had been trying to use the 'mouse jump' to catch earthworms, but without success, ran across to the vixen to investigate the earthworm she had captured. The cub continued with its unsuccessful attempts until the vixen caught a third earthworm, which she did not pull out completely from its burrow. She kept it taut till the cub took it from her, and repeated the process shortly thereafter. The cub now started using its mother's technique.

Although this instance of social learning in the red fox seems to be the only one recorded, tutoring may be more prevalent in this species. Henry (1986) mentions that cubs often accompany their parents presumably on foraging trips, but that he never observed them on such forays. The red fox can also be very sociable on occasions (Macdonald 1987). On the other hand, the wide geographical distribution spanning a great variety of habitats, differing in food availability and identity, probably explains why selection for social learning in this species has not been more pervasive – the extent of 'instinctive' behaviour is such that survival is assured even if foxes operate singly.

The dingo, *Canis lupus dingo*

When dingo pups are small, pack members provision them with rabbits caught far from the den (Corbett 1995). The rabbits around the den are left alone, and allow growing pups to approach them closely. However, as pups start to mature, provisioning decreases, and pups start going further afield, and for longer periods, by themselves. At about the same time the rabbits near the den become more jumpy and bolt down holes at the approach of a dingo; pups now stalk rabbits and only rush the last few metres if they seem to feel they have a chance of success. One female was seen to coach the pups in the art of stalking (Corbett 1995): she would run for several metres, then stop and wait for the pups to catch up, then chase after the rabbits again. The pups copied her movements, stopped when she stopped and ran when she ran.

After 6 months of age pups also frequently accompany males, following their moves, thus suggesting that young dingoes refine their hunting and other behaviours by following the behaviour of their elders (Corbett 1995).

Bat-eared foxes, *Otocyon megalotis*

After bat-eared fox pups are weaned, or towards the end of the pre-weaning period, the male initiates them into foraging (Nel 1978, Pauw 1997).

In a semi-desert area such as the southwestern Kalahari a lactating female forages for long periods ($> 8\,\text{h night}^{-1}$) while the male guards the young, gradually leading them further and further from the den as they grow older. During this period males capture invertebrate prey, e.g. sunspiders (Solifugae),

Figure 15.1. Bat-eared fox.

and the cubs are then called to the male with a low whistle; the sunspider is released, whereupon the cubs try to capture it. If they are unsuccesful, the male repeats the process. After weaning and initiation into foraging, cubs remain with their parents as a nuclear family for at least a further 6–7 months, i.e. from birth in October–December until the family breaks up in June–July of the following year. During this period of close association with the parents the family forages as a group; when foraging/feeding on harvester termites, if one of the group locates a food patch (feeding colony of termites) the others are called to it, again with a low whistle. In this manner, both the prey source (e.g. sunspiders or termites) and foraging tactics are demonstrated to the pups (Nel 1996).

Wild dogs, *Lycaon pictus*

Wild dogs hunt cooperatively in packs, utilising a range of predominantly larger antelope (Van Lawick and van Lawick-Goodall 1970, Kruuk 1972, Schaller 1972). In the Serengeti plain some packs also hunt zebra (Malcolm and van Lawick 1975, Malcolm 1984a); these packs are large (eight or more adult pack members) but pack size does not seem to be the crucial factor – similar-sized packs ignore zebra, and not all members of zebra-hunting packs join in the kill.

As individual wild dogs can be recognised by the colour patterns, packs can also be identified. In two out of 10 packs in the Serengeti in East Africa hunting zebra was the norm; this 'tradition' in at least one pack was learned by each generation from its predecessor (Malcolm 1984a). In the space of 10 years at least three generations of wild dogs were involved in this pack. Similarly, knowledge of the location of water sources, prey concentrations and range boundaries may be passed on by tradition (Malcolm 1984a).

Absence of the opportunity to learn from experienced adult pack members may also be detrimental to the survival of young wild dogs. On three occasions captive reared wild dogs were reintroduced to the Etosha National Park in Namibia. All reintroductions failed because packs were unskilled at capturing large prey; they gradually starved, or were killed by lions. In addition, apart from apparently lacking an adequate escape response to lions, a pack of seven dogs did not displace cheetahs at a kill, and also tolerated black-backed jackal at their own kills (Scheepers and Venzke 1995). Efficient hunting is crucial to survival of wild dog packs; where the risk of kleptoparasitism by e.g. spotted hyaenas is high, increased metabolic needs cannot be met and wild dog populations decline (Gorman *et al.* 1998).

Hunting efficiency, and appropriate behaviour to secure kills by other species, or chasing away consumers of own kills involve socially acquired behaviours in the company of experienced adults.

Black-backed jackal, *Canis mesomelas*

Along the Skeleton Coast of the Namib Desert in Namibia black-backed jackals carry prey captured or scavenged on the beach to low coastal hummocks that back onto the beach. Utilising such specific feeding sites, probably for thermoregulatory reasons, is unique to groups whose territories include such hummocks (Dreyer and Nel 1990). In addition, when chasing cormorants on the beaches and causing them to take flight, these coastal jackal cut across the semi-circle executed by the cormorants as they turn to gain altitude, jump and pluck them from the air, at heights up to 2 m. Other groups that have cormorant roosting or breeding sites in their ranges will bark underneath the sites, causing the agitated birds to regurgitate their meals onto the ground, dislodge their chicks from the nest, or leave the roosting site. In the latter case, birds usually dive to gain speed and, if going too low, can also be plucked from the air. Some of these adaptations for increasing feeding efficiency are unique to this population of black-backed jackal, and are probably perpetuated through social learning rather than by learning as individuals.

Figure 15.2. Black-backed jackal.

There is a lack of studies on the occurrence of social learning in wild canids, but at least now we may ask relevant questions, and the existing observations should lead to further research. A few experimental studies have also been undertaken that allow much closer examination of social influences involved, even if they do not address ecologically valid conditions for social learning. The following are given as examples of experimental studies of socially mediated learning. They are also important because they demonstrate clear practical applications.

One study involves domestic dogs. *C. lupus familiaris*, and the other a study of black-backed jackals, *C. mesomelas*.

Seeking and retrieval of narcotics by dogs

Slabbert and Rasa (1997) studied the performance of juvenile German Shepherd dogs in retrieving sachets with narcotics, under different experimental conditions. Young dogs were either separated from their mothers at a young age (6 weeks), or had extended maternal contact (till age 12 weeks), while the mothers were either untrained, or trained to locate and retrieve hidden sachets. The trained bitches located the sachets solely on the basis of scent; visual stimuli were absent.

Young dogs that had extended maternal contact (12 weeks) and the opportunity to observe their mothers perform the learned skill (seeking and retrieval of sachets filled with narcotics) acquired this skill, and retained it for an extended period of time, i.e. when tested at 6 months of age. Pups separated from their trained mothers at 6 weeks age, or those left with their untrained mothers for 12 weeks, had significantly lower performance rates at testing. Slabbert and Rasa hypothesise that the skill acquired by the juveniles who spent an extended period with their mothers and observed them searching for hidden sachets, and their mothers being praised for their successes by the handler, came about through social learning. Klopfer (1961) suggested that social learning explains the retention of the effect of an observation after an interval (here 3 months).

Early separation from the mother also had negative effects on both physical condition and weight gain of the pups, and such pups had a higher susceptibility to disease. All these factors were related to symptoms of separation stress, as measured by behavioural differences between separated and nonseparated pups at 6 weeks age. Slabbert and Rasa (1997) suggest that training of working dogs, through observation of trained mothers during a 6-week period of extended maternal contact, would provide a viable alternative to conventional training methods which depends on positive reinforcement of learned responses, without impairment of dog/human relations and with improved pup health and survival.

Aversion learning by black-backed jackal

In many parts of southern Africa black-backed jackal prey heavily on small livestock, and their numbers are therefore controlled. The commonest methods are cyanide guns ('coyote getters'), foothold traps ('gin traps') and hunting. However, in areas where cyanide guns or traps were set during control operations their success rate steadily decreased, as jackals seem to progressively learn to avoid these devices. It was postulated that both inherent behavioural abilities, specifically stimulus discrimination, and learnt behavioural patterns could explain the avoidance of novel objects, in this case control devices (Brand 1993).

Avoidance of poisoned bait by black-backed jackal was suggested by Van der Merwe (1953), while avoidance learning has been shown to occur in coyotes, *Canis latrans* (Gustavson *et al.* 1974, Olsen and Lehner 1978, Horn 1983), kit foxes, *Vulpes macrotis* (Zoellick and Smith 1986) and dogs (Black 1959, Cornwell and Fuller 1961). However, the possible role of social learning processes were not discussed.

Brand and Nel (1997) investigated both stimulus discrimination (reaction towards novel objects) and the acquisition of avoidance behaviour in jackal under experimental conditions. Especially important for control operations was the way in which avoidance behaviour could spread through a population. In addition to determining whether the response time of jackal to bait (control object) was shorter in the absence than in the presence of novel objects, e.g. cyanide guns, and whether response to novel objects differs between inexperienced and experienced jackal, the difference in response to novel objects (control devices) by each member of a twosome was also investigated.

It was found that the animals showed an increased response time when exposed to novel objects, as opposed to being exposed to meat bait only. This indicates an inherent cautiousness towards unknown or novel stimuli.

Testing for avoidance of set cyanide guns (but armed with bitter juice) and padded foothold traps showed that there was no significant difference in the ratio of pairs activating or not activating cyanide guns when tested again after 3 or 6 months, but significantly more inexperienced than experienced jackal were 'captured'. In addition, mates of experienced jackal learned to avoid cyanide guns.

Brand (1993) also found that when naive juveniles were tested together with their experienced parents (i.e. previously exposed to experiments) the pups showed no significant difference in their response to cyanide guns, i.e. following their parents' example, they avoided them. When tested 3 months later on their own, these pups still did not pull the cyanide guns. When experienced pups and naive pups, tested in pairs, were exposed to functional devices 3 months later it was found that a significant difference in negative response (i.e. avoidance to cyanide guns) was evident between the experienced and naive pairs – the experienced pups avoided the devices, the naive pups did not.

Concluding remarks

Young canids, as with most other carnivores, are typically exposed to an extended period of parental care. The social environment in which the cubs grow up is initially dominated by their mother, who comforts, suckles and grooms them, but later on can include also paternal influences and even those of other adults and older siblings, as pack members or helpers, depending on the social structure of the species. Variables would include the extent and nature of interactions with adults and older siblings, the number of adults or siblings involved, and the duration of the period from the end of suckling to dispersal of young. The latter could prove to be the most important, as it is during this period that the young are becoming involved in group activities, e.g. hunting. During this period of exposure to adults and maturing siblings, opportunities for observing the behaviour of other group members is to be expected, even if the extent and nature of the observed behaviours vary. Socially learnt behaviours could be locally adaptive (Galef 1995) and stable if reinforced (Laland 1996), e.g. in family groups or larger aggregations. Although difficult to demonstrate empirically, a predisposition to learn from observing conspecifics could be a strong selective force affecting survival, as in carnivores, where mortality of subadults can be very high. An individual carnivore can, by trial and error, improve its prey capture techniques. In many cases, the nature of the prey would entail no risk to a young, inept canid, trying to subdue the prey on its own, even if it were unsuccessful at first. But where so many species, especially in the family Canidae, show a group, family or even pack social structure, learning by observing group members may play a major role in perfecting such behaviours, especially when potentially dangerous, large prey are captured. The extent to which canids will do so depends upon selection for social learning being operative, that is, upon such learning being adaptive. The general constraints of adaptation to local conditions have been discussed, e.g. by Galef (1995). What would be the case in canids? Most species are catholic in their prey usage, but learning from parents which prey items are most advantageous would favour the quick learner, especially as mortality during the period from weaning to dispersal can be substantial, e.g. in black-backed jackal, *Canis mesomelas* (Ginsberg and Macdonald 1990). Similarly, where prey availability is ephemeral, multiple users would not decrease individual intake. For example, patches of termites foraging above ground are time-depleted, once these are being harvested by bat-eared foxes. Acquiring the habit of calling group members (in this case all close relatives) will increase fitness of the particular individual. Learning the boundaries of one's group territory can also be adaptive, in that prior access to food or water resources results. For example, in the Kalahari black-backed jackal pups are dominant even to adults from other groups in their home territory as regards access to waterholes or carcasses (unpublished data).

The examples that have been given here are intended to open up questions for research in this area. There is a great deal to do, but even with the limited data at our disposal it is clear that social learning may play an important role during the maturation of cubs in many canid species, and that individual differences in the amount learned could well be decisive to survival. But these are hypotheses, that need to be tested.

Acknowledgements

Initial research for this paper was done while visiting to the Department of Biology, University of Haifa at Oranim, Israel. I wish to thank my host, Abraham Haim, for many courtesies and help; he and his colleagues for many stimulating discussions; and Yoram Yom-Tov of Tel-Aviv University for putting their library at my disposal. Our Departmental librarian at Stellenbosch University, Mrs C van der Westhuÿsen, was exceptionally helpful in locating and obtaining references. Barbara Cook, the editors and referees suggested much appreciated improvements to the manuscript.

References

Bekoff, M. (1975). Social behavior and ecology of African Canidae: a review. In *The Wild Canids. Their Systematics, Behavioral Ecology and Evolution*, ed. M. W. Fox, pp. 120–42. New York: Van Nostrand Reinhold.

Bekoff, M. (1977). *Canis latrans* Say. *Mammal. Species*, **79**, 1–9.

Bekoff, M. and Wells, M. C. (1980). The social ecology of coyotes. *Sci. Am.*, **242**(4), 112–20.

Bekoff, M., Diamond, J. and Mitton, J. B. (1981). Life-history patterns and sociality in canids: body size, reproduction, and behaviour. *Oecologia* (Berl.), **50**, 386–90.

Bekoff, M., Daniels, J. and Gittleman, J. L. (1984). Life history patterns and the comparative social ecology of carnivores. *Annu. Rev. Ecol. Syst.*, **15**, 191–232.

Bester, J. L. (1982). The behavioral ecology and management of the Cape fox *Vulpes chama* (A. Smith) with special reference to the Orange Free State. MSc thesis, University of Pretoria. (In Afrikaans.)

Biben, M. (1983). Comparative ontogeny of social behaviour in three South American canids, the maned wolf, crab-eating fox and bush dog: implications for sociality. *Anim. Behav.*, **31**, 814–26.

Black, A. H. (1959). Heart rate changes during avoidance learning in dogs. *Can. J. Psychol.*, **13**, 229–42.

Bowen, W. D. (1981). Variation in coyote social organization: the influence of prey size. *Can. J. Zool.*, **59**, 639–52.

Bradbury, J. W. and Vehrencamp, S. L. (1976). Social organization and foraging in Emballonurid bats. 1. Field studies. *Behav. Ecol. Sociobiol.*, **1**, 337–81.

Brand, D. J. (1993). The influence of behaviour on the management of black-backed jackal. PhD thesis, University of Stellenbosch.

Brand, D. J. and Nel, J. A. J. (1997) Avoidance of cyanide guns by black-backed jackal. *Appl. Anim. Behav. Sci.*, **55**, 177–82.

Camenzind, F. J. (1978). Behavioral ecology of coyotes in the National Elk Refuge, Jackson, Wyoming. In *Coyotes: Biology, Behavior and Management*, ed. M. Bekoff, pp. 267–94. New York: Academic Press.

Caro, T. M. and Hauser, M. D. (1992). Is there teaching in non-human animals? *Q. Rev. Biol.*, **67**, 151–74.

Clutton-Brock, T. H. and Harvey, P. H. (1977). Primate ecology and social organization. *J. Zool., Lond.*, **183**, 1–39.

Corbett, L. K. (1995). *The Dingo in Australia and Asia*. Ithaca and London: Comstock/Cornell.

Cornwell, A. C. and Fuller, J. L. (1961). *Conditioned Operant – Pavlovian Interactions*. London: Wiley.

Dietz, J. M. (1984). Ecology and social organization of the maned wolf (*Chrysocyon brachycurus*). *Smithsonian*

Contrib. Zool., No. **392**, 1–5.

Dietz, J. M. (1985). *Chrysocyon brachyurus. Mammal. Species*, **234**, 1–4.

Dreyer, H. van A. and Nel, J. A. J. (1990). Feeding-site selection by black-backed jackals on the Namib Desert coast. *J. Arid Environ.*, **19**, 217–24.

Galef, B. G., Jr (1993). Functions of social learning about food: a causal analysis of effects of diet novelty on preference transmission. *Anim. Behav.*, **46**, 257–63.

Galef, B. G., Jr. (1995). Why behaviour patterns that animals learn socially are locally adaptive. *Anim. Behav.*, **49**, 1325–34.

Galef, B. G., Jr. and Whiskin, E. L. (1994). Passage of time reduces effects of familiarity on social learning: functional implications. *Anim. Behav.*, **48**, 1057–62.

Geffen, E., Gompper, M. E., Gittleman, J. L., Luha, H. R., Macdonald, D. W. and Wayne, R. K. (1996). Size, life history traits, and social organization in the Canidae: a reevaluation. *Am. Nat.*, **147**, 140–60.

Ginsberg, J. R. and Macdonald, D. W. (1990). *Foxes, Wolves, Jackals and Dogs. An Action Plan for the Conservation of Canids.* Gland, Switzerland: IUCN/SSC Canid Specialist Group.

Giraldeau, L-A. (1997). The ecology of information use. In *Behavioral Ecology*, 4th edn, ed. J. R. Krebs and N. B. Davies, pp. 42–68. Oxford: Blackwell Science.

Giraldeau, L-A, Caraco, T. and Valone, T. J. (1994). Social foraging: individual learning and cultural transmission of innovations. *Behav. Ecol.*, **5**, 35–43.

Gittleman, J. L. (1985). Functions of communal care in mammals. In *Evolution – Essays in Honour of John Maynard Smith*, ed. P. J. Greenwood, P. H. Harvey and M. Slatkin, pp. 187–205. Cambridge: Cambridge University Press.

Gorman, M. L., Mills, M. G., Raath, J. P. and Speakman, J. R. (1998). High hunting costs make African wild dogs vulnerable to kleptoparasitism by hyaenas. *Nature*, **391**, 479–81.

Gustavson, C. R., Garcia, J., Hankins, W. G. and Rusiniak, K. W. (1974). Coyote aversion control by aversive conditioning. *Science*, **184**, 581–3.

Harrington, F. G., Paquet, P. C., Ryon, J. and Fentress, J. C. (1982). Monogamy in wolves: a review of the evidence. In *Wolves of the World*, ed. F. H. Harrington and P. C. Paquet, pp. 209–22. Park Ridge: Noyes Publications.

Henry, J. D. (1986). *Red Fox – The Catlike Canid.* Washington, DC: Smithsonian Institution Press.

Hersteinsson, P. (1984). The behavioural ecology of the Arctic fox (*Alopex lagopus*) in Iceland. PhD thesis, Oxford University.

Heyes, C. M. (1993). Imitation, culture and cognition. *Anim. Behav.*, **46**, 999–1010.

Horn, S. W. (1983). An evaluation of predatory suppression in coyotes using lithium chlor-ide-induced illness. *J. Wildl. Manage.*, **47**, 999–1009.

Jarman, P. J. (1974). The social organisation of antelope in relation to their ecology. *Behaviour*, **48**, 215–67.

Johnsingh, A. J. T. (1982). Reproductive and social behaviour of the Dhole, *Cuon alpinus* (Canidae). *J. Zool., Lond.*, **198**, 443–63.

Kleiman, D. G. (1977). Monogamy in mammals. *Q. Rev. Biol.*, **52**, 39–69.

Kleiman, D. G. and Brady, C. A. (1978). Coyote behavior in the context of recent canid research: problems and perspectives. In *Coyotes: Biology, Behavior and Management*, ed. M. Bekoff, pp. 163–88. New York: Academic Press.

Kleiman, D. G. and Eisenberg, J. F. (1973). Comparisons of canid and felid social systems from an evolutionary perspective. *Anim. Behav.*, **21**, 637–59.

Klopfer, P. H. (1961). Observational learning in birds: the establishment of behavioural modes. *Behaviour*, **17**, 71–80.

Kruuk, H. (1972). *The Spotted Hyena.* Chicago: University of Chicago Press.

Kruuk, H. (1975). Functional aspects of social hunting by carnivores. In *Function and Evolution in Behaviour*, ed. G. Baerends, C. Beer and A. Manning, pp. 119–41. Oxford: Clarendon Press.

Laland, K. N. (1996). Is social learning always locally adaptive? *Anim. Behav.*, **52**, 637–40.

Lamprecht, J. (1978a). On diet, foraging behaviour and interspecific food competition of jackals in the Serengeti National Park, East Africa. *Z. Säugetierkunde*, **43**, 210–23.

Lamprecht, J. (1978b). The relationship between food competition and foraging group size in some larger carnivores. *Z. Tierpsychol.*, **46**, 337–43.

Macdonald, D. W. (1979). The flexible social system of the golden jackal, *Canis aureus*. *Behav. Ecol. Sociobiol.*, **5**, 17–38.

Macdonald, D. W. (1980). The red fox, *Vulpes vulpes*, as a predator upon earthworms, *Lumbricus terrestris*. *Z. Tierpsychol.*, **52**, 171–200.

Macdonald, D. W. (1981). Resource dispersion and the social organisation of the red fox, *Vulpes vulpes*. In *Proceedings of the World-wide Furbearer Conference*, ed. J. A. Chapman and D. Ursley, pp. 918–49. Frostburg, Maryland.

Macdonald, D. W. (1984a). The dog family. In *The Encyclopaedia of Mammals*, vol. 1, ed. D. W. Macdonald, pp. 56–7. London: George Allen and Unwin.

Macdonald, D. W. (1984b). Foxes. In *The Encyclopaedia of Mammals*, vol. 1, ed. D. W. Macdonald, pp. 68–73. London: George Allen and Unwin.

Macdonald, D. W. (1987). *Running with the Fox*. London: Unwin Hyman.

Macdonald, D. W. (1996). Social behaviour of captive bush dogs (*Speothos vernaticus*). *J. Zool., Lond.*, **239**, 525–43.

Macdonald, D. W. and Courtenay, O. (1996). Enduring social relationships in a popula-

tion of crab-eating zorros, *Cerdocyon thous*, in Amazonian Brazil (Carnivora, Canidae). *J. Zool. Lond.*, **239**, 329–55.

MacPherson, A. H. (1969). *The Dynamics of Canadian Arctic Fox Populations. Canadian Wildlife Service* No. 8: 1–52. Ottawa: Dept. of Indian Affairs and Northern Development.

Malcolm, J. R. (1984). African wild dog. In *The Encyclopaedia of Mammals*, vol. 1, ed. D. W. Macdonald, pp. 76–9. London: George Allen and Unwin.

Malcolm, J. R. and van Lawick, H. (1975). Notes on wild dogs (*Lycaon pictus*) hunting zebras. *Mammalia*, **39**, 231–40.

McGrew, J. C. (1979). *Vulpes macrotis. Mammal. Species*, **123**, 1–6.

Mech, L. D. (1970). *The Wolf*. New York: Natural History Press.

Mills, M. G. L. (1978). The comparative socio-ecology of the Hyaenidae. *Carnivore*, **1**, 1–7.

Mills, M. G. L. (1990). *Kalahari Hyaenas. The Comparative Behavioural Ecology of Two Species*. London: Unwin Hyman.

Moehlman, P. D. (1986). Ecology of cooperation in canids. In *Ecological Aspects of Social Evolution. Birds and Mammals*, ed. D. I. Rubenstein and R. W. Wrangham, pp. 64–86. Princeton: Princeton University Press.

Moehlman, P. D. (1989). Intraspecific variation in canid social systems. In *Carnivore Behavior, Ecology and Evolution*, ed. J. L. Gittleman, pp. 143–63. London: Chapman and Hall.

Nel, J. A. J. (1978). Notes on the

food and foraging behavior of the bat-eared fox, *Otocyon megalotis. Bull. Carnegie Mus. Nat. Hist.*, **6**, 13–27.

Nel, J. A. J. (1996). The termitenator. *BBC Wildl. Mag.*, **14**(5), 24–30.

Olsen, A. and Lehner, P. N. (1978). Conditioned avoidance of prey in coyotes. *J. Wildl. Manage.*, **42**, 676–9.

Pauw, A. (1997). Bat-ears. Africa's insect-eating fox. *Africa Environ. Wildl.*, **5**(3), 30–9.

Reich, A. (1981). The behaviour and ecology of African wild dogs in Kruger National Park. PhD thesis, Yale University.

Schaller, G. B. (1972). *The Serengeti Lion*. Chicago: University of Chicago Press.

Scheepers, J. L. and Venzke, K. A. E. (1995). Attempts to reintroduce African wild dogs *Lycaon pictus* into Etosha National Park, Namibia. *S. Afr. J. Wildl. Res.*, **25**, 138–40.

Sheldon, J. W. (1992). *Wild Dogs. The Natural History of the Nondomestic Canidae*. London: Academic Press.

Slabbert, J. M. and Rasa, O. A. E. (1997) Observational learning of an acquired maternal behaviour pattern by working dog pups: an alternative training method? *Appl. Anim. Behav. Sci.*, **53**, 309–16.

Van der Merwe, N.J. (1953). The jackal. *Fauna Flora*, **4**, 1–83.

Van Lawick, H. (1973). *Solo*. London: Collins.

Van Lawick, H. and van Lawick-Goodall, J. (1970). *Innocent Killers*. London: Collins.

Vincent, L. and Bekoff, M. (1978). Quantitative analyses of the ontogeny of predatory behaviour in coyotes, *Canis latrans.* *Anim. Behav.*, **26**, 225–31.

Wyman, J. (1967). The jackals of the Serengeti. *Animals,* **10**, 79–83.

Zoellick, B. W. and Smith, N. S. (1986). Capturing desert kit foxes at dens with box traps. *Wildl. Soc. Bull.,* **74**, 284–6.

Part 5

Dolphins and whales: communication and foraging in aquatic environments

Editors' comments

KRG and HOB

This section focuses on the social transmission of foraging, vocal and social skills in cetaceans in relationship to the challenges of capturing fish and mammalian prey, communicating through an aquatic medium, and meeting the demands of complex social structure.

The prey capture techniques of several species of whales and dolphins are described by James Boran and Sara Heimlich (chapter 16). They include the cooperative herding of fish, stranding fish and sea mammals on beaches, stunning fish with tail slaps, emission of bubbles to confuse and congregate schools of fish and/or combinations of these and other behaviours. To some extent foraging techniques vary according to species. Some cetacean species, however, use a number of techniques. Bottlenose dolphins, in particular, are able to exploit a wide range of dietary resources, because of their ability to use a diversity of foraging procedures. For example, some bottlenose dolphins have been observed carrying sponges on the tips of their snouts while searching for food, a behaviour that may dislodge fish from the bottom (Smolker *et al.* 1997), and, off the coast of Brazil, others herd fish into nets held by fishermen (Pryor *et al.* 1990). Killer whales are also noted for the diversity of their diets that may include invertebrates, fish, and marine mammals including other whales.

Many cetaceans resemble great apes, humans, and elephants in that they have large bodies, large brains, and long lifespans, and they give birth to a single young that is nursed for a year or more. These life-history strategies reflect considerable parental investment of time and energy during periods of pregnancy and lactation. Killer whales and, perhaps other cetaceans, also resemble tool-using apes and humans in that they use foraging procedures whose mastery requires years of practice on the part of the young and, hence, potentially further prolongs the period of dependency.

The ability to socially acquire and transmit foraging procedures is likely to benefit individual members of any species that uses highly varied foraging techniques. Social learning will be especially valuable to members of those species whose foraging procedures are sufficiently complex that individual animals would be unlikely to discover them on their own (Gibson, chapter 19, this volume, Shennan and Steele, chapter 20, this volume). In species whose foraging procedures can be mastered only after years of practice, parents will also benefit reproductively from any behaviours, such as teaching, that accelerate learning, hence, foraging independence, on the part of the young. (Shennan and Steele, chapter 20, this volume). In animals that hunt cooperatively, as do many killer whales and bottlenose dolphins, active teaching may also assure that juveniles contribute to group foraging success at the earliest possible age, and, hence, benefit all members of the cooperative group. Given these considerations, we would expect to find the social acquisition of foraging techniques, perhaps, even the teaching of foraging techniques, in some cetacean groups.

As described by Boran and Heimlich in chapter 16, some of the best-studied cetacean species may meet these theoretical expectations. One well-documented instance of the social acquisition of foraging techniques occurred in humpbacked whales. These animals use a bubble cloud method of foraging: i.e. they dive beneath schools of fish and blow bubbles that result in a concentration of fish within a bubble circle. They then swallow large numbers of fish simultaneously. One population of humpbacked whales devised a variant of this technique called lobtailing. These whales slap the water with their tails, then dive and emit bubbles. The lobtailing behaviour evidently stuns the fish. The frequency of lobtailing behaviour increased in this population with time as young humpbacked whales born into the group acquired it, evidently by observing older members of the group, although not necessarily their own parents. The sponge using and net-driving techniques of bottlenose dolphins are also thought to be socially acquired.

Several examples of possible teaching have also been reported in dolphins and killer whales. Dolphins and killer whales both live in complex social groups and engage in cooperative behaviours. Young animals must learn the social structure of their groups and cooperative behaviour. Some observations reported by Boran and Heimlich suggest that active teaching of social behaviours may occur in killer whales. Older animals, for example, encourage younger ones to engage in mock battles and appear to demonstrate appropriate fighting techniques and postures.

Other examples of probable teaching relate to foraging endeavours. Atlantic spotted nosed dolphins, for instance, forage on bottom dwelling fish. Mother dolphins sometimes drive fish from the bottom and then permit their offspring to practise pursuit and capture techniques (Herzing 1996, cited in chapter 16).

Some bottlenose dolphin and killer whale populations practice foraging techniques that involve driving fish (by dolphins) or marine mammals (by killer whales) onto beaches. In the process, these cetaceans beach themselves and will die unless they succeed in reentering the water. In Punte Norte, Argentina, young killer whales actively watch as older animals first drive sea lions onto the beach and then strand and free themselves. In 2/3 of the observed episodes of juveniles practicing these behaviours in Argentina, adult males stood by and watched. The adults charged the beach when juveniles were unsuccessful, and, then, the juveniles would try again. One killer whale female in the Crozet Islands helped her juvenile strand by pushing it towards the beach.

In addition to complex foraging techniques and social structure, many cetaceans possess well-developed vocal skills. Vincent Janik (chapter 17) details vocal capacities of bottlenose dolphins in relationship to the communicative challenges of an aquatic environment, and Boran and Heimlich summarise the vocal capacities of killer whales and the song learning behaviours of humpbacked whales. All of these species modify vocal output in response to social input. Killer whales and dolphins also possess definitive vocal imitation capacities and pod-specific vocal dialects. Dolphins and killer whales use socially modified vocalisations for individual identification and for group cohesion. As Janik notes, individual identification by the vocal–auditory channel may be an essential adaptation for dolphin mothers and calves who can become widely separated while the mother hunts. Whether cetaceans use socially acquired or socially modified vocalisations for the transmission of environmental information remains an open question. Some laboratory reports cited by Janik, however, do suggest that socially acquired dolphin vocalisations can be used to refer to specific objects, and as Boran and Heimlich point out, group cohesion and individual recognition theories cannot explain the full diversity of killer whale vocalisations.

References

Herzing, D. (1996). Vocalizations and associated underwater behaviour of free-ranging Atlantic spotted dolphins, *Stenella frontalis*, and bottlenose dolphins, *Tursiops truncatus*. *Aquat. Mammal.*, **22**, 61–79.

Pryor, K., Lindberg, J., Lindberg, S. and Milano, R. (1990). A dolphin-human fishing cooperative in Brazil. *Mar. Mammal Sci.*, **6**, 77–82.

Smolker, R. A., Richards, A. F., Conner, R. C., Mann, J. and Berggren, P. (1997). Sponge carrying by dolphins (Delphinidae, *Tursiops* sp.): a foraging specialization involving tool use? *Ethology*, **103**, 454–65.

16

Social learning in cetaceans: hunting, hearing and hierarchies

James R. Boran and Sara L. Heimlich

Introduction

The order *Cetacea* is made up of approximately 78 species which vary substantially in body size, behavioural ecology, social organisation and sociality (Leatherwood and Reeves 1983). Body lengths range from 1.9 m for the harbour porpoise, *Phocoena phocoena*, to 31 m for the blue whale, *Balaenoptera musculus*. Whereas some species are relatively solitary, with assemblages of tens or more occurring only when food is especially clumped (e.g. blue whales: Calambokidis and Steiger 1997), groups of up to 1000 are typical of some of the oceanic dolphins (Würsig 1979). The degree and permanence of movements, site fidelity and mixing between age/sex groups also vary (Tyack 1986). However, the interaction between morphology, ecology and social features is still not well understood for the majority of species.

This is hardly surprising because their aquatic habitat presents unique challenges for long-term observations of the same groups; only a fraction of a cetacean's life is conducted at the surface; their migratory patterns can cover thousands of miles; and their occurrence in study areas can be short. Consequently, relationships among individuals can be difficult to document.

Cetaceans are distributed in all oceans, from polar to equatorial waters. They are also found in a number of large river systems (e.g. Amazon, Ganges, Yangtzee). They occur in enclosed seas (e.g. the Black Sea), coastal waters and deep, mid-ocean regions. However, for the majority of species only basic information is available on their natural histories.

The species which have been best studied live in coastal waters, such as some bottlenose dolphin, *Tursiops truncatus*, and killer whale, *Orcinus orca*, populations (Wells 1991, Ford *et al.* 1994). Migratory species which exhibit some degree of site fidelity, such as humpback, *Megaptera novaeangliae*, blue, right, *Eubalaena* spp. and grey, *Eschrichtius robustus*, whales, which return to the same environs year after year, have also been successfully studied (e.g. Darling 1984, Weinrich 1991, Payne 1995).

Most cetaceans are long-lived and many (predominantly the odontocetes:

Figure 16.1. Killer whales © J. Boran/S. Heimlich.

Brodie 1969) have long periods of sexual and social maturation, on par with large, social terrestrial mammals, particularly elephants (Moss 1988), wild dogs (Frame *et al.* 1979) and some primate species (Harvey and Clutton-Brock 1985). Long maturation in cetaceans probably improves survival by extending the period for learning the complexities of 'ocean lore', such as habitat parameters, feeding and navigational strategies, and social requirements. Social learning may facilitate this prolonged maturation and acquisition of the skills and information relevant to cetacean lifestyles.

Despite a lack of wide ranging and detailed information in this area, cetaceans present interesting examples and hypotheses for socially facilitated acquisition of fundamental survival skills, i.e. social learning. It has been proposed that social learning is more likely to occur in species which are opportunists (reviewed in Lefebvre and Giraldeau 1996). The cetacean species reviewed here have had to adapt to exploiting the patchy, clumped resources of the oceanic environment, and meeting the challenges presented by life in a three-dimensional, vast and often featureless medium. Living in groups improves foraging success and protection from predators (Norris and Dohl 1980, Wells *et al.* 1980, Norris and Schilt 1988). Undoubtedly, improvements are gained, in part, because groups also enable development of new methods for resource exploitation, through learned behaviours which are transmitted through the group.

The most probable candidates for social learning are species which live in consistent groupings, where repeated interactions between the same individuals occur (as reviewed for cetaceans which appear to exhibit reciprocal altruism: Connor and Norris 1982). For these species, group synchrony is especially important for meeting the challenges of living in the three-dimensional ocean, where resources are sparsely distributed.

The aim of this chapter is to present examples of socially mediated learning from long-term studies of free-ranging populations in which individual animals were identified within resident groups, where information obtained in a social context from relatively experienced individuals is advantageous for the relatively inexperienced.

We take examples from a variety of studies that provide detailed observations of the same individuals over time and contribute interesting examples of unique, group-specific and site-specific behaviours in which socially mediated learning is strongly implicated. The examples include observations from our own work on killer whales and pilot whales.

Our review is organised under three broad topics: feeding specialisations ('hunting'), vocal traditions ('hearing') and species variability in social structure ('hierarchies'). The majority of the examples are of feeding specialisations because these have been most clearly documented, but vocal traditions are also clearly exhibited in the few species that have been studied. There are fewer examples of variability in social structure, possibly because the detailed, long-term studies required for this are few and just beginning to yield data.

Feeding specialisations

Cetaceans must have the ability to both locate prey and track changes in the environment over large distances and, frequently, long time-spans. Sound, either ambient, prey-generated or cetacean-generated (e.g. echolocation in odontocetes) is of paramount importance to cetacean foraging success, and allows them to extend foraging beyond the limitations of vision. However, memories of previous feeding success are probably equally integral to overall foraging success in cetaceans (Würsig 1986). Such memories are most likely to be best retained by the older, more experienced members of the group, and may be transferred to other members through social learning.

More particularly however, most cetaceans have specialised foraging methods, of which the following are interesting in the context of socially mediated learning.

The subsurface behaviour of Atlantic spotted dolphins, *Stenella frontalis*, has been studied in the clear waters on the offshore, shallow sand banks in the

Figure 16.2. Killer whale catching a sea lion © R. Hoelzel.

Bahamas since 1985 (Herzing 1996). Underwater video and sound recordings document behavioural interactions, including foraging techniques.

The primary prey are bottom-dwelling fish, which are frequently buried in the sand and which dolphins often locate with echolocation (Herzing 1996). Social learning appears to be an important element in detection and capture of these fish. Young calves (less than one year old) regularly travel behind and beneath their mother as she scans the sea bottom. Often, the mother drives out the fish but allows the calf to pursue the chase and attempt capture. The mother's deferral of feeding appears to be an attempt to teach the calf how to catch its own food (Herzing pers. commun.). Juveniles also appear to watch other juveniles during foraging, or even during play when chasing and harassing stingray. The older juveniles repeatedly demonstrate their technique and then allow the younger animals to try it themselves (Herzing pers. commun.).

Bottlenose dolphins, *Tursiops truncatus*, are highly adaptable feeders (Shane *et al.* 1986). They specialise on locally abundant food types, and many elements of their feeding specialisations are transmitted from generation to generation through social learning. For example, dolphins feeding in shallow salt marshes often chase fish up onto the beach and then push themselves back into the water, and adults have been observed demonstrating this to young dolphins (Hoese 1971). A review of intentional beaching behaviour in bottlenose dolphins found that it has developed in geographically separated populations (Silber and Fertl 1995), but the learning processes involved in its acquisition have not been well studied. However, suggestions for the ways in which social learning may be important in this behaviour are provided by killer whales, which also exhibit intentional beaching. We discuss this in a later section of the paper.

A unique and on-going cooperative interaction between bottlenose dolphins and human fishermen off the coast of Brazil (Pryor *et al.* 1990) is interesting in various ways. The dolphins help bring fish (mullet, *Mugil cephalus*) to the nets. The cooperation provides benefits to both the fishermen and the dolphins. During one half-hour observation, the dolphins drove fish to the net six times and the four successful catches totalled over 80 kg of fish. Although the fishermen never give the dolphins any of the fish they catch, the dolphins feed on the fish which are stunned or missed by the net. The benefit for the dolphins may be a reduction in the energy expenditure required to catch their estimated 10–12 kg daily food requirement; the effort of dolphins foraging for mullet on their own has been described as a 'strenuous pursuit' (Pryor *et al.* 1990).

The complexity of the behaviour and the strict protocol to the sequence of events suggests there is learning by both dolphins and humans (Pryor *et al.* loc. cit.). The men stand in water one metre deep in a line parallel to shore, holding their net. The dolphins position themselves several metres away and eventually dive and swim away from the net. They then turn back towards the shore and swim rapidly towards the net, quickly stop and dive about 5–7 metres away from the net, creating a small wave in the process. If the dolphins perform the complete 'submerge–depart–return' sequence, the fishermen cast their net, trapping the fish underneath it. The dolphins dive in the area offshore of the net and bring fish to the surface to feed on them (Pryor *et al.* loc. cit.). The energy level of the dolphin's final rush to the net appears to indicate the size of the fish school, showing that the dolphin cues are graded.

The 25–30 dolphins which participate in this activity are termed 'good' dolphins by the fishermen. They appear to be resident to the area, based on repeated sightings of recognisable individuals, and the same individuals participate regularly. There are also 'bad' dolphins which do not help the fishermen and often try to disrupt the activity. These dolphins do not cooperate with conspecifics and feed using other methods (Pryor *et al.* loc. cit.).

The 'good' dolphins participate as part of an interacting group with mixed age classes. One group was composed of three generations: an adult female with two adult offspring, one of whom had a calf. It is unclear whether the young of these dolphins are recruited into this behaviour after their segregation into juvenile bands. If they are, the process of social learning might be available for documentation. However, the specialisation of this cooperative

Figure 16.3. Bottlenose dolphin © J. Boran/S. Heimlich.

behaviour in a specific social group implicates social learning as a mechanism for its development.

Humpback whales, *Megaptera novaeangliae*, also provide an example of a feeding specialisation which appears to be transmitted through social learning. Humpbacks undertake annual migrations between tropical, winter breeding grounds and high latitude feeding grounds. Because of the low productivity of tropical waters, humpback whales do not feed on the breeding grounds. All of their feeding must take place during a six-to-eight month period in productive, high latitude waters every year (Hain *et al.* 1982, Weinrich *et al.* 1992, Clapham 1996).

Humpback whales feed on schooling fish and are generally classified as 'swallowers' (Nemoto 1959), which engulf a single mouthful of prey at a time, as opposed to 'skimmers' such as right whales, which filter food by swimming with an open mouth. However, humpbacks have also developed a number of specialised techniques for concentrating the prey before they eat it. One such technique has been called 'bubble cloud' feeding (Hain *et al.* 1982), in which a whale emits bubbles from its blowhole beneath a school of fish while gradually ascending to the surface. The bubbles are thought to entrap the fish. The whale subsequently dives under the bubble cloud and lunges up through the centre of the circle of bubbles with its mouth open and takes in a big gulp of fish.

Weinrich *et al.* (1992) documented the acquisition and spread of a novel adaptation of bubble feeding, called 'lobtail feeding', during the course of their long-term research programme in which individual animals were identified. The behaviour involves an animal first slapping the surface of the water with its tail one or more times to create an effervescence; it then immediately dives under this area, emitting a bubble cloud, and emerges in the centre of the area to feed. The benefits of this feeding method over simple bubble feeding are still unclear. The splash of the lobtail may serve to startle the fish and cause them to clump before the whale emits its bubble cloud (Weinrich *et al.* loc. cit).

The percentage of animals using the lobtail feeding method steadily increased over time. In 1980, out of a sample size of 31 feeding animals, no animals were observed using this behaviour. In the following year, 2 of 51 animals (3.9%) were lobtail feeding and this percentage continued to increase until 1989, when 42 of 83 animals (50.6%) were using the technique. Of the total study population of 250 animals, 95 (38%) were observed to use lobtail feeding at some point. Eight animals who were first observed using the behaviour during the initial three years of the study continued to engage in it for the remainder of the study (Weinrich *et al.* loc. cit).

The spread of this behaviour was primarily due to acquisition by young animals. The youngest whales observed lobtail feeding were 2 years old (sexual maturity is 5 years old), but the behaviour was incomplete; the whales lobtailed then resurfaced without a feeding lunge as if the animals were practising (Weinrich *et al.* loc. cit). Animals first identified before 1982 and presumed to be older, (although exact ages were not always known) were significantly less common lobtail feeders (13 of 104: 13%) than were those animals first identified after 1982 (82 of 146 animals: 56%). Of those animals documented as calves during the study, 51% (20 of 38 animals) were lobtail feeders. As only three mothers of these 20 calves were lobtail feeders, assessment of maternal training remains inconclusive.

The main increase in lobtail feeding appeared to be from animals being born into the population, and not from lobtail feeding immigrants. It is likely that this behaviour was acquired through socially mediated learning, including behavioural tradition. The relative extent to which independent, individual experience and social mediation are involved in the acquisition of the behaviour clearly deserves more attention.

Killer whales exhibit manifold specialisations in feeding behaviours, but the common and indeed universal element observed world-wide, is group cooperation. The overall high degree of within-group organisation observed in most killer whale populations, especially during feeding, implicates social learning at some level. Exceptional behavioural inventiveness and self-generated innovations in group repertoires are the hallmarks of killer whales in captivity; these traits underlie the great popularity for captive displays of killer whales (Defran and Pryor 1980). This supports the implication that some behaviours in killer whales are acquired through social learning.

The killer whale is a cosmopolitan and highly adaptable member of the family *Delphinidae* whose large body size (4–6 m) and large teeth suggest specialisation on large prey. However, prey is diverse (Hoyt 1990, Jefferson *et al.* 1991), ranging from small fish to large whales, for example, blue whales (Tarpy 1979) and sperm whales (Arnbolm *et al.* 1987).

Feeding specialisations of killer whales have been examined since the early 1970s, in the inland waters of British Columbia and Washington, USA (Bigg *et al.* 1990, Ford *et al.* 1994). One of the major discoveries has been that there are three distinct 'forms' of killer whales, of which two live sympatrically in the region, these are named 'residents' and 'transients'. The third form, 'offshores' (Ford *et al.* 1994), will not be dealt with in this review. The residents and transients differ in distribution, seasonal occurrence, social organisation, acoustic dialects, behavioural repertoires and prey choice (Ford and Fisher 1983, J.R. Heimlich-Boran 1986, 1988, S. L. Heimlich-Boran 1986, 1988, Ford 1989, 1991, Morton 1990, Olesiuk *et al.* 1990, Felleman *et al.* 1991, Baird *et al.* 1992, Hoelzel 1993, Baird and Dill 1995, 1996, Nichol and Shackleton 1996). They also differ in the genetic aspects of pigmentation patterns, dorsal fin morphology and mitochondrial DNA, which indicate they are reproductively isolated (Duffield 1986, Baird and Stacey 1988, Hoelzel and Dover 1990).

Resident whales favour the five species of Pacific salmon, *Oncorhynchus* spp., as prey. They have been seen with salmon in their mouths (Ford *et al.* 1994), and collection of fish scales in the vicinity of feeding whales indicates that 95% of all feeding is on salmon (Nichol and Shackleton 1996).

Resident whale movement patterns follow major tidal streams used by migratory salmon, and suggest foraging specialisation. Although they are seen during all months, resident whales occur with much greater frequency during July through October, which is the height of the salmon migration from the Pacific Ocean to natal rivers for spawning (J. R. Heimlich-Boran 1986, Nichol and Shackleton 1996). The timing of killer whale occurrence in specific sub-areas of their range has also been correlated with the occurrence of salmon caught by recreational fishermen (J. R. Heimlich-Boran 1986, 1988) and with regular salmon estimates using data from commercial catches and direct counts at river mouths (Nichol and Shackleton 1996). In fact, there are indications that different resident pods may specialise on individual salmon species (Nichol and Shackleton 1996). Such fine-tuning is indicative of some degree of social learning.

Residents exhibit highly coordinated foraging behaviours which result in a collaborative hunting effort. For the system to work, some degree of social learning must be in place for the acquisition of the behaviours. During foraging, group-specific vocalisations are numerous and echolocation is heavily used (Hoelzel and Osborne 1986, Ford 1989). Calls are recognisable by all group members and undoubtedly help coordinate the group. Group synchrony is important to locate dispersed fish schools and to concentrate them into larger groups for capture and feeding.

However, this socially mediated group coordination has its limits. Once a school of fish is located and concentrated, synchrony breaks down and feeding appears to be solitary (J. R. Heimlich-Boran 1986, 1988, Felleman *et al.* 1991, Hoelzel 1993). Prey sharing is rare and has only been observed between females and calves (Baird pers. commun.).

Transients also exhibit foraging strategies which imply some socially mediated learning functions during their acquisition. Pod specific variability in techniques has been observed in transients, which appear to specialise on marine mammal prey rather than fish. The most common prey is the harbour seal, *Phoca vitulina*, but sea lion (unknown spp.), elephant seal, *Mirounga angustirotris*, harbour porpoise, *Phocoena phocoena*, and Dall's porpoise, *Phocoenoides dalli*, have also been taken (Morton 1990, Baird and Dill 1995). Although seen sporadically during all months, transients are most common in the region during the August/September harbour seal pupping and weaning season (Baird and Dill 1995). While some transient pods tend to forage nearshore, focusing on harbour seal haul-out areas, others favour offshore feeding, capturing prey in open water (Baird and Dill 1995).

Transients also hunt cooperatively, with group members pursuing individual prey. (Morton 1990, Baird and Dill 1995, 1996). Once a seal is captured, all group members converge and often work together in tearing apart the prey, indicating a large degree of prey sharing (Baird and Dill 1995). Transient whales primarily hunt by stealth, with little or no vocalisations or echolocation (Morton 1990, Felleman *et al.* 1991, Baird and Dill 1995).

Killer whale populations in other locations show similar behavioural adaptability and foraging specialisations. In the fjords of northwest Norway, October to January is the season when over-wintering herring, *Clupea harengus*, occur (Similä *et al.* 1996). Pod-specific differences in seasonal occurrence have been identified among the 39 pods which use the region. Most pods occur in both summer and autumn and seem to follow herring migrations, but some pods (24%) occur only in summer (Similä *et al.* 1996). These differences may relate to different prey preferences; one of the summer pods was observed feeding on harbour seal as well as herring (Similä *et al.* 1996).

The whales feeding on herring often exhibit a unique strategy termed 'carousel feeding' (Similä and Ugarte 1993, after Bel'kovitch *et al.* 1991) in which they swim in coordinated circles around and under a loosely clumped school. As they slowly reduce the size of the circle, they turn, flash their white bellies at the school and emit large bubbles, which help to drive the fish toward the surface. This is accompanied by vocalising and lob-tailing at the surface. When the school is concentrated at the surface, the circling whales begin to slap at it underwater with their tails, hitting fish to stun or kill them. They then turn back and eat the immobilised fish. A loud sound (either due to cavitation or actual contact with fish) can be heard with these tail slaps, and it may also help to stun fish (T. Similä pers. commun.).

This highly coordinated behaviour requires the members of the hunting group constantly to adapt their own behaviour to that of other members of the group. There is also an aspect of altruism (Hamilton 1964); some whales may defer feeding by continuing to encircle the school while others feed (Christensen 1978). This may then be reciprocated later (see discussion of the potential for reciprocal altruism in dolphins in Connor and Norris 1982).

Learning is integral to acquiring skill in 'carousel feeding'. It is suggested by observations of juvenile whales, who exhibit a higher percentage of unsuccessful tail slaps than adults, either due to missing the school or not slapping hard enough (T.Similä pers. commun.). It may be that young animals become adept through experience, and perhaps by observing more successful individuals, although some proficiency could come simply from growing in size.

A unique killer whale feeding method in which socially mediated learning has been clearly observed is that of 'intentional stranding'. Other dolphin species have also been observed to intentionally come onto the beach when chasing fish prey (e.g. bottlenose dolphins: Hoese 1971, Silber and Fertl 1995, and humpback dolphins: Peddermors and Thompson 1994).

Off Punta Norte, Argentina, and in the Crozet Islands of the southern Indian Ocean, killer whales strand themselves on the beach to capture southern sea lions, *Otaria flavescens*, and southern elephant seals, *Mirounga leonina* (Lopez and Lopez 1985, Hoelzel 1991, Guinet 1991, Guinet and Bouvier 1995). This behaviour is risky: one juvenile female was found permanently stranded above the water line in the Crozet Islands and would have died if researchers had not helped push her back into the water (Guinet 1991). Females were the most predominant stranders in the Crozet Islands, where beaches are shallowly sloped and a large, bulky adult male would have difficulty getting close enough to capture a seal. Both males and females stranded themselves off the steeper slopes of the Argentinean beaches.

Intentional stranding as a predation technique appears to be reasonably cost effective. Killer whales in Argentina successfully caught prey in approximately one out of every three strandings (Lopez and Lopez 1985, Hoelzel 1991). In each of three pods, there was one adult whale who was responsible for most of the hunting. However, in one pod of two adult males, there were a few observations of synchronous stranding and in these cases the males had a slightly higher success rate (75% together versus 31% solitarily when attacking pups: Hoelzel 1991), because one whale could capture pups the other missed. However, it is unknown whether each actually obtained more food than they would have by hunting individually.

Food sharing has been observed at both sites (Hoelzel 1991, Guinet 1991) and it could possibly play an instructive role in the demonstration and acquisition of this predation technique. Typically, after a pup is captured on the beach (usually alive and restrained by the hind flippers), the whales back off the beach and bring the pup into deeper water where other pod members wait. Occasionally, the prey is released and actively pursued by other whales, often resulting in the seal pup being hit into the air by a head toss or slap from the tail; sometimes, there is simply active milling in the vicinity of the whale with its prey, and the prey is never seen again. The prey is eventually dismembered. Hoelzel (1991) calculated that in 44% of his observations a whale other than the capturing whale was observed to eat part of the prey. This rate varied between pods: the pod with two adult males shared only 27% of kills while the pods with subadults shared 78% and 86% of all kills (although this could have been because the pods with subadults were more likely to release the prey which would have made observations of sharing easier to see: Hoelzel 1991).

Learning, through active observation by juveniles of adults, may be another mechanism through which skilled intentional stranding develops. In two-thirds of the observations of juveniles stranding in Argentina, a nearby adult male remained stationary at the surface, oriented towards the juvenile. On seven of these sessions, the adult male charged the beach after the juvenile had made a number of unsuccessful attempts, but the male did not capture any pinnipeds. Following this, the juvenile would repeat its behaviour (Lopez and Lopez 1985). In the Crozet Islands, a female was observed helping her juvenile calf to strand, actively pushing it in towards the beach. Then the female would position herself further onshore and help the calf return to the water (Guinet 1991). On another occasion, a mother had to rescue her stranded calf by swimming rapidly towards shore and creating a large wave which gave the calf room to push itself off the beach.

Hoelzel (1991) reported that only two adult females made the kills and that juveniles often observed this at close range, stranding alongside and slightly behind the adults. In Crozet, two of the juveniles always stranded with their mothers, while another stranded more commonly with a female who was not its mother (Guinet 1991). Inclusion or tolerance of inexperienced juveniles by adults, despite the potential for increased capture loss, is indicated by Hoelzel (1991); the capture success of one female decreased when juveniles were present (82% solitarily versus 31% when accompanied), yet nearly half of her hunts were with juveniles (Hoelzel 1991).

Individual practice plays an important part in the development of intentional stranding. In Argentina, immature whales, observed during 27 hunting sessions, performed numerous unsuccessful attempts. They '... swam directly towards pinnipeds on the coast, became stranded only centimetres from the pinnipeds, and generally remained next to them for 5–10 seconds. These beachings occurred in areas where an adult killer whale could have been successful in trapping a pinniped' (p. 182, Lopez and Lopez 1985). Prey capture practice opportunities are also provided when prey captured by adults is released offshore (Hoelzel 1991).

Guinet and Bouvier (1995) also identified 'social beaching play' as a mechanism for learning predation techniques, where intentional stranding occurs even though prey is absent from the beach; they consider its practice by juveniles to be fundamental in the development of skilled intentional stranding. They observed that most juvenile strandings occurred in the absence of prey (81 of 88 events). They also noted the rate of juvenile stranding increased with age. The first successful intentional stranding occurred at an estimated age of 6 years, although the juvenile required assistance from its mother to re-enter the water, suggesting it was still perfecting its technique. (Guinet and Bouvier 1995).

The risks associated with intentional stranding suggest that parental investment is of critical importance in ensuring the safety of the young during the development or acquisition of this behaviour (Guinet and Bouvier 1995). Usually it is only a few individuals who appear to do most of the stranding, suggesting that it is a learned skill which is acquired through socially mediated learning, and this includes elements of active instruction (teaching).

The observations of apparent teaching by experienced individuals of inexperienced individuals, in the acquisition of intentional stranding skills, implies that both the behaviour, and the exploitation of prey which requires it, is important in the repertoire of killer whales in Argentina and the Crozet Islands. Its importance is reflected in the investment given to training young animals. In addition to naive individuals observing skilled individuals to learn behaviours, skilled individuals appear to be actively teaching naive individuals.

Caro and Hauser (1992) have proposed a definition for teaching in non-human animals:

> *An individual actor **A** can be said to teach if it modifies its behaviour only in the presence of a naive observer, **B**, at some cost or at least without obtaining an immediate benefit for itself. **A**'s behaviour thereby either encourages or punishes **B**'s behaviour, or provides **B** with experience, or sets an example for **B**. As a result, **B** acquires knowledge or learns a skill earlier in life or more rapidly or efficiently than it might otherwise do, or that it would not learn at all. (p. 153)*

The intentional stranding behaviour of killer whales appears to fit most of these criteria. Adults (usually the mothers) alter their behaviour in the presence of juveniles. They either simply stop their own activities and increase their vigilance towards the young or actively push their calves in towards the beach. In the event of problems, they actively rescue the young from getting permanently stranded on the beach. The potential for the mother to strand herself permanently involves some cost, especially when she positions herself even higher on the beach than her calf (Guinet 1991). Pushing her calf onto the beach could also involve the cost of stranding her calf to a point where the calf got stuck and died.

However, to completely fulfil Caro and Hauser's (1992) definition of teaching it will have to be shown that juveniles who receive no instruction in intentional stranding have a lower success rate at capturing prey than those who do. An anecdotal suggestion that training alone was important, came from observation of one juvenile in Crozet who stopped practising its intentional stranding during the first year in which its previous 'teacher' (not its mother) had disappeared and was presumed to have died (Guinet 1991). The following year the juvenile resumed stranding with a different female. However, as the juvenile was first observed at an age before it had completely acquired the skill it wasn't possible to evaluate its capture success rate. Such an evaluation requires observation of an adult who had never received training when it was young. These don't seem to exist because all juveniles receive training.

Vocal traditions

Vocal learning in mammals is difficult to document, but appears to be relatively rare (Janik and Slater 1997). However, cetaceans, along with bats, are extremely dependent on their highly specialised acoustic senses (Jerison 1986) and many have elaborate vocal repertoires. Vocal learning has been documented in a handful of cetacean species.

One of the most striking examples of social learning in the development of vocal traditions is the song of the humpback whale. Male humpbacks sing long and complex songs when on their tropical winter breeding grounds (Payne and McVay 1971). Songs are only rarely heard on the higher latitude feeding grounds during the summer (McSweeney *et al.* 1989). Hence, songs are hypothesised to be sexual displays indicative of male fitness (Tyack 1981). One idea is that since whales breathe at the same place in each song cycle, song length could be indicative of breath-holding capability (Chu and Harcourt 1986). This could be used by females to choose mates. Although copulation has only rarely been observed in humpbacks, male singers stop singing when they join a female/calf pair and courtship behaviour ensues (Tyack 1981). Agonistic interactions can occur between animals known to be singers (presumably males) when both escort the same female–calf pair. Song has also been hypothesised to be a male secondary sexual characteristic acting as a ritualised form of male–male competition (Darling 1983).

Songs are composed of discrete notes, or units of sound, which are combined to form phrases of relatively uniform duration. Variable numbers of phrases are combined to form themes which occur in a predictable, cyclical order to form the song. The song, when repeated in an unbroken sequence (occasionally over 10 hr in duration), is termed a song session (Payne and McVay 1971, Payne *et al.* 1983).

All whales in the same ocean basin sing the same song during any given year (Payne and Guinee 1983), but the song progressively changes from year to year (Payne *et al.* 1983). Themes are always sung in the same fixed order, but some of these are occasionally dropped and phrases are varied in frequency, duration, spacing or number of units (Payne *et al.* 1983). This change occurs in the songs of individual humpbacks and is not due to changing membership of breeding groups over time (Guinee *et al.* 1983). These complex changes are incorporated by each singer so that at any given time all singers on the same breeding ground sing the same song.

A form of group consensus appears to dictate an active process of progressive changes in song from season to season. However, changes are not due to a simple forgetting of the previous breeding season's song. Song is not always at its most fixed at the end of a singing season, by which time all the singers have been regularly listening to each other (Payne *et al.* 1983). All evidence suggests that the song is held in memory over the summer and that changes only occur gradually during the singing season, when the singers are together on the breeding grounds (Payne *et al.* loc. cit., Guinee *et al.* 1983).

Through the singing season, themes gain or lose popularity. If a theme is very popular (i.e. used by nearly 100% of the population), it may become a permanent part of the song as quickly as within two months. Interestingly, a theme which is only used by 50% of the population seems doomed to die out (Payne *et al.* 1983). There are certain periods when the song is in a relatively stable state and other times when the song appears to be undergoing change. These periods are indicated by the presence of 'transitional phrases' (Payne and McVay 1971) These are phrases which occur between two themes and which contain elements of both themes. The relative abundance of transitional phrases do not follow any fixed pattern of occurrence within the course of a singing season. In fact, some of these progressive changes appear to continue the same trends from year to year, ruling out the effect of environmental parameters, such as temperature or day length, which fluctuate on an annual or seasonal basis (Payne *et al.* 1983).

The progressive change of humpback whale song, where all males change their song in the same way, involves a large degree of socially mediated modification that engages a constant feedback among all the singers.

Killer whales have also been shown to use group-specific call repertoires, or dialects, in two locations: along the coast of British Columbia, Canada (Ford and Fisher 1983, Ford 1989, 1991), and off northern Norway (Strager 1995). These dialects probably serve to maintain the cohesion and integrity of killer whale communities. As with vocal dialects in birds, there should be a large degree of social learning in the development of this behaviour, although detailed studies have not been done to date.

Killer whales emit three types of vocalisations: echolocation clicks, whistles and pulsed calls (Ford and Fisher 1983). Echolocation clicks are sound emissions of short, broad band pulses. The returning echo provides information on the location and target density of surrounding objects. Whistles are narrow band, pure tone vocalisations which are common during low activity level behaviours (Ford 1989). Pulsed calls are the most common and are composed of high repetition pulses (up to 5000/s) which create 'scream-like' sounds. Most pulsed calls are highly stereotyped and discrete, and have been described in nonoverlapping categories (Ford and Fisher 1983, Ford 1989). There are also some pulsed calls, termed variable calls, which cannot be categorised (Ford and Fisher 1983). Discrete calls (whistles and pulsed calls, but not echolocation clicks) compose 90% of all communication sounds.

Each killer whale group (pod) in British Columbia has been photographically identified, and each has been found to emit a limited number of call types (Ford *et al.* 1994). Sixty-nine call types have been identified, but the resident pods have average repertoires of only 10.7 discrete calls (range: 7–17 calls: Ford 1991). Each pod has its most common calls, which are emitted in repetitive series by more than one animal at a time. Pod repertoires have remained stable

over decades; recordings made during the 1960s during early capture operations revealed discrete calls which were still used by the same pods in 1983
(Ford 1991).

The call repertoires of the two nonoverlapping resident communities are
completely distinct. However, pods in the same community which associate
together often share a few calls. In some cases, pod-specific minor variations in
shared calls have been detected, but the call is still recognisable as the same call
(Ford and Fisher 1983). The proportion of shared calls between pods has been
used to generate association matrices and define 'acoustic clans'. Four clans
have been identified in the two resident, nonoverlapping communities and it
has been hypothesised that members of the same acoustic clan share a common
ancestry (Ford 1991). Subsequent divergence of acoustic repertoires into
group-specific dialects could result from 'errors in the copying of calls across
generations and the vocal idiosyncrasies of individual animals' (p. 743, Ford
1989).

A unique situation exists in the northern community, which has three
distinct acoustic clans (Ford 1991). Killer whales appear to be the only
nonhuman animal with unique group-specific dialects in communities which
inhabit the same range and occasionally associate (Ford 1991).

The benefits of learning the group-specific call repertoire may help to
ensure acceptance by the group and allow continued access to the benefits of
group living, such as improved food-finding (Ford 1989, 1991). Pod-specific
call repertoires may also help to notify all group members of the presence and
location of their fellow pod members and thus help to coordinate and synchronise group activities. This was especially evident when sporadic calling
during foraging seemed to incite chorusing from distant group members (Ford
1989). In general, the discrete, pod-specific calls comprised nearly 100% of
calls during foraging and travelling, while during socialising, variable calls
comprised over 30% of all calls (Ford 1989). However, the call repertoires are
likely to serve more than just a intra-group contact function: only one or two
calls would be needed for that. Ford (1989) identified a few behavioural
correlations with the relative production of various calls in the repertoire. For
example, five of 16 calls for one pod comprised 78% of all calls during foraging.
During the unique situation of two pods meeting, one pod used a specific call
significantly more than during any other behaviour (Ford 1989). These are
only the beginning steps on the way to an understanding of the killer whale
communication system.

The role of learning and socially mediated influences upon vocal learning is
further supported because killer whales are known to be capable of vocal
learning through stimulus–response training (van Heel *et al.* 1982). Pods have

been heard giving poor imitations of another pods' calls, but this is rare (Ford 1991). In captivity, when animals from different dialect groups are housed together, some mimicry of each others calls has been noted (Bain 1986).

Perhaps the best evidence for vocal learning comes from studies of the development of vocalisations of a captive-born calf (Bowles *et al.* 1988). The calf's early vocalisations were completely different from the mother's call repertoire, but by the end of its first year, the calf was using four of the mother's most common calls. The calf was not simply copying all it heard: it never reproduced the calls of a highly vocal (but unrelated) pool mate, but selectively learned the calls of its mother (Bowles *et al.* loc. cit.). On two occasions, when the calf was only 14 days old, the mother gave her call and the calf immediately gave a call which contained some of the components of the mother's call, suggesting that the calf was trying to copy its mother, or perhaps even that the mother was demonstrating to her calf (Bowles *et al.* 1988).

Although vocal learning is evident in killer whales, details of the role of social learning *per se* are still unknown. This is especially true for the development and maintenance of killer whale dialects in the wild.

Recent work on bottlenose dolphins suggests that they also may have regional dialects. Wang *et al.* (1995) compared acoustic parameters of whistle vocalisations among three adjacent populations, and compared these with other geographically isolated populations. Differences were greatest between isolated populations and similarities between adjacent populations corresponded to the known rates of mixing of specific individuals between those populations.

Sayigh *et al.* (1990) suggests dolphin infants appear to learn whistles from their mothers. However, the role of vocal learning and social learning has not been carefully studied. Similarities in some vocalisations from individuals from adjacent populations indicates that bottlenose dolphins whistle repertoires may be influenced by cross-exposure (Wang *et al.* 1995). This implies that some degree of social learning may occur.

The role of vocal learning in bottlenose dolphins is examined more completely by Janik (chapter 17, this volume).

At its simplest level, vocal learning probably functions to maintain group identity in cetaceans, which can disperse over large distances and potentially have a high degree of exposure to unrelated conspecifics. As we come to understand cetacean communication systems better, vocal learning may also be found to help coordinate specific group activities, such as foraging specialisations for local prey and environmental conditions.

Social organisation

Many cetacean species have adapted to life in the ocean by developing a highly structured social system which serves to optimise food acquisition, access to mates and protection from predators in the vast, three-dimensional, aquatic environment (Norris and Dohl 1980). At a basic level, cetacean schools often behave similarly to schools of fish, e.g. reacting to danger by closing ranks, becoming polarised and repressing individual behaviour (Norris and Schilt 1998). However, cetaceans have also developed elements of social complexity which might be expected of large-brained, group-living mammals, such as intricate social relationships with individual variations in behaviour (S. L. Heimlich-Boran, 1986). This may be facilitated by the adaptive advantage which echo-location sensory abilities give many species. In any case, social cetaceans appear to have overcome some of the difficulties of living in this environment through social specialisations which allow individuals to learn from each other.

A good example of social learning in the development of social organisation is provided by observational studies on Atlantic spotted dolphins (Herzing 1996). These studies have provided insights on the role of aggressive behaviours and dominance relationships. In many cases, these interactions occur vocally, but since our understanding of dolphin communication is so limited, they are difficult to interpret. A more obvious form of social interaction has been through direct body contact or visual displays. Underwater observations of other dolphin species have suggested that body posture is an important feature in signalling emotional state and intent (Norris *et al.* 1994).

Social learning is undoubtedly involved in a young animal's ongoing understanding of its role in the society. For example, juveniles are often encouraged by older dolphins to engage in mock battles involving head-to-head charges and occasional body contact (Herzing pers. commun.). The most interesting feature of these interactions is that the older animals seem to be demonstrating the social signals of body contact and charging, shoving the younger animals gently to help them develop the proper posturing. These interactions never seem to escalate; it appears that the older animals are simply instructing the younger in the 'rules' of aggressive interaction (Herzing pers. commun.).

This observational example shows one way in which social learning (and perhaps even teaching) may help to maintain social cohesion within groups.

Further, killer whales appear to have two alternative strategies of social organisation. These strategies, while probably initially based on the foraging specialisations described earlier, have extended into most aspects of their lives. We briefly review the differences, and suggest ways in which social learning may be involved.

Resident whales in British Columbia and Washington State live in large, stable groups (mean pod size = 11.6) which occur throughout the year in restricted community areas along the 800 km coast (Bigg *et al.* 1990, Ford *et al.* 1994). All 305 whales (as of 1993: Ford *et al.* 1994) have been identified, and genealogies have been established on the basis of known births and female associations with animals that were immature at the start of the study (1973). There has never been any mixing of pod membership; animals that have disappeared are assumed to have died (Bigg *et al.* 1990). The result is that both male and female juveniles remain in their mother's pod into adulthood. This lack of dispersal is very rare among mammals and implies a strong benefit of being a permanent member of one's family group (S. L. Heimlich-Boran 1986, 1988, J. R. Heimlich-Boran 1993).

In contrast, transient killer whales live in small, somewhat fluid groups (mean pod size = 2.7) which occur sporadically throughout the ranges of both resident communities, and which specialise on pinniped prey. Transient pods occasionally associate together and there have been suggestions that there may be some dispersal from the natal group based on observed mixing between pods (Baird *et al.* 1992).

These different distributions based on distinct feeding preferences result in small-scale spatial segregation between residents and transients, although they have often been observed within a few kilometres of each other (Morton 1990). There does appear to be some avoidance of residents by transients, and there is one instance of an aggressive interaction where a resident appeared to attack a transient (Baird and Dill 1995).

While the habits of transients more closely parallel the habits of other killer whales around the world, the resident form has only been conclusively described off the west coast of North America and off northern Norway (Similä and Ugarte 1993, Strager 1995). It could be that the resident form can only develop under special conditions where resources are sufficiently abundant to support the larger group sizes. Social learning could be involved in fine-tuning the adaptation to these local conditions. Of course, other factors could also favour the development and maintenance of a 'resident'-type social organisation. Being a member of a kin-related group should help support the cooperative behaviour required to locate and concentrate dispersed prey (Hamilton 1964). Adult males even appear to defer from the typical mammalian pattern of male dispersal to find mating opportunities. Inbreeding avoidance must be occurring by mating between individuals from different pods (S. L. Heimlich-Boran 1986, 1988).

In some large delphinids (*Orcinus* and *Globicephala*), females appear to live beyond reproductive age into a 'post-reproductive' period (Marsh and Kasuya

1986, Olesiuk *et al.* 1990), or at least have increasingly longer calving intervals such that they produce no calves in their later years (Martin and Rothery 1993). This may be seen as a form of terminal investment (Trivers 1974, Clutton-Brock 1984), where a female increases her parental investment in her last offspring. In short-finned pilot whales, although females may live up to 80 years, no females were found to still be ovulating beyond the age of 40 (Kasuya and Marsh 1984, Marsh and Kasuya 1984).

These whales also have an extended maturation period of the young (15–20 years for males), suggesting that parental investment is critically important for these whales. We suggest that one of the main forms of parental investment could be the cross-generational transfer of information through social learning (after S. L. Heimlich-Boran 1986, 1988).

The existence of an age class living beyond its reproductive lifespan is uncommon in mammals (Hrdy and Whitten 1987). The question arises as to what role the old animals play. The extended care of the last calf is one important role. But it could be that the older members help inexperienced individuals to adapt to the local environment through their prolonged experience (S. L. Heimlich-Boran, 1986), for example by improving group survival through memories of previous feeding success (Würsig 1986).

In summary, social learning plays an important role in the ontogeny of social organisation (e.g. in Atlantic spotted dolphins, where adults appear to train adolescents in the details of social interaction), as well as in the development of social specialisations. Killer whale specialisations in social organisation, possibly related to feeding specialisations, are shown in variability of group size and fluidity between sympatric groups. Although there is genetic differentiation between these groups, there is likely to be an element of social learning involved in the founding development and ongoing maintenance of these distinct social structures.

Conclusions

There are good examples of research that indicate social learning in the behaviour of a variety of cetacean species. We have given cases for free-ranging species in these areas, namely feeding specialisation ('hunting'), learned acoustic repertoires ('hearing') and learning and specialisation in social organisation ('hierarchies').

Feeding specialisations have been found in a number of cetacean groups. Sympatric killer whale, *Orcinus orca*, communities have specialised for either fish or marine mammal prey. Some humpback whales, *Megaptera novaean-*

gliae, have developed a novel method of bubble-feeding which has spread through the population. One group of Brazilian bottlenose dolphins, *Tursiops truncatus,* has learned to cooperate with human fishermen, demonstrating how social learning has allowed exploitation of a novel food source. Atlantic spotted dolphins, *Stenella frontalis,* train their young in fish-catching and killer whales in Argentina and the Crozet Islands appear to train immatures in intentional beaching techniques to catch seals.

Learned acoustic repertoires have been demonstrated in only a few species, but represent some of the few examples of mammalian vocal learning (Janik and Slater 1997). Male humpback whales use elaborate 'songs' as breeding displays which show annual, progressive changes, implying that social learning may be operating through group consensus. Bottlenose dolphins have population differences in vocalisations which show graded variation with geographic distance. Killer whales have stable, group-specific dialects which are unique to adjacent communities, while some calls are shared between temporary associates.

Social organisation forms the basis for all social learning in the sense that animals must live together to learn from each other. However, strictly learned variations in social structure have been difficult to document in cetaceans. Interestingly, the one study where detailed underwater observations have been regularly made (Herzing's work on Atlantic spotted dolphins in the Bahamas) has given examples of young learning (or being taught) the rules of social interaction. Long-term observations of killer whales have revealed a number of group-specific variations in social structure which may have resulted from social learning. Additional studies on other species using similarly detailed methods will likely reveal more ways in which social learning plays an important role in the fine-tuning of social structure to fit local conditions.

Interesting and functionally significant as these examples are, it is important to note that the bulk of the information from cetaceans so far is observational rather than quantitative.

Future field work must focus on the specific roles of individuals in social interactions. Focal animal studies, such as those that are being used in investigations of bottlenose dolphins in Australia (Connor *et al.* 1992) and spotted dolphins in the Bahamas (Herzing 1996), will be highly informative. The use of underwater video and sound recording (e.g. Herzing 1996) will allow these focal interactions to be carefully documented. Moreover, results from captive studies may help to identify potential areas for research.

Additionally, comparative studies between cetaceans and other mammals are highly important. Cetaceans, as mammals who have adapted to a completely different environment from that of terrestrial mammals, can provide a useful addition to an understanding of the ways in which habitat affects mammalian social systems (Würsig 1989).

Acknowledgements

This paper couldn't have been written without the fine research carried out by the scientists whose work is reviewed here. The contents were also improved immensely with comments from Robin Baird, John Ford, Denise Herzing, Victor Janik, Karen Pryor, Martin Rosen, Tiu Similä, Mason Weinrich and Bernd Würsig. We would also like to thank the help and patience of our editors, Hilary Box and Kathleen Gibson.

References

Arnbolm, T., Papastavrou, V., Weilgart, L. S. and Whitehead, H. (1987). Sperm whales react to an attack by killer whales. *J. Mammal.*, **68**, 450–3.

Bain, D. E. (1986). Acoustic behaviour of *Orcinus*: sequences, periodicity, behavioral correlates and an automated technique for call classification. In *Behavioral Biology of Killer Whales*, ed. B. Kirkevold and J. Lockard, pp. 335–71. New York: A. R. Liss.

Baird, R. W., Abrams, P. A. and Dill, L. M. (1992). Possible indirect interactions between transient and resident killer whales: implications for the evolution of foraging specialization in the genus *Orcinus*. *Oecologia*, **89**, 125–32.

Baird, R. W. and Dill, L. M. (1995). Occurrence and behaviour of transient killer whales: seasonal and pod-specific variability, foraging behaviour, and prey handling. *Can. J.*

Zool., **73**, 1300–11.

Baird, R. W. and Dill, L. M. (1996). Ecological and social determinants of group size in transient killer whales. *Behav. Ecol.*, **7**, 408–16.

Baird, R. W. and Stacey, P. J. (1988). Variation in saddle patch pigmentation in populations of killer whales (*Orcinus orca*) from British Columbia, Washington, and Vancouver Island. *Can. J. Zool.*, **66**, 2582–5.

Bel'kovich, V. M., Ivanova, E. E., Yefremenkova, O. V., Kozarovitsky, L. B. and Kharitonov, S. B. (1991). Searching and hunting behaviour in the bottlenose dolphin (*Tursiops truncatus*) in the Black Sea. In *Dolphin Societies*, ed. K. Pryor and K. S. Norris, pp. 38–67. Berkeley, CA: University of California Press.

Bigg, M. A., Olesiuk, P. F., Ellis, G. M., Ford, J. K. B. and Balcomb, K. C. (1990). Social or-

ganization and genealogy of resident killer whales (*Orcinus orca*) in the coastal waters of British Columbia and Washington State. *Rep. Int. Whal. Commn.*, **Special Issue 12**, 383–405.

Bowles, A. E., Young, W. G. and Asper, E. D. (1988). Ontogeny of stereotyped calling of a killer whale calf, *Orcinus orca*, during her first year. *Rit Fiskideildar*, **11**, 251–75.

Brodie, P. F. (1969). Duration of lactation in cetaceans: an indicator of required learning. *Am. Midl. Nat.*, **82**, 312–14.

Calambokidis, J. and Steiger, G. (1997). *Blue Whales*. Grantown-on-Spey, Scotland: Colin Baxter Photography.

Caro, T. M. and Hauser, M. D. (1992). Is there teaching in non-human animals? *Q. Rev. Biol.*, **67**, 151–74.

Christensen, I. (1978). The killer whale (*Orcinus orca*) in the northeast Atlantic. *Fisken Hav.*,

1978(1), 23–31.

Chu, K. and Harcourt, P. (1986). Behavioral correlations with aberrant patterns in humpback whale songs. *Behav. Ecol. Sociobiol.*, **19**, 309–312.

Clapham, P. J. (1996). The social and reproductive biology of humpback whales: an ecological perspective. *Mammal Rev.*, **26**, 27–49.

Clutton-Brock, T. H. (1984). Reproductive effort and terminal investment in iteroparous animals. *Am. Nat.*, **123**, 212–29.

Connor, R. C. and Norris, K. S. (1982). Are dolphins reciprocal altruists? *Am. Nat.*, **119**, 358–74.

Connor, R. C., Smolker, R. A. and Richards, A. F. (1992). Two levels of alliance formation among male bottlenose dolphins (*Tursiops* sp.). *Proc. Natl. Acad. Sci. USA*, **89**, 987–90.

Darling, J. D. (1983). Migrations, abundance and behaviour of Hawaiian humpback whales *Megaptera novacangliae* (Borowski). PhD thesis, University of California, Santa Cruz.

Darling, J. D. (1984). Gray whales off Vancouver Island, British Columbia. In *The Gray Whale*, ed. M. L. Jones, S. L. Swartz and S. Leatherwood, pp. 267–87. Orlando, FL: Academic Press.

Defran, R. H. and Pryor, K. (1980). The behaviour and training of cetaceans in captivity. In *Cetacean Behaviour: Mechanisms and Functions*, ed. L. Herman, pp. 319–62. New York: J. Wiley-Interscience.

Duffield, D. A. (1986). *Orcinus orca*: taxonomy, evolution, cytogenetics and population structure. In *Behavioral Biology of Killer Whales*, ed. B. Kirkevold and J. S. Lockard, pp. 19–33. New York: A. R. Liss.

Felleman, F. L., Heimlich-Boran, J. R. and Osborne, R. W. (1991). Feeding ecology of the killer whale (*Orcinus orca*). In *Dolphin Societies*, ed. K. Pryor and K. S. Norris, pp. 113–47. Berkeley, CA: University of California Press.

Ford, J. K. B. (1989). Acoustic behaviour of resident killer whales (*Orcinus orca*) off Vancouver Island, British Columbia. *Can. J. Zool.*, **67**, 727–45.

Ford, J. K. B. (1991). Vocal traditions among resident killer whales (*Orcinus orca*) in coastal waters of British Columbia. *Can. J. Zool.*, **69**, 1454–83.

Ford, J. K. B. and Fisher, H. D. (1983). Group-specific dialects of killer whales (*Orcinus orca*) in British Columbia. In *Communication and Behaviour of Whales*, ed. R. S. Payne, pp. 129–61. Boulder, CO: Westview Press.

Ford, J. K. B., Ellis, G. M. and Balcomb, K. C. (1994). *Killer Whales*. Vancouver: UBC Press.

Frame, L. H., Malcolm, J. R., Frame, G. W. and van Lawick, H. (1979). Social organization of African wild dogs (*Lycaon pictus*) on the Serengeti plains, Tanzania 1967–1978. *Z. Tierpsychol.*, **50**, 225–49.

Guinee, L., Chu, K. and Dorsey, E. M. (1983). Changes over time in the songs of known individual humpback whales (*Megaptera novaeangliae*). In *Com-munication and Behaviour of Whales*, ed. R. Payne, pp. 59–80. Boulder, CO: Westview Press.

Guinet, C. (1991). Intentional stranding apprenticeship and social play in killer whales (*Orcinus orca*). *Can. J. Zool.*, **69**, 2712–16.

Guinet, C. and Bouvier, J. (1995). Development of intentional stranding hunting techniques in killer whale (*Orcinus orca*) calves at Crozet Archipelago. *Can. J. Zool.*, **73**, 27–33.

Hain, J. H. W., Carter, G., Kraus, S., Mayo, C. and Winn, H. (1982). Feeding behaviour of the humpback whale in the western North Atlantic. *Fishery Bull.*, **80**, 259–68.

Hamilton, W. D. (1964). The genetical theory of social behaviour. *J. Theor. Biol.*, **7**, 1–52.

Harvey, P. H. and Clutton-Brock, T. H. (1985). Life history variation in primates. *Evolution*, **39**, 559–81.

Heimlich-Boran, J. R. (1986). Fishery correlations with the occurrence of killer whales in greater Puget Sound. In *Behavioral Biology of Killer Whales*, ed. B. C. Kirkevold and J. S. Lockard, pp. 113–31. New York: A. R. Liss.

Heimlich-Boran, J. R. (1988). Behavioral ecology of killer whales (*Orcinus orca*) in the Pacific Northwest. *Can. J. Zool.*, **66**, 565–78.

Heimlich-Boran, J. R. (1993). Social organisation of the short-finned pilot whale, *Globicephala macrorhynchus*, with special reference to the comparative social ecology of delphinids. PhD thesis, University of Cambridge.

Heimlich-Boran, S. L. (1986). Cohesive relationships among Puget Sound killer whales. In *Behavioral Biology of Killer Whales*, ed. B. Kirkevold and J. S. Lockard, pp. 251–84. New York: A. R. Liss.

Heimlich-Boran, S. L. (1988). Association patterns and social dynamics of killer whales (*Orcinus orca*) in Greater Puget Sound. MA thesis: Moss Landing Marine Laboratories, San Jose State University.

Herzing, D. (1996). Vocalizations and associated underwater behaviour of free-ranging Atlantic spotted dolphins, *Stenella frontalis*, and bottlenose dolphins, *Tursiops truncatus*. *Aquat. Mamm.*, **22**, 61–79.

Hoelzel, A. R. (1991). Killer whale predation on marine mammals at Punta Norte, Argentina: food sharing, provisioning and foraging strategy. *Behav. Ecol. Sociobiol.*, **129**, 1–8.

Hoelzel, A. R. (1993). Foraging behaviour and social group dynamics in Puget Sound killer whales. *Anim. Behav.*, **45**, 581–91.

Hoelzel, A. R. and Dover, G. A. (1990). Genetic differentiation between sympatric killer whale populations. *Heredity*, **66**, 191–5.

Hoelzel, A. R. and Osborne, R. W. (1986). Killer whale call characteristics; implications for cooperative foraging strategies. In *Behavioral Biology of Killer Whales*, ed. B. Kirkevold and J. S. Lockard, pp. 373–406. New York: A. R. Liss.

Hoese, A. D. (1971). Dolphin feeding out of water in a salt marsh. *J. Mammal.*, **52**, 222–3.

Hoyt, E. (1990). *Orca: The Whale Called Killer*. London: Hale.

Hrdy, S. B. and Whitten, P. L. (1987). Patterning of sexual activity. In *Primate Societies*, ed. B. B. Smuts, D. L. Cheney, R. M. Seyfarth, R. W. Wrangham and T. T. Struhsaker, pp. 370–84. Chicago: Chicago University Press.

Janik, V. M. and Slater, P. J. B. (1997). Vocal learning in mammals. *Adv. Stud. Behav.*, **26**, 59–99.

Jefferson, T. A., Stacey, P. J. and Baird, R. W. (1991). A review of killer whale interactions with other marine mammals: predation to coexistence. *Mammal. Rev.*, **22**, 35–47.

Jerison, H. J. (1986). The perceptual world of dolphins. In *Dolphin Cognition and Behavior: a Comparative Approach*, ed. R. J. Schusterman, J. A. Thomas and F. G. Wood, pp. 141–66. Hillsdale, NJ: Erlbaum Assoc.

Kasuya, T. and Marsh, H. (1984). Life history and reproductive biology of the short-finned pilot whale, *Globicephala macrorhynchus*, off the Pacific coast of Japan. *Rep. Int. Whal. Comm.*, **Special Issue 6**, 259–310.

Leatherwood, S. and Reeves, R. R. (1983). *The Sierra Club Handbook of Whales and Dolphins*. San Francisco: Sierra Club Books.

Lefebvre, L. and Giraldeau, L-A. (1996). Is social learning an adaptive specialization? In *Social Learning in Animals: The Roots of Culture*, ed. C. M. Heyes and B. G. J. Galef, pp. 107–28. San Diego, CA: Academic Press.

Lopez, J. C. and Lopez, D. (1985). Killer whales (*Orcinus orca*) of Patagonia, and their behaviour of intentional stranding while hunting nearshore. *J. Mammal.*, **66**, 181–3.

Marsh, H. and Kasuya, T. (1984). Ovarian changes in the short-finned pilot whale, *Globicephala macrorhynchus*. *Rep. Int. Whal. Comm.*, **Special Issue 6**, 311–35.

Marsh, H. and Kasuya, T. (1986). Evidence for reproductive senescence in female cetaceans. *Rep. Int. Whal. Comm.*, **Special Issue 8**, 57–74.

Martin, A. R. and Rothery, P. (1993). Reproductive parameters of female long-finned pilot whales (*Globicephala melas*) around the Faroe Islands. *Rep. Int. Whal. Comm.*, **Special Issue 14**, 263–304.

McSweeney, D. J., Chu, K. C., Dolphin, W. F. and Guinee, L. N. (1989). North Pacific humpback whale songs: a comparison of southeast Alaskan feeding ground songs with Hawaiian wintering ground songs. *Mar. Mammal Sci.*, **5**, 139–48.

Morton, A. (1990). A quantitative comparison of the behaviour of resident and transient forms of the killer whale off the central British Columbia coast. *Rep. Int. Whal. Commn.*, **Special Issue 12**, 245–8.

Moss, C. W. (1988). *Elephant Memories: Thirteen Years in the Life of an Elephant Family*. New

York: W. Morrow and Co.

Nemoto, T. (1959). Food of baleen whales with reference to whale movements. *Sci. Rep. Whales Res. Inst.*, **16**, 149–290.

Nichol, L. M. and Shackleton, D. M. (1996). Seasonal movements and foraging behaviour of northern resident killer whales (*Orcinus orca*) in relation to the inshore distribution of salmon (*Oncorhynchus* spp.) in British Columbia. *Can. J. Zool.*, **74**, 983–91.

Norris, K. S. and Dohl, T. P. (1980). The structure and function of cetacean schools. In *Cetacean Behaviour*, ed. L. M. Herman, pp. 211–61. New York: Wiley and Sons.

Norris, K. S. and Schilt, C. R. (1988). Cooperative societies in three-dimensional space: on the origins of aggregations, flocks, and schools, with special reference to dolphins and fish. *Ethol. Sociobiol.*, **9**, 149–79.

Norris, K. S., Würsig, B., Wells, R. S. and Würsig, M. (1994). *The Hawaiian Spinner Dolphin*. Berkeley, CA: University of California Press.

Olesiuk, P. F., Bigg, M. A. and Ellis, G. M. (1990). Life history and population dynamics of resident killer whales (*Orcinus orca*) in the coastal waters of British Columbia and Washington State. *Rep. Int. Whal. Comm*, **Special Issue 12**, 209–43.

Payne, R. (1995). *Among Whales*. New York: Scribner.

Payne, K., Tyack, P. and Payne, R. (1983). Progressive changes in the songs of humpback whales (*Megaptera novaeangliae*): a detailed analysis of two seasons in Hawaii. In *Communication and Behaviour of Whales*, ed. R. Payne, pp. 9–57. Boulder, CO: Westview Press.

Payne, R. and Guinee, L. (1983). Humpback whale (*Megaptera novaeangliae*) songs as an indicator of 'stocks'. In *Communication and Behaviour of Whales*, ed. R. Payne, pp. 333–58. Boulder, CO: Westview Press.

Payne, R. and McVay, S. (1971). Songs of humpback whales. *Science*, **173**, 585–97.

Peddermors, V. M. and Thompson, G. (1994). Beaching behaviour during shallow water feeding by humpback dolphins *Sousa plumbea*. *Aquat. Mammal.*, **20**, 65–7.

Pryor, K., Lindberg, J., Lindberg, S. and Milano, R. (1990). A dolphin–human fishing cooperative in Brazil. *Mar. Mammal Sci.*, **6**, 77–82.

Sayigh, L. S., Tyack, P. L., Wells, R. S. and Scott, M. D. (1990). Signature whistles of free-ranging bottlenose dolphins (*Tursiops truncatus*): stability and mother-offspring comparisons. *Behav. Ecol. Sociobiol.*, **26**, 247–60.

Shane, S. H., Wells, R. S. and Würsig, B. (1986). Ecology, behaviour, and social organization of the bottlenose dolphin: a review. *Mar. Mammal. Sci.*, **2**, 34–63.

Silber, G. K. and Fertl, D. (1995). Intentional beaching by bottlenose dolphins (*Tursiops truncatus*) in the Colorado River Delta. *Aquat. Mammal.*, **21**, 183–6.

Similä, T., Holst, J. C. and Christensen, I. (1996). Occurrence and diet of killer whales in northern Norway: seasonal patterns relative to the distribution and abundance of Norwegian spring-spawning herring. *Can. J. Fish. Aquat. Sci.*, **53**, 769–79.

Similä, T. and Ugarte, F. (1993). Surface and underwater observations of cooperatively feeding killer whales in northern Norway. *Can. J. Zool.*, **71**, 1494–9.

Strager, H. (1995). Pod-specific call repertoires and compound calls of killer whales, *Orcinus orca* Linnaeus, 1758, in the waters of northern Norway. *Can. J. Zool.*, **73**, 1037–47.

Tarpy, C. (1979). Killer whale attack. *Nat. Geogr.*, **155**, 542–5.

Trivers, R. L. (1974). Parent-offspring conflict. *Am. Zool.*, **14**, 249–64.

Tyack, P. (1981). Interactions between singing Hawaiian humpback whales and conspecifics nearby. *Behav. Ecol. Sociobiol.*, **8**, 105–16.

Tyack, P. (1986). Population biology, social behavior and communication in whales and dolphins. *Trends Ecol. Evol.*, **1**, 144–50.

van Heel, W. H. D., Kamminga, C. and van der Toorn, J. D. (1982). An experiment in two way communication in *Orcinus orca* L. *Aquat. Mammal.*, **9**, 69–82.

Wang, D., Würsig, B. and Evans, W. E. (1995). Whistles of bottlenose dolphins: comparisons among populations. *Aquat. Mammal.*, **21**, 65–77.

Weinrich, M. T. (1991). Stable social associations among humpback whales (*Megaptera novaeangliae*) in the southern Gulf of Maine. *Can. J. Zool.*, **69**, 3012–19.

Weinrich, M. T., Schilling, M. R. and Belt, C. R. (1992). Evidence for acquisition of a novel feeding behaviour: lobtail feeding in humpback whales, *Megaptera novaeangliae. Anim. Behav.*, **44**, 1059–72.

Wells, R. S. (1991). The role of long-term study in understanding the social structure of a bottlenose dolphin community. In *Dolphin Societies: Discoveries and Puzzles*, ed. K. Pryor and K. S. Norris, pp. 199–225. Berkeley, CA: University of California Press.

Wells, R. S., Irvine, A. B. and Scott, M. D. (1980). The social ecology of inshore odontocetes. In *Cetacean Behavior: Mechanisms and Processes*, ed. L. M. Herman, pp. 263–317. New York: Wiley and Sons.

Würsig, B. (1979). Dolphins. *Sci. Am.*, **240**, 136–48.

Würsig, B. (1986). Delphinid foraging strategies. In *Dolphin Cognition and Behaviour: A Comparative Approach*, ed. R. J. Schusterman, J. A. Thomas, and F. G. Wood, pp. 347–59. Hillsdale, NJ: Erlbaum Assoc.

Würsig, B. (1989). Cetaceans. *Science*, **244**, 1550–7.

17

Origins and implications of vocal learning in bottlenose dolphins

Vincent M. Janik

Introduction

Before we can study functional aspects of socially mediated learning it must be established first that a behaviour pattern is learned and second that it is mediated socially. The investigation of socially mediated learning in communication systems requires only the first step. If learning takes place it can only be social since communication can only occur between individuals. Socially mediated learning can affect animal communication in three different ways. The two more common ones are through changes in comprehension (i.e. learning to understand the contexts with which a particular signal is associated) and changes in usage (i.e. learning when to use a signal). These forms of learning have also been described as contextual learning (Janik and Slater 1997). A rarer form of social learning in acoustic communication is vocal learning. This term refers to cases in which an individual modifies the form of its signals as result of experience with those of another individual. It can result either in signals becoming matched or in distinct differences arising between individuals. Contextual learning that is related to communication has been found in many mammalian species (Salzinger and Waller 1962, Molliver 1963, Lilly 1965, Myers *et al.* 1965, Schusterman and Feinstein 1965, Burnstein and Wolff 1967, Lal 1976), but only a few studies have shown vocal learning.

In mammals, primates have been the main focus of research on vocal learning (Cheney and Seyfarth 1990). This is partly because vocal learning is a prerequisite for the acquisition of spoken languages in humans. Comparative studies have tried to investigate the evolution of vocal learning by looking at nonhuman primates, but only found evidence for learning of temporal and amplitude parameters. Despite extensive research on nonhuman primates, no convincing evidence for vocal learning in the frequency domain has been found (Janik and Slater 1997). Hence, some authors assume that among mammals vocal learning is unique to humans (Jürgens 1992, Meltzoff 1996). However, studies from as far back as 1972 have demonstrated clearly that some mammals are capable of vocal learning of frequencies. Caldwell and Caldwell

Figure 17.1. Bottlenose dolphins in Western Australia (above) and the Moray Firth, Scotland (below).

(1972) described such learning in a captive bottlenose dolphin, *Tursiops truncatus.* These results were subsequently confirmed in a more detailed study by Richards *et al.* (1984). Further, all individuals within a humpback whale, *Megaptera novaeangliae,* population changed their song in synchrony in a way that could only be explained if vocal learning was involved (Guinee *et al.* 1983, Payne and Payne 1985). Harbour seals, *Phoca vitulina*, have been found to copy human speech (Ralls *et al.* 1985). Infants of greater horseshoe bats, *Rhinolophus ferrumequinum*, adjusted the main frequency of their echolocation calls to those of their mothers by vocal learning (Jones and Ransome 1993). Similarly, greater spear-nosed bats, *Phyllostomus hastatus*, have group-specific calls in groups of unrelated individuals which suggested that vocal learning was involved (Boughman 1997). Unfortunately, we know very little about the adaptive significance of vocal learning in these cases. In bats, vocal learning seems to be important for individual recognition, but it is not clear what other contexts might be affected by their vocal learning skills. Humpback whales and harbour seals show singing behaviour similar to that of birds, in which males produce repetitive sequences of sounds during the breeding season. This suggests a convergent evolutionary pathway to that of songbirds.

Of those mammals that show vocal learning dolphins are the most similar to primates in their social behaviour. Experimental studies have also shown that their cognitive abilities are similar to those of the great apes (Herman *et al.* 1993). They are capable of generalisation of rules, formation of abstract concepts such as same/different (Herman *et al.* 1994), object labelling (Richards *et al.* 1984), referential reporting the presence or absence of named objects (Herman and Forestell 1985), and semantic and syntactic understanding in artificial languages including the understanding of novel sentences (Herman *et al.* 1984). However, unlike nonhuman primates, dolphins rely almost entirely on the acoustic channel for their communication and they are capable of vocal learning. They do not produce songs but use their sounds extensively in social interactions within complex social groups. In this chapter I will evaluate what role vocal learning played in the evolution of cetaceans and what implications it had once it evolved. To consider hypotheses about the origins and implications of vocal learning, however, we need first to look at current knowledge about the social structure and vocal communication of bottlenose dolphins.

The social structure of bottlenose dolphins

Bottlenose dolphins can be found in groups of one to several hundred animals. The average group size lies between 3 and 30 individuals (reviewed in Wells *et al.* 1980). In some areas group size is slightly larger in deep water (Würsig 1978), but average group size does not seem to vary significantly between coastal and pelagic areas. However, groups of several hundred individuals can only be found in the open ocean (Scott and Chivers 1990). The waters off southern Africa present a notable exception, as average group size there was 140.3 animals in coastal areas (Saayman *et al.* 1973). Differences in group size may be related to environmental factors that influence group structure. Dusky dolphins, *Lagenorhynchus obscurus*, for example, show different group sizes in different coastal habitats (Würsig *et al.* 1991). However, differences in average group sizes or in individual association coefficients between studies can also be caused by different definitions of what a group is, and one has to be careful in comparing data from different studies.

Studies on the structure of bottlenose dolphin societies have commonly used association coefficients in which the total number of sightings of two animals together is multiplied by two and then divided by the sum of the total sightings of each individual. These studies that were carried out in Sarasota, USA (Wells *et al.* 1987, Wells 1991), and Shark Bay, Australia (Smolker *et al.* 1992), have shown patterns of association among individuals that are charac-teristic of a fission–fusion society. Members of a community form groups of different sizes but in most cases the individual composition of these groups changes frequently. Individual male dolphins can be found associating with different female groups on different days. However, males also tend to associ-ate in dyads or triplets. These male alliances herd and consort with females for mating purposes (Connor *et al.* 1992a, 1996). Further, separate alliances have been observed to fuse into a second-order alliance of four or more animals that replaced dyadic alliances as female consorts (Connor *et al.* 1992a). Alliances between specific males can be extremely stable and can have association coefficients that are as high as those of mothers and their calves. In some cases males have not been seen without their ally for over two years (Smolker *et al.* 1992).

Females, on the other hand, tend to associate in matrilineal groups. How-ever, this is a pattern that can only be recognised in long-term studies. Group composition is much more fluid on a daily basis leading to the impression of a female network rather than an organisation of entirely separate groups. Fe-males give birth to one calf after 12 months of gestation. In Sarasota, mothers and calves were almost always seen together until a calf reached 3–4 years of age. Association coefficients between mothers and calves then decreased grad-ually in the following years (Wells *et al.* 1987). Once they have been weaned, bottlenose dolphins associate in subadult groups that are composed of both sexes. This period lasts from weaning to approximately 8–12 years of age in females and to 10–15 years in males (Wells 1991). After that time males and females show adult association patterns with females returning to their mat-rilineal group and males starting to form alliances with other males.

The vocal repertoire of bottlenose dolphins

Bottlenose dolphins produce various different kinds of sounds. Among these, clicks, burst-pulsed sounds and whistles have received the most attention. Clicks are very brief, broad-band signals used for echolocation. The communicative value of clicks has not yet been identified. Certain dolphin species, such as the Hector's dolphin, *Cephalorhynchus hectori*, do not produce whistles and use only click sounds to communicate (Dawson 1991). Bastian (1967) found that a bottlenose dolphin was capable of copying the lever pressing behaviour of a concealed conspecific that produced elaborate click trains during its performance. The animals were only rewarded if both pressed the same of two levers, but only one individual had a light in its pool that indicated which lever was the rewarded one in a particular trial. The associated click trains could have been communication or echolocation signals. Bottlenose dolphins are able to extract information from the echoes of echolocation clicks produced by conspecifics purely by listening to them (Xitco and Roitblat 1996). In this case the original signaller might not even notice that another individual gained any information. Thus, it will be a challenging task for researchers to distinguish between clicks used for echolocation that can be intercepted by others to gather information and those emitted specifically to communicate.

Burst-pulsed sounds consist of click trains with a very short interclick interval, so that they appear somewhat tonal to the human ear. We do not know how a dolphin perceives such a sound and whether the distinction between clicks and burst-pulsed sounds has any relevance to the animal itself. There are many cases in which these sounds start with a slower click train which then accelerates to become a burst-pulsed sound. The reverse process can be found at the end of burst-pulsed sounds. However, they are usually composed of low frequency clicks which are unusual in echolocation. It has been shown that burst-pulsed sounds are a common vocalisation form in agonistic interactions (Overstrom 1983).

The best studied communication signals of bottlenose dolphins are their whistles. These are narrow-band, tonal signals that lie between 1 and 24 kHz. The sound repertoire of an individual bottlenose dolphin comprises several whistle types, one of which is highly stereotyped (Caldwell *et al.* 1990, Janik *et al.* 1994). Caldwell and Caldwell (1965) described this stereotyped whistle for the first time and found that it was consistent within individuals. They called these signals signature whistles and hypothesised that they facilitate individual recognition (Caldwell and Caldwell 1968). Studies on wild individuals that were frequently captured for recordings and then released again have shown that the signature whistle of an individual can remain stable for at least 12 years

(Sayigh *et al.* 1990). In all studies in which individuals were isolated for recording, the signature whistle was the primary whistle type, often accounting for almost 100% of all whistles (Caldwell and Caldwell 1965, Caldwell *et al.* 1990, Sayigh *et al.* 1990, 1995, Janik *et al.* 1994). However, isolation was the only context studied. Janik and Slater (1998) have shown that undisturbed bottlenose dolphins that are well habituated to captivity do not use signature whistles while they are together but only when out of sight of each other. In studies where bottlenose dolphins have been found to produce stereotyped whistles even if swimming together (Caldwell and Caldwell 1965, Tyack 1986b, Caldwell *et al.* 1990), individuals were recorded in unusual situations, mostly shortly after capture or medical procedures. If social animals are exposed to novel situations, cohesion calls can be expected at higher rates since they would ensure that close associates are informed about each other's position and, thus, facilitate cooperation in, for example, defensive actions. Thus, the contexts in which signature whistles are used and their stability over time support the hypothesis that they are primarily group cohesion calls.

Infants develop their own signature whistle in the first few months of life (Caldwell and Caldwell 1979) and there is accumulating evidence that the development of signature whistles is strongly influenced by vocal learning. Captive bottlenose dolphin calves often develop whistles that are similar to the whistle used by the human trainer or similar to whistles of their pool mates but not to those of their mothers (Tyack 1997). One stranded calf that was raised by a female that was not her mother developed a whistle similar to that of the care-giving female (Tyack and Sayigh 1997). Finally, a bottlenose dolphin calf that was housed with two Pacific white-sided dolphins, *Lagenorhynchus ob-liquidens*, and seven bottlenose dolphins developed a whistle similar to one of the white-sided dolphins (Caldwell and Caldwell 1979). Vocal learning in adult bottlenose dolphins has been reported from experimental studies of clicks (Moore and Pawloski 1990), burst-pulsed sounds (Caldwell and Caldwell 1972) and whistles (Caldwell and Caldwell 1972, Richards *et al.* 1984). Dolphins are extremely versatile in copying new sounds and have been shown to be capable of producing copies of novel sounds at the first attempt (Richards *et al.* 1984). However, we do not know yet how genetic parameters limit the flexibility of call ontogeny of cetaceans.

It is important to note that several other calls do not seem to fit into the three categories mentioned so far. Examples are brays (dos Santos *et al.* 1995), low-frequency, narrow-band (LFN) sounds (Schultz *et al.* 1995), and pops (Connor and Smolker 1996). Like LFN sounds, brays are low frequency and narrow band but much longer in duration and with many harmonics, while pops are very short, low frequency, broad-band signals. We know very little about the functional significance of these call types and further research on them is badly needed.

Origins of vocal learning in dolphins

Several hypotheses have been brought forward to explain the evolution of vocal learning. Three hypotheses for the evolution of vocal learning in birds relate to features of their environment. The population identity idea assumes vocal learning to be important to maintain local adaptations to specific habitats (Nottebohm 1972). The habitat matching hypothesis suggests that vocal learning is important in matching the acoustic transmission characteristics of different habitats (Hansen 1979). Finally, the intense speciation hypothesis suggests that vocal learning helps to maintain species recognition in a habitat where species density is high (Nottebohm 1972). Janik and Slater (1997) have argued that these hypotheses seem to be unlikely for cetaceans in view of the marine environment in which they live. The sea is a vast, relatively homogeneous habitat in comparison with most terrestrial environments (Spiesberger and Fristrup 1990). Environmental parameters that influence sound transmission in air such as air temperature, turbulence, and foliage or obstacle density do not vary as much in the sea. Temperature does not fluctuate as much, water currents are slow compared with wind speed in air, and obstacles are rare. Therefore, differences in sound transmission characteristics between marine environments are a lot smaller than between terrestrial ones. Thus, local adaptations to specific habitats or an ability to adapt vocalisations to different sound transmission characteristics are less likely to be of importance in the sea. Similarly, the intense speciation hypothesis was developed for a tropical rainforest situation with hundreds of different bird species. In dolphins, those species that were not restricted to coastal areas did not face any spatial limits to dispersal in the open sea. Furthermore, dolphins are predators and not primary consumers like many birds. Therefore, their density cannot be as high as those of birds if the amount of resources is comparable and it is unlikely that dolphin species that were restricted to coastal areas would have found enough resources for intense speciation. Thus, all three hypotheses for the evolution of vocal learning that refer to habitat structure seem unlikely for cetaceans.

The most plausible reasons for the evolution of vocal learning in cetaceans can be found in the sexual selection hypothesis and the individual recognition hypothesis. If the complexity of the call repertoire of an individual relates to its fitness this can be used to assess quality in mate choice or intrasexual competition. Here, sexual selection would greatly favour the evolution of vocal learning to enlarge an animal's repertoire and therefore increase its reproductive success. Many birds, for example, use song in mate attraction and territory defence. In cetaceans both toothed whales and baleen whales show vocal learning. Singing behaviour can be found in baleen whales, and sexual selection could have favoured the evolution of vocal learning in this group. Singing male humpback whales, for example, are on average further apart than non-singers, suggesting a spacing function of song (Frankel *et al.* 1995). However, toothed whales such as dolphins do not produce song but use vocal learning in other contexts.

Vocal learning can also facilitate individual recognition (Janik and Slater 1997). In most animals individual recognition is made possible by individual differences in voice characteristics. These are caused by differences in vocal tract morphology or in the neural control of sound production. Thus, they affect the production of all calls emitted by an individual. In some mammals individuals produce isolation calls. These are usually better suited for long distance communication than other calls but identity is encoded in voice characteristics (Lieblich *et al.* 1980) and not in the call type that is used. Such voice characteristics tend to have a higher variability in species that live in noisy environments than in those living in quieter ones (Beecher 1991, Loesche *et al.* 1991). An individual recognition system becomes more effective as the ratio between inter-individual variability and intra-individual variability in the recognition parameter gets larger (Beecher 1991). However, at a certain background noise level it becomes difficult to maintain individual recognition with variants of the same call type even if the intra-individual variability is very small. Noise can mask significant differences between calls. Masking of recognition signals could either be caused by non-specific noise in the same frequency band or by conspecifics calling at the same time. Another problem for a voice based recognition system arises if voice characteristics of individuals can change. Vocal learning would allow an individual to solve such problems by incorporating a new call type specifically for individual recognition.

The fission–fusion structure of dolphin societies is similar to those of several primate species. In such societies communication is needed to maintain bonds and alliances. Additionally, dolphins need an effective communication system simply to stay in touch with members of their current group. This becomes most obvious in the case of mothers and calves. In Shark Bay, Australia, non-weaned calves spend a considerable amount of time away from their mother, either socialising or on their own, and separation distances of more than 100 metres have been found (Smolker *et al.* 1993). Similar observations have been made in the Moray Firth, Scotland (Janik, unpublished data).

Separations that leave the calf on its own are most likely related to the mother's feeding strategies. Adult dolphins can reach swimming speeds of up to 10 metres per second (Lang and Pryor 1966) and dispersal is possible in all three dimensions. Infants can probably not keep up with their mothers when they are hunting, and they also save energy by staying behind. However, calves are also vulnerable to predation and need the mother for nursing. Dolphins cannot carry their infant or deposit it at a den. Therefore, an effective communication system to keep track of each other is the only way to maintain mother–infant contact.

Only certain communication channels have the required properties to ensure long distance contact. Dolphins have a good sense of vision, but underwater visibility is very limited, especially in turbid, nutrient rich waters. Olfaction is a very important communication channel for mother–infant recognition in mammals (reviewed in Halpin 1991). Dolphins can detect different solutions of chemicals by taste, but it seems that their olfactory sense is rather limited (Nachtigall 1986). Furthermore, communication by chemical substances is relatively slow at long distances. Thus, only the acoustic channel is suited to individual recognition and group maintenance at longer distances.

Dolphins could keep track of each other acoustically in three different ways. First, they might be able to locate group members by using their elaborate echolocation system. However, there are several reasons why this is an unlikely solution. Echolocation is very directional, so that the animal would have to scan all possible directions to locate a group member. Furthermore, the blubber and skin of bottlenose dolphins have anechoic properties (Au 1996). The acoustic target strength (i.e. the amount of sound that is reflected off a target) of a dolphin is much lower than that of a fish, especially at the high sound frequencies that could give enough detail to distinguish between individuals. In the Moray Firth, Scotland, vocalising bottlenose dolphins can be localised at more than 500 m distance from the nearest microphone by using a passive acoustic localisation method (Janik 1998). It seems unlikely that a bottlenose dolphin would be capable of discriminating individuals over that distance purely by echolocating. Furthermore, the air-filled cavities of dolphins that reflect sound best change shape under different water pressures. In a diving mammal these would not be good features for use in individual recognition.

The second possibility is simply listening to the echolocation sounds produced by other group members. If these clicks are individually distinctive they could be used to keep track of group members. However, similar problems apply. Echolocation clicks are very directional. In order to intercept a signal the animal would have to be in line with the echolocation target of the relevant group member. Furthermore, dolphins do not produce clicks continuously, and this would make position monitoring difficult. High frequencies also get attenuated quickly, so that this system would not have a very wide range.

Finally, individuals could make their position and identity known by emitting individually specific recognition signals. But why do dolphins not simply recognise each other by their voices without the need to learn a special signal? We have already seen that background noise can be a problem. The sea is a high background noise environment (Spiesberger and Fristrup 1990) that can mask individual differences especially over the long separation distances that can be found between communicating dolphins. But there is another potential problem that is related to diving. Individual voice characteristics are influenced by head morphology. Dolphins have several air sacs that are used to recycle air for sound production. It is likely that the shape of these air filled cavities in the dolphin's forehead influences its voice characteristics. Such cavities change shape if an animal is diving because of the increasing water pressure outside. Thus, not only the shape of these cavities but also voice characteristics might change with swimming depth (Tyack 1991). Indeed, Ridgway *et al.* (1997) showed that the power spectra of beluga, *Delphinapterus leucas*, sounds were very different depending on the depth at which the sound was produced. Therefore, dolphin voice characteristics do not seem to provide reliable information on the identity of a caller. Instead bottlenose dolphins seem to ensure individual recognition over long distances by encoding identity in a specific whistle type. Vocal learning gives them the opportunity to develop such new whistle types that can be used in this context. This is further supported by the fact mentioned above that signature whistle development is influenced by vocal learning.

This leaves us with two possible reasons for the evolution of vocal learning in cetaceans if we assume that it is still used in its original context. Baleen whales seem to use vocal learning in the context of mate attraction and/or intra-sexual competition (Tyack 1986a). Toothed whales on the other hand most probably use it in individual recognition. It is difficult to decide which of these was the original context in which vocal learning evolved or whether vocal learning evolved twice within the cetaceans. One argument can be made to support the primacy of individual recognition. Baleen whales only branched off from the older toothed whales around 10–40 million years ago (Milin-kovitch *et al.*1993). Thus, since none of the toothed whales has been found to show singing behaviour it seems likely that individual recognition was the significant factor in the evolution of vocal learning. However, further studies on individual recognition, especially in baleen whales, are needed to come to a conclusion about its role in the evolution of vocal learning in cetaceans.

Implications of vocal learning

One of the main problems for a recognition system based on learned signals is that others can copy the recognition signal. Even though one bottlenose dolphin produces its specific signature whistle consistently if it is isolated from its group members, other individuals are capable of producing the same whistle. This can introduce a serious problem if learned signals are crucial for individual recognition. How does the receiver know which individual produced a call if it cannot use voice characteristics? Birds, for example, find it more difficult to distinguish between individuals that sing the same song type than between individuals that sing different song types (McGregor and Avery 1986, Stoddard *et al.* 1992, Beecher *et al.* 1994). However, bottlenose dolphins nevertheless seem to use learned signals in individual recognition.

Two questions are critical to the impact of whistle copying: do bottlenose dolphins copy the signature whistles of others, and why would they do so? Tyack (1986b) was the first to investigate signature whistle copying. To identify which of the two individuals in a pool was producing a call he employed so-called vocalights. Vocalights consist of a microphone and amplitude sensitive light emitting diodes (LEDs). These devices were attached to a dolphin's head so that whenever an individual produced a call its LEDs lit up. With this set-up, Tyack found that each individual produced mainly one of the two stereotyped signature whistles he recorded, but that approximately 25% of each individual's whistles were copies of the other animal's signature whistle. Janik (1998) found that whistle copying also occurs between wild dolphins. Thus, strictly speaking signature whistles are not individually specific.

There are several possible reasons why dolphins copy signature whistles. Feekes (1977) hypothesised that shared, learned calls can act as a password to exclude non-members from a group. This hypothesis assumes that an individual cannot produce an exact copy of a call on first exposure. Richards *et al.* (1984), however, have shown that individual dolphins can produce high fidelity copies of novel calls immediately. It is, of course, difficult to assess whether another dolphin could recognise that a copy was emitted by an animal with no practice in producing that call. However, the cues present for such recognition could only lie in small parameter variations. In the previous section on the origins of vocal learning I discussed problems imposed on the communication system by background noise and changing water pressure. That discussion showed that it seems unlikely that an individual could discriminate between genuine and copied calls of the same type at a distance by listening to slight parameter differences.

This would make whistle copying a useful tool to deceive others. Unfortunately, dolphin whistles can carry over several kilometres (Janik 1998) and the animals are difficult to observe in their natural habitat. Therefore, there are no behavioural studies that have the data required to look at deception. However, dolphins could use copying for deception in several contexts. An individual could lure a mother away from a food patch by copying her infant's whistle. In agonistic or territorial encounters, a dolphin could pretend to be more than one individual by producing calls of several animals as in the Beau Geste hypothesis proposed for birds (Krebs 1977). Deception is likely to occur if the costs involved in being detected are lower than the benefits that can be achieved by performing it or if detection is highly unlikely. Considering the cognitive abilities of dolphins even tactical deception (Whiten and Byrne 1988) seems possible. However, false signals can only be advantageous if they are relatively rare (Wiley 1994). If not, specific adaptations to counteract deception would be expected, since signature whistles still seem to be important in individual recognition and group cohesion. These issues have so far received little research attention in dolphins.

However, the common occurrence of signature whistle matching, i.e. the production of the same signature whistle type by two individuals in rapid succession, found in dolphins (Janik 1998) suggests that deception is not one of the main functions of whistle copying. In these cases it is immediately clear to all receivers that one of the callers is not the individual the specific signature whistle belongs to. A more likely explanation is that copying another individual's whistle is used to address a particular animal as suggested by Tyack (1986b). In the highly fluid social system of bottlenose dolphins this interpretation is particularly convincing. Since individuals often swim within acoustic range of each other, the ability to address specific individuals could be used to maintain group cohesion even in large feeding aggregations. Also if individuals are trying to coordinate their movements at a distance a labelling system would be useful. Bottlenose dolphins have been found to hunt cooperatively (Hoese 1971) and adult males sometimes synchronise their swimming in the context of mating (Connor et al. 1992a). Tyack (1993) described an instance in which a female reacted selectively to the copying of her whistle by another female while other animals did not. Gwinner and Kneutgen (1962) observed similar cases in which ravens, Corvus corax, and white-rumped shamas, Copsychus malabaricus, used specific calls of their mates, which caused the partner to return to the caller. Matching another animal's calls can also occur in agonistic contexts and can elicit an aggressive response without cooperation being involved. Many songbirds use song matching in such contexts (reviewed in Catchpole and Slater 1995). Bottlenose dolphins do not seem to use whistles much in aggressive interactions at close range (Overstrom 1983), but they could be important over longer distances.

Richards *et al.* (1984) trained a bottlenose dolphin to imitate specific computer sounds in response to different objects. They called this behaviour vocal labelling and suggested that dolphins use signature whistles as labels for other dolphins. While copying is a descriptive term, labelling suggests that the individual has a learned, internal representation that connects the label and the corresponding object independent of the context. In observational studies it is often difficult to recognise whether the animal has made such a connection or whether it only formed a context-specific association, i.e. that the production of a specific call leads to a specific result. Context-specific associative learning is very common in animals. Gwinner and Kneutgen (1962) in their study on call copying in birds, for example, argued that the female bird had learned to associate her copying with a return response of the male. This does not require the internal connection between the male and its call that is required for labelling. In human language, on the other hand, new labels are created spontaneously and are used to communicate about the environment. Terrace (1985) and Wilkins and Wakefield (1995) have suggested that labelling behaviour may have been a first step in the evolution of human language. They pointed out that several animal species can be taught to use new labels for objects, but none shows that behaviour in its own communication system or uses new labels without being rewarded. Humans, on the other hand, develop or copy novel sounds spontaneously to label objects often without any reward being involved.

Very little is known about the mechanisms of whistle copying or possible labelling in dolphins. Reiss and McCowan (1993) investigated labelling behaviour in dolphins experimentally. They reported that bottlenose dolphins spontaneously produced novel whistles that have been presented to them during presentations of different objects. These whistles were later also heard when the animals were allowed to manipulate these objects in their tank without any further reward being given by the experimenters. However, in studies such as this it can be difficult to determine whether a call is newly learned or was in an animal's repertoire before. Reiss and McCowan (1993) reported that the whistles used were novel to the animals. However, the spontaneous whistles produced by their study animals while manipulating objects were remarkably similar to contours in their baseline data from the same individuals before onset of their vocal mimicry study (compare figures in Reiss and McCowan 1993, and Fig. 1 in McCowan and Reiss 1995). Bottlenose dolphins can shift the frequency band in which a whistle contour lies (Richards *et al.* 1984), so that differences in the absolute frequency of contours seems less important. Therefore, it is still not clear whether bottlenose dolphins use novel sounds spontaneously to label objects. The social system of bottlenose dolphins and the large number of individuals each interacts with would make a labelling system advantageous. However, further studies on the contexts of call copying are needed to decide what mechanisms might be involved.

Conclusions

We have seen that individual recognition was a likely selection pressure for the evolution of vocal learning in dolphins. Since dolphins are limited to acoustic signals for communication by the properties of their environment and by their high mobility, this puts special selection pressures on their communication system. It may have led not only to the development of learned recognition calls, but possibly also to the usage of copies of these calls to address specific individuals.

The evolution of vocal learning seems to be a crucial event in the phylogeny of odontocetes and probably greatly facilitated their development into the social, highly mobile, aquatic predators they are today. Inevitable separations of mothers and infants while feeding on fast fish are not a problem if identity and position information can be transmitted reliably between them. This allows for a prolonged nursing phase and frees the mother from providing a massive amount of resources to her offspring in a very short time as is the case in other marine mammals such as many phocid seals (reviewed in Trillmich 1996).

A long nursing phase also gives the infant time to acquire many of its abilities by learning. In societies where cooperation and/or social learning occur, it is possible to invade new habitats rapidly by exploiting food sources that otherwise would not be available (e.g. Terkel 1996, Byrne chapter 18, this volume). Bottlenose dolphins seem to have made extensive use of this possibility. They can be found world-wide in almost all marine habitats and have adopted a generalist feeding strategy displaying a wide variety of feeding methods, e.g. using a sponge as a tool to chase fish from the bottom (Smolker *et al.* 1997), herding fish on to a beach and then feeding on them by beaching themselves (Hoese 1971), digging for fish in the bottom by burying themselves (Rossbach and Herzing 1997) and possibly, in association with false killer whales, *Pseudorca crassidens*, causing sperm whales, *Physeter macrocephalus*, to defaecate or regurgitate and then feeding on half digested food items (Palacidos and Mate 1996). These techniques usually occur in only one population or group of dolphins and are apparently not part of the whole species' repertoire. There also is some evidence for teaching of feeding strategies in the largest dolphin species, the killer whale, *Orcinus orca* (Guinet and Bouvier 1995). A long nursing phase allows the young to gather information on feeding techniques as well as about the social structure of the community.

We still know very little about how dolphins use sounds socially. It is likely that vocal learning affects many other aspects of communication in addition to the individual recognition system. Maintaining complex social relationships requires complex communication. Any form of cooperation or coordinated behaviour would require extensive transmission of information. Again, due to the lack of visual contact position information is probably one of the most important, but other referential information is required as well if a specific behaviour has to be performed in synchrony. A good example is in cooperative hunting strategies. Lions (Stander 1992) and chimpanzees (Boesch 1994) rely heavily on visual information to coordinate their behaviour during cooperative hunts. In dolphins all this information would have to be transmitted acoustically. They show coordinated behaviours in feeding contexts (Hoese 1971) and between males in alliances (Connor *et al.* 1992b). However, we know little about how such coordination is achieved. To understand the role that vocal learning might have played in the evolution of sociality we need to investigate, first, the extent to which non-communicative sounds such as swimming noises or sounds produced for echolocation are used to acquire information that is not available through vision and, second, how communication sounds are used to transmit information that is needed to co-operate in a particular behaviour.

Acknowledgements

I would like to thank Richard Connor, Jeff Graves, Peter McGregor, Laela Sayigh, Peter Slater and the editors for comments on a draft of this paper. Funding was provided by a DAAD-Doktorandenstipendium aus Mitteln des zweiten Hochschulsonderprogramms.

References

Au, W. W. L. (1996). Acoustic reflectivity of a dolphin. *J. Acoustical Soc. Am.*, **99**(6), 3844–8.

Bastian, J. (1967). The transmission of arbitrary environmental information between bottlenose dolphins. In *Animal Sonar Systems – Biology and Bionics*, ed. R. G. Busnel, pp. 803–73. Jovey-en-Josas, France: Laboratoire de Physiologie Acoustique.

Beecher, M. D. (1991). Successes and failures of parent-offspring recognition in animals. In *Kin Recognition*, ed. P. G. Hepper, pp. 94–124. Cambridge: Cambridge University Press.

Beecher, M. D., Campbell, S. E. and Burt, J. M. (1994). Song perception in the song sparrow: birds classify by song type but not by singer. *Anim. Behav.*, **47**, 1343–51.

Boesch, C. (1994). Chimpanzees-red colobus monkeys: a predator-prey system. *Anim. Behav.*, **47**, 1135–48.

Boughman, J. W. (1997). Greater spear-nosed bats give group-distinctive calls. *Behav. Ecol. Sociobiol.*, **40**, 61–70.

Burnstein, D. D. and Wolff, P. C. (1967). Vocal conditioning in the guinea pig. *Psychonom. Sci.*, **8**, 39–40.

Caldwell, M. C. and Caldwell, D. K. (1965). Individual whistle contours in bottlenose dolphins (*Tursiops truncatus*). *Nature*, **207**, 434–5.

Caldwell, M. C. and Caldwell, D. K. (1968). Volcalization of naive captive dolphins in small groups. *Science*, **159**, 1121–3.

Caldwell, M. C. and Caldwell, D. K. (1972). Vocal mimicry in the whistle mode by an Atlantic bottlenosed dolphin. *Cetology*, **9**, 1–8.

Caldwell, M. C. and Caldwell, D. K. (1979). The whistle of the Atlantic bootlenosed dolphin (*Tursiops truncatus*) – Ontogeny. In *Behavior of Marine Animals: Current Perspectives in Research, Vol. 3, Cetaceans*, ed. H. E. Winn and B. L. Olla, pp. 369–401. New York: Plenum Press.

Caldwell, M. C., Caldwell, D. K.

and Tyack, P. L. (1990). Review of the signature-whistle-hypothesis for the Atlantic bootlenose dolphin. In *The Bottlenosed Dolphin*, ed. S. Leatherwood and R. R. Reeves, pp. 199–234. San Diego, CA: Academic Press.

Catchpole, C. K. and Slater, P. J. B. (1995). *Bird Song: Biological Themes and Variations*. Cambridge: Cambridge University Press.

Cheney, D. L. and Seyfarth, R. M. (1990). *How Monkeys See the World*. Chicago: University of Chicago Press.

Connor, R. C. and Smolker, R. A. (1996). 'Pop' goes the dolphin: a vocalization male bottlenosed dolphins produce during consortships. *Behaviour*, **133**, 643–62.

Connor, R. C., Smolker, R. and Richards, A. F. (1992a). Two levels of alliance formation among male bottlenosed dolphins (*Tursiops* sp.). *Proc. Natl. Acad. Sci. USA*, **89**, 987–90.

Connor, R. C., Smolker, R. A. and Richards, A. F. (1992b). Dolphin alliances and coalitions. In *Coalitions and Alliances in Humans and Other Animals*, ed. A. H. Harcourt and F. B. M. De Waal, pp. 415–43. Oxford: Oxford University Press.

Connor, R. C., Richards, A. F., Smolker, R. A. and Mann, J. (1996). Patterns of female attractiveness in Indian Ocean bottlenose dolphins. *Behaviour*, **133**, 37–69.

Dawson, S. M. (1991). Clicks and communication: the behavioural and social contexts of Hector's dolphin vocalizations.

Ethology, **88**, 265–76.

dos Santos, M. E., Ferreira, A. J. and Harzen, S. (1995). Rhythmic sound sequences by aroused bottlenose dolphins in the Sado estuary, Portugal. In *Sensory Systems of Aquatic Mammals*, ed. R. A. Kastelein, J. A. Thomas and P. E. Nachtigall, pp. 325–34. Woerden, Netherlands: De Spil Publishers.

Feekes, F. (1997). Colony specific song in *Cacicus cela* (Icteridae, Aves): the pass-word hypothesis. *Ardea*, **65**, 197–202.

Frankel, A. S., Clark, C. W., Herman, L. M. and Gabriele, C. M. (1995). Spatial distribution, habitat utilization, and social interactions of humpback whales, *Megaptera novaeangliae*, off Hawai'i, determined using acoustic and visual techniques. *Can. J. Zool.*, **73**, 1134–46.

Guinee, L. N., Chu, K. and Dorsey, E. M. (1983). Changes over time in the songs of known individual humpback whales (*Megaptera novaeangliae*). In *Communication and Behavior of Whales*, ed. R. Payne, pp. 59–80. Boulder, CO: Westview Press.

Guinet, C. and Bouvier, J. (1995). Development of intentional stranding hunting techniques in killer whale (*Orcinus orca*) calves at Corzet Archipelago. *Can. J. Zool.*, **73**, 27–33.

Gwinner, E. and Kneutgen, J. (1962). Über die biologische Bedeutung der 'zweckdienlichen' Anwendung erlernter Laute bei Vögeln. *Z. Tierpsychol.*, **19**(6), 692–6.

Halpin, Z. T. (1991). Kin recognition cues of vertebrates. In *Kin Recognition*, ed. P. G. Hepper, pp. 220–58. Cambridge: Cambridge University Press.

Hansen, P. (1979). Vocal learning: its role in adapting sound structures to long-distance propagation and a hypothesis on its evolution. *Anim. Behav.*, 27, 1270–1.

Herman, L. M. and Forestell, P. H. (1985). Reporting presence or absence of named objects by a language-trained dolphin. *Neurosci. Biobehav. Rev.*, 9, 667–81.

Herman, L. M., Richards, D. G. and Wolz, J. P. (1984). Comprehension of sentences by bottlenosed dolphins. *Cognition*, 16, 129–219.

Herman, L. M., Pack, A. A. and Morrel-Samuels, P. (1993). Representational and conceptual skills of dolphins. In *Language and Communication: Comparative Perspectives*, ed. H. L. Roitblat, L. M. Herman and P. E. Nachtigall, pp. 403–42. Hillsdale, NJ: Lawrence Erlbaum Associates.

Herman, L. M., Pack, A. A. and Wood, A. M. (1994). Bottlenose dolphins can generalize rules and develop abstract concepts. *Mar. Mammal Sci.*, 10, 70–80.

Hoese, H. D. (1971). Dolphin feeding out of water in a salt marsh. *J. Mammal.*, 52(1), 222–3.

Janik, V. M. (1998). Functional and organizational aspects of vocal repertoires in bottlenose dolphins (*Tursiops truncatus*), PhD thesis, School of Environmental and Evolutionary Biology, University of St Andrews.

Janik, V. M. and Slater, P. J. B. (1997). Vocal learning in mammals. *Adv. Stud. Behav.*, 26, 59–99.

Janik, V. M. and Slater, P. J. B. (1998). Context-specific use suggests that bottlenose dolphin signature whistles are cohesion calls. *Anim. Behav.*, 56, 829–38.

Janik, V. M., Dehnhardt, G. and Todt, D. (1994). Signature whistle variations in a bottlenosed dolphin, *Tursiops truncatus*. *Behav. Ecol. Sociobiol.*, 35(4), 243–8.

Jones, G. and Ransome, R. D. (1993). Echolocation calls of bats are influenced by maternal effects and change over a lifetime. *Proc. R. Soc. Lond. B*, 252, 125–8.

Jürgens, U. (1992). Introduction and review. In *Nonverbal Vocal Communication: Comparative and Developmental Approaches*, ed. H. Papousek, U. Jürgens and M. Papousek, pp. 3–5. Cambridge: Cambridge University Press and Editions de la Maison des Sciences de l'Homme.

Krebs, J. R. (1997). The significance of song repertoires: the Beau Geste hypothesis. *Anim. Behav.*, 25, 475–8.

Lal, H. (1967). Operant control of vocal responding in rats. *Psychonom. Sci.*, 8, 35–6.

Lang, T. G. and Pryor, K. (1966). Hydrodynamic performance of porpoises (*Stenella attenuata*). *Science*, 152, 531–3.

Lieblich, A. K., Symmes, D., Newman, J. D. and Shapiro, M. (1980). Development of the isolation peep in laboratory-bred squirrel monkeys. *Anim. Behav.*, 28, 1–9.

Lilly, J. C. (1965). Vocal mimicry in *Tursiops*: ability to match numbers and durations of human vocal bursts. *Science*, 147, 300–1.

Loesche, P., Stoddard, P. K., Higgins, B. J. and Beecher, M. D. (1991). Signature versus perceptual adaptations for individual vocal recognition in swallows. *Behaviour*, 118, 15–25.

McCowan, B. and Reiss, D. (1995). Whistle contour development in captive-born infant bottlenose dolphins (*Tursiops truncatus*): role of learning. *J. Comp. Psychol.*, 109(3), 242–60.

McGregor, P. K. and Avery, M. I. (1986). The unsung song of great tits (*Parus major*): learning neighbors' songs for discrimination. *Behav. Ecol. Sociobiol.*, 18, 311–16.

Meltzoff, A. N. (1996). The human infant as imitative generalist: a 20-year progress report on infant imitation with implications for comparative psychology. In *Social Learning in Animals: The Roots of Culture*, ed. C. M. Heyes and B. G. Galef, pp. 347–70. San Diego, CA: Academic Press.

Milinkovitch, M. C., Orti, G. and Meyer, A. (1993). Revised phylogeny of whales suggested by mitochondrial ribosomal DNA sequences. *Nature*, 361, 346–8.

Molliver, M. E. (1963). Operant control of vocal behavior in the cat. *J. Exper. Anal. Behav.*, **6**(2), 197–202.

Moore, P. W. B. and Pawloski, D. A. (1990). Investigations on the control of echolocation pulses in the dolphin (*Tursiops truncatus*). In *Sensory Abilities of Cetaceans*, ed. J. Thomas and R. A. Kastelein, pp. 305–16. New York: Plenum Press.

Myers, S. A., Horrel, J. A. and Pennypacker, H. S. (1965). Operant control of vocal behavior in the monkey *Cebus albifrons*. *Psychonom. Sci.*, **3**, 389–90.

Nachtigall, P. E. (1986). Vision, audition, and chemoreception in dolphins and other marine mammals. In *Dolphin Cognition and Behavior: A Comparative Approach*, ed. R. J. Schusterman, J. A. Thomas and F. G. Wood, pp. 79–113. Hillsdale, NJ: Lawrence Erlbaum Associates.

Nottebohm, F. (1972). The origins of vocal learning. *Am. Nat.*, **106**, 116–40.

Overstrom, N. A. (1983). Association between burst-pulse sounds and aggressive behavior in captive Atlantic bottlenosed dolphins (*Tursiops truncatus*). *Zoo Biol.*, **2**, 93–103.

Palacidos, D. M. and Mate, B. R. (1996). Attack by false killer whales (*Pseudorca crassidens*) on sperm whales (*Physeter macrocephalus*) in the Galapagos Islands. *Mar. Mammal Sci.*, **12**(4), 582–7.

Payne, K. and Payne, R. (1985). Large scale changes over 19 years in songs of humpback whales in Bermuda. *Z. Tierpsychol.*, **68**, 89–114.

Ralls, K., Fiorelli, P. and Gish, S.

(1985). Vocalizations and vocal mimicry in captive harbor seals, *Phoca vitulina. Can. J. Zool.*, **63**, 1050–6.

Reiss, D. and McCowan, B. (1993). Spontaneous vocal mimicry and production by bottlenose dolphins (*Tursiops truncatus*): evidence for vocal learning. *J. Comp. Psychol.*, **107**(3), 301–12.

Richards, D. G., Wolz, J. P. and Herman, L. M. (1984). Vocal mimicry of computer-generated sounds and vocal labeling of objects by a bottlenosed dolphin, *Tursiops truncatus. J. Comp. Psychol.*, **98**, 10–28.

Ridgway, S., Carder, D., Smith, R., Kamolnick, T. and Elsberry, W. (1997). First audiogram for marine mammals in the open ocean and at depth: hearing and whistling by two white whales down to 30 atmospheres. *J. Acoustic. Soc. Am.*, **101**(5), 3136.

Rossbach, K. A. and Herzing, D. L. (1997). Underwater observations of benthic-feeding bottlenose dolphins (*Tursiops truncatus*) near grand Bahama Island, Bahamas. *Mar. Mammal Sci.*, **13**(3), 498–504.

Saayman, G. S., Taylor, C. K. and Bower, D. (1973). Diurnal activity cycles in captive and free-ranging Indian Ocean bottlenose dolphins (*Tursiops truncatus* Ehrenburg). *Behaviour*, **44**, 212–33.

Salzinger, K. and Waller, M. B. (1962). The operant control of vocalization in the dog. *J. Exp. Anal. Behav.*, **5**, 383–9.

Sayigh, L. S., Tyack, P. L., Wells, R. S. and Scott, M. D. (1990).

Signature whistles of free-ranging bottlenose dolphins, *Tursiops truncatus*: mother–offspring comparisons. *Behav. Ecol. Sociobiol.*, **26**, 247–60.

Sayigh, L. S., Tyack, P. L., Wells, R. S., Scott, M. D. and Irvine, A. B. (1995). Sex differences in signature whistle production of free-ranging bottlenose dolphins, *Tursiops truncatus. Behav. Ecol. Sociobiol.*, **36**, 171–7.

Schultz, K. W., Cato, D. H., Corkeron, P. J. and Bryden, M. M. (1995). Low frequency narrow-band sounds produced by bottlenose dolphins. *Mar. Mammal Sci.*, **11**(4), 503–9.

Schusterman, R. J. and Feinstein, S. H. (1965). Shaping and discriminative control of underwater click vocalizations in a California sea lion. *Science*, **150**, 1743–4.

Scott, M. D. and Chivers, S. J. (1990). Distribution and herd structure of bottlenose dolphins in the eastern tropical Pacific Ocean. In *The Bottlenose Dolphin*, ed. S. Leatherwood and R. R. Reeves, pp. 387–402. San Diego, CA: Academic Press.

Smolker, R. A., Richards, A. F., Connor, R. C. and Pepper, J. W. (1992). Sex differences in patterns of association among Indian Ocean bottlenose dolphins. *Behaviour*, **123**, 38–69.

Smolker, R. A., Mann, J. and Smuts, B. B. (1993). Use of signature whistles during separation and reunions by wild bottlenose dolphin mothers and infants. *Behav. Ecol. Sociobiol.*, **33**, 393–402.

Smolker, R. A., Richards, A. F.,

Connor, R. C., Mann, J. and Berggren, P. (1997). Sponge carrying by dolphins (Delphinidae, *Tursiops* sp.): a foraging specialization involving tool use? *Ethology*, **103**, 454–65.

Spiesberger, J. L. and Fristrup, K. M. (1990). Passive localization of calling animals and sensing of their acoustic environment using acoustic tomography. *Am. Nat.*, **135**(1), 107–53.

Stander, P. E. (1992). Foraging dynamics of lions in a semi-arid environment. *Can. J. Zool.*, **70**, 8–21.

Stoddard, P. K., Beecher, M. D., Loesche, P. and Campbell, S. E. (1992). Memory does not constrain individual recognition in a bird with song repertoires. *Behaviour*, **122**, 274–87.

Terkel, J. (1996). Cultural transmission of feeding behavior in the black rat (*Rattus ratus*). In *Social Learning in Animals: The Roots of Culture*, ed. C. M. Heyes and B. G. Galef, pp. 17–47. San Diego, CA: Academic Press.

Terrace, H. S. (1985). In the beginning was the 'name'. *Am. Psychol.*, **40**(9), 1011–28.

Trillmich, F. (1996). Parental investment in pinnipeds. *Adv. Stud. Behav.*, **25**, 533–77.

Tyack, P. (1986a). Population biology, social behaviour, and communication in whales and dolphins. *Trends Ecol. Evol.*, **1**, 144–50.

Tyack, P. (1986b). Whistle repertoires of two bottlenosed dolphins, *Tursiops truncatus*:

mimicry of signature whistles? *Behav. Ecol. Sociobiol.*, **18**, 251–7.

Tyack, P. (1991). If you need me, whistle. *Nat. Hist.*, **8/91**, 60–1.

Tyack, P. L. (1993). Animal language research needs a broader comparative and evolutionary framework. In *Language and Communication: Comparative Perspectives*, ed. H. L. Roitblat, L. M. Herman and P. E. Nachtigall, pp. 115–52. Hillsdale, NJ: Lawrence Erlbaum Associates.

Tyack, P. L. (1997). Development and social functions of signature whistles in bottlenose dolphins *Tursiops truncatus*. *Bioacoustics*, **8**, 21–46.

Tyack, P. L. and Sayigh, L. S. (1997). Vocal learning in cetaceans. In *Social Influences on Vocal Development*, ed. C. Snowdon and M. Hausberger, pp. 208–33. Cambridge: Cambridge University Press.

Wells, R. S. (1991). The role of long-term study in understanding the social structure of a bottlenose dolphin community. In *Dolphin Societies*, ed. K. Pryor and K. S. Norris, pp. 199–235. Berkeley, CA: University of California Press.

Wells, R. S., Irvine, A. B. and Scott, M. D. (1980). The social ecology of inshore odontocetes. In *Cetacean Behavior: Mechanisms and Functions*, ed. L. M. Herman, pp. 263–317. New York: John Wiley and Sons.

Wells, R. S., Scott, M. D. and Irvine, A. B. (1987). The social

structure of free-ranging bottlenose dolphins. In *Current Mammalogy*, ed. H. H. Genoways, pp. 247–305. New York: Plenum Press.

Whiten, A. and Byrne, R. W. (1988). Tactical deception in primates. *Behav. Brain Sci.*, **11**, 233–44.

Wiley, R. H. (1994). Errors, exaggeration, and deception in animal communication. In *Behavioral Mechanisms in Evolutionary Ecology*, ed. L. A. Real, pp. 157–98. Chicago: The University of Chicago Press.

Wilkins, W. K. and Wakefield, J. (1995). Brain evolution and neurolinguistic preconditions. *Behav. Brain Sci.*, **18**, 161–226.

Würsig, B. (1978). Occurrence and group organization of Atlantic bottlenose porpoises (*Tursiops truncatus*) in an Argentine bay. *Biol. Bull.*, **154**, 348–59.

Würsig, B., Cipriano, F. and Würsig, M. (1991). Dolphin movement patterns: information from radio and theodolite tracking studies. In *Dolphin Societies: Discoveries and Puzzles*, ed. K. Pryor and K. S. Norris, pp. 79–111. Berkeley, CA: University of California Press.

Xitco, M. J. and Roitblat, H. L. (1996). Object recognition through eavesdropping: passive echolocation in bottlenose dolphins. *Anim. Learning Behav.*, **24**(4), 355–65.

Part 6

The great ape – human adaptation: culture and the cognitive niche

Editors' comments

KRG and HOB

Anthropologists have long defined humans as the cultural animal and assumed that learned traditions occur in no other animals (White 1959). This anthropocentric perspective is clearly erroneous. Nonetheless, humans certainly exceed other animals in the shear quantity of culturally transmitted information and skills that are possessed by many human societies. Many humans societies also differ from those of other animals in that they exhibit a cultural 'ratcheting effect', that is, human cultural innovations build and expand upon each other, especially in the technological realm. (Tomasello *et al.* 1993, Boyd and Richerson 1996, Gibson 1993, Visalberghi 1993). Mithen chapter 21, this volume). Chapters in this section of the volume focus on the cognitive capacities that make human culture possible, in comparison with the cognitive and social learning capacities of our closest phylogenetic kin, the great apes. Great apes and humans share 98 to 99% of their genes and common life-history strategies involving large body and brain sizes, long lifespans, prolonged maturational periods, and intense maternal investment characterised by normally giving birth to a single young that is nursed for several years.

Given the close phylogenetic relationships between great apes and humans it is not surprising that some of the clearest examples of animal social traditions come from the great apes. Chimpanzee populations, for example, differ in tool-using, foraging and other traditions in ways that cannot be explained by ecological conditions and that have been termed cultural (McGrew 1992, Boesch 1996). Further, both chimpanzees (Hayes and Hayes 1952) and orangutans (Russon and Galdikas 1993) have been observed imitating human tool-using behaviours, e.g. weeding, sweeping, using boats and hanging hammocks. One of the best-documented examples of animal teaching behaviour also comes from chimpanzees (Boesch 1996). In the Taï Forest of the Ivory Coast, chimpanzees crack hard nuts using anvils and wooden or stone hammers. Young chimpanzees master these techniques only after approximately 8 years of practice. During this period, mother chimpanzees share nuts, a high quality food, with their young. As Shennan and Steele (chapter 20, this volume) note, teaching is likely to be an adaptive strategy when the mastery of foraging or other techniques requires years of practice on the part of the young. For example, any behaviour that accelerates the mastery of the nut-cracking task by young chimpanzees will increase their chances of survival in the advent of maternal death and, by decreasing the period of maternal investment, potentially increase the reproductive capacity of the mother. Consequently, it is perhaps not surprising that Taï chimpanzee mothers have been observed demonstrating the appropriate use of hammers to their young (Boesch 1993).

Tool use, however, is not the only foraging strategy that may require complex manipulative techniques. In this section, Byrne (chapter 18) notes that even those species or populations of great apes that do not use tools, feed on foods that require considerable food processing prior to ingestion. In his view, great ape foraging tasks are, in general, more complex than those of monkeys in terms of the food processing required, and, hence, in terms of their demands on animal manipulative abilities (see also King chapter 2, this volume). He gives a detailed description of some gorilla plant foraging techniques to illustrate his point. Gorillas are herbivorous animals who easily find food. They do, however, feed on plants that require considerable processing prior to ingestion, because they are defended by spines and stings or encased in hard woody sheaths. Gorillas process such foods by methods that involve a sequence of manipulative stages that must be performed in the correct sequence in order to yield an appropriate result. Byrne suggests that young gorillas learn effective methods of eating these foods by a process that he calls 'programme level imitation'. This involves the imitation of the outcomes of a series of manipulative steps (e.g. folding, stripping), but not the imitation of the actual hand movements used. That great apes possess the ability to imitate the outcomes of a series of manipulative actions on objects is also evident from laboratory experiments and from the imitative capacities of chimpanzees and orangutans that have been kept in captivity (Russon and Galdikas 1993, Byrne and Russon 1998, Whiten 1998). Little evidence, however, suggests that apes can actually mimic motor actions.

Byrne's descriptions of gorilla food processing techniques are of interest with respect to an existing hypothesis of the selective pressures that led to the evolution of advanced cognitive capacities in great apes and humans. Specifically, it has been proposed that a major selective agent for advanced sensorimotor intelligence, tool-using abilities, imitation, information donation, teaching and self-awareness in the common ancestor of great apes and humans was extractive foraging on a diversity of foods that must first be removed from outer matrices in which they are enclosed such as shells, rinds or bark (Gibson 1993, King 1994, Parker and Gibson 1977, 1979, Parker 1996). Chimpanzee nut-cracking, termiting and ant-dipping behaviours qualify as extractive foraging. Most gorilla plant processing behaviours described by Byrne do not, but they do, nonetheless, require some imitative capacity and the intellectual capacity to organise a sequence of acts hierarchically. Moreover, not all primate extractive foragers can imitate. Capuchins, for example, forage extractively on both animal and plant foods and use tools in captivity, but appear to lack imitative capacities (Gibson 1990, Visalberghi and Fragaszy 1990).

One difference between the tool-assisted extractive foraging behaviours of wild chimpanzees and those of captive capuchins lies in the complexity of the tool-using tasks. Capuchin nut-cracking behaviour appears to involve only repetitive banging behaviours. In contrast, chimpanzee termiting and nut-cracking behaviours require the appropriate organisation of a series of manipulative tasks. In the wild, young chimpanzees master the individual motor acts essential for cracking nuts with stone hammers long prior to mastering the ability to organise them in an appropriate cracking sequence. During the years prior to mastery, they repeatedly observe the nut-cracking behaviours of adults (Inoue-Nakamura and Matsuzawa 1997). Hence, rather than extractive foraging *per se*, the primary foraging advantage provided by great ape abilities to acquire object manipulation behaviours socially may be the ability to utilise foods whose acquisition or preparation requires serially organised actions on objects. The tool using behaviours of chimpanzees and the plant processing behaviours of gorillas both meet these criteria.

As noted above, great apes can imitate an appropriate series of actions to yield particular results on objects, but there is little evidence that they can imitate motor actions or vocalisations. Humans can imitate motor actions and vocalisations. They also possess information transmission capacities not possessed by great apes, most notably language, but also music, dance, mime and art. Gibson (chapter 19) suggests that these and other human cognitive and social learning capacities reflect the interactions of many neural areas that have expanded in size in human evolution. These neural expansions provide humans with enhanced procedural and declarative learning capacities and with information processing and mental constructional capacities that exceed those of other primates. Mental constructional capacities, in particular, serve as essential neurological underpinnings of human linguistic, imitative and technological capacities. Human culture and technology depend upon human social learning capacities, but also facilitate them and make them both possible and advantageous. Technology, for example, enhances human art, music and linguistic communication and provides human adults with foraging capacities that permit sexual divisions of labour, food sharing between males and females, and long-term maternal and paternal provisioning of the young. Such provisioning permits a prolonged period of childhood dependency in which children can master foraging and other technical skills in the presence of adults of both sexes.

Shennan and Steele (chapter 20) elaborate the theme of human technology by providing a cost–benefit analysis of the social transmission of human tool-making and other crafts. They hypothesise that the time requirements of mastering the production of crafts are such that most instruction should be provided by genetically related kin, i.e. parents. They further suggest that both parents and children will benefit by instruction and practice at the earliest possible age. A survey of traditional societies supports this model. In such societies, children usually learn sex-appropriate crafts from the same sexed parent, usually prior to adolescence. They note that while, in great apes, vertical transmission of tool-making and other foraging techniques primarily occurs within the mother–young unit, in humans both parents contribute to the transmission of craft traditions. The vertical transmission of human craft techniques suggests that throughout most of human history cultural innovations would have spread with the dispersal of human lineages, rather than by primarily horizontal transmission. Ultimately, the survival of particular skills and techniques would have depended on the survival and reproductive success of the lineages that used them.

In the final chapter (chapter 21), Mithen extends the focus on human technology into prehistorical times. For approximately 1 000 000 years, premodern human forms in Africa, Western Asia and Europe produced a specific tool type – the Acheulian hand axe. Hand axes were bilaterally symmetrical tools with a broad base and a sharp tip. They were used for animal butchery and may also have been used for hunting and for the processing of plant foods. Hand axe manufacture required cognitive capacities not thought to be possessed by the great apes (Wynn 1979). The appropriate manufacture of hand axes also requires several processing steps and is, thus, unlikely to have been repeatedly reinvented. Mithen argues persuasively that the production of hand axes over a period of one million years indicates that the hominids who made them possessed imitation capacities.

Despite the possession of imitative abilities, no cultural evolution occurred among the hand axe makers, at least not in the technological sense. The ratcheting effect (Tomaselto *et al.* 1993) in which technological innovations have built upon each other and resulted in cultural evolution in many modern human societies simply did not exist. This finding is important in light of hypotheses that imitation is the key factor distinguishing humans from apes and permitting cultural evolution (Boyd and Richerson 1996). Imitation capacities may be essential for cultural evolution to occur, but clearly they are not sufficient.

We first begin to see rapid cultural evolution in Europe approximately 50 000 years ago. Many theories have been proposed to explain the sudden emergence of rapid cultural change at this time period. Most of them postulate a sudden emergence of language or other cognitive ability. It is important to note, however, that humans achieved modern anatomical form and brain size more than 100 000 years ago – at least 50 000 years before the onset of rapid cultural change. Moreover, despite the tendency to assume that the cultural ratcheting effect is a human cultural universal, it is not. Many cultures continued to live a stone age existence well into the nineteenth century. Hence, it is possible that the sudden appearance of rapid cultural change in Europe approximately 50 000 years ago represents new ecological strategies and/or the influence of a very few key inventions that were, perhaps, as suggested by Shennan and Steele, spread by a few lineages.

References

Boyd, R and Richerson, P. (1996). Why culture is common, but cultural evolution is rare. In *Evolution of Social Behavior Patterns in Primates and Man*, ed. W. G. Runciman, J. Maynard-Smith and R. I. M. Dunbar, pp. 77–93. Oxford: Oxford University Press.

Boesch, C. (1993). Aspects of transmission of tool use in wild chimpanzees. In *Tools, Language and Cognition in Human Evolution*, ed. K. R. Gibson and T. Ingold, pp. 171–183, Cambridge: Cambridge University Press.

Boesch, C. (1996). The emergence of cultures among wild chimpanzees. In *Evolution of Social Behaviour Patterns in Primates and Man*, ed. W. G. Runciman, J. Maynard Smith and R. I. M.

Dunbar, pp. 251–68. Oxford: Oxford University Press.

Byrne, R. and Russon, A. E. (1998). Learning by imitation: a hierarchical approach. *Behav. Brain Sci.*, **21**, 667–72.

Gibson, K. R. (1990). Tool use, imitation, and deception in a captive cebus monkey. In *Language and Intelligence in Monkeys and Apes: Comparative Developmental Perspectives*, ed. S. T. Parker and K. R. Gibson, pp. 205–18, Cambridge: Cambridge University Press.

Gibson, K. R. (1993). Generative interplay between technical capacities, social relations, imitation, and cognition. In *Tools, Language and Cognition in Human Evolution*, ed. K. R. Gibson and T. Ingold, pp. 131–7, Cambridge: Cambridge Uni-

versity Press.

Hayes, K. J. and Hayes, C. (1952). Imitation in a home-reared chimpanzee. *J. Comp. Physiol. Psychol.*, **45**, 450–9.

King, B. J. (1994). *The Information Continuum: Evolution of Social Information Transfer in Monkeys, Apes, and Hominids*. Santa Fe, CA: School of American Research Press.

McGrew, W. C. (1992). *Chimpanzee Material Cuture: Implications for Human Evolution*. Cambridge: Cambridge University Press.

Inoue-Nakamura, N. and Matsuzawa, T. (1997). Development of stone tool use by wild chimpanzees (*Pan troglodytes*). *J. Comp. Psychol.*, **111**, 159–73.

Parker, S. T. (1996). Apprenticeship in tool-mediated extractive foraging: the origins of imitation, teaching, and self-awareness in great apes. In *Reaching into Thought: The Minds of Great Apes*, ed. A. E. Russon, K. A. Bard and S. T. Parker, pp. 348–70. Cambridge: Cambridge University Press.

Parker, S. T. and Gibson, K. R. (1977). Object manipulation, tool-use and sensorimotor intelligence as feeding adaptations in cebus monkeys and great apes. *J. Human Evol.*, **6**, 623–41.

Parker, S. T. and Gibson, K. R. (1979). A developmental model for the evolution of language and intelligence in early hominids. *Behav. Brain Sci.*, **2**, 367–408.

Russon, A. E. and Galdikas, B. M. F. (1993). Imitation in free-ranging rehabilitant orang-utans (*Pongo pygmaeus*) *J. Comp. Psychol.*, **107**, 147–61.

Tomasello, M., Kruger, A. C and Ratner, H. H. (1993). Cultural learning. *Behav. Brain Sci*, **16**, 495–552.

Visalberghi, E. (1993). Capuchin monkeys: a window into tool use in apes and humans. In *Tools, Language and Cognition in Human Evolution.* ed. K. R. Gibson and T. Ingold, pp. 138–50. Cambridge University Press.

Visalberghi, E. and Fragaszy, D. (1990). Do monkeys ape? In *Language and Intelligence in Monkeys and Apes: Comparative Developmental Perspectives*, ed. S. T. Parker and K. R. Gibson, pp. 247–73. Cambridge: Cambridge University Press.

White, L. A. (1959). *The Evolution of Culture: The Development of Civilization to the Fall of Rome.* Cambridge: Cambridge University Press.

Whiten, A. (1998). Imitation of the sequential structure of actions by chimpanzees (*Pan troglodytes*). *J. Comp. Psychol.*, **112**(3), 270–81.

Wynn, T. G. (1979). The intelligence of later Acheulean hominids. *Man (N.S.)*, **14**, 371–91.

18

Cognition in great ape ecology: skill-learning ability opens up foraging opportunities

Richard W. Byrne

'By their fruits ye shall know them'

(St. Matthew vii, 20)

There is increasing acceptance among primatologists that great apes may have mechanisms of social learning in addition to those of monkeys. In particular, great apes may possess the ability to learn new manual procedures by observation, although this is a matter of some controversy (Custance *et al.* 1995, Heyes 1993, Miles *et al.* 1996, Tomasello 1996). What is less often considered is what *difference* this putative ability might make to the natural lives of the ape species. Do apes do things differently, or do different things? Yet, taking that approach immediately provides evidence relevant to the controversy; if the answer to both questions is 'no', it becomes most unlikely that any cognitive difference exists between apes and monkeys. Therefore, by concentrating on natural behaviour we attain a double end, gaining evidence with theoretical import as well as illuminating the comparative ecology of the primates. The domain in which learning new manual procedures most directly affects the behavioural ecology of apes is that of *feeding*. This chapter, therefore, first identifies differences in characteristic food gathering behaviour between monkeys and apes, then asks whether these differences might reflect more powerful, cognitively complex mechanisms of social learning in great apes, and finally considers whether there are any indications of these mechanisms in the acquisition of natural feeding behaviour.

Monkeys and apes are haplorhine primates, a more derived group than the strepsirhines (lemurs and lorises), which retain primitive mammalian features such as reflective tapetum and wet rhinarium. Numerous species of haplorhines are found throughout the tropics and in some more temperate climates, but most of these are monkeys, and the great apes form only a small taxon of 4–7 species – depending on the taxonomy adopted – restricted to central zones of Africa (chimpanzees and gorillas) and to the islands of Sumatra and Borneo (the orangutan). In addition, of course, the human is a great ape, most closely related to the chimpanzees (Goodman *et al.* 1994, Kim and Takenaka 1996, Waddell and Penny 1996), which increases the interest of great ape cognition to the human sciences. Monkeys and apes are most often diurnal and found living in semi-permanent groups, although species vary considerably in social organisation and behavioural ecology. This variation is perhaps most dramatic among the great apes, which include species that are solitary, monogamous, harem-forming, or living in large, fission–fusion groups of both sexes. Long lifespan, prolonged maternal care and often highly social behaviour present great opportunities for social learning in monkeys and apes, but despite extensive long-term field studies in the last 40 years surprisingly little is known of the significance of social learning for their survival.

Obtaining an adequate diet is a much greater problem for a monkey or an ape than it is for most mammals. Most carnivores consume a diet largely of flesh, ensuring balanced nutrient composition. Most herbivores obtain a substantial proportion of their energy budget with the aid of bacterial fermentation, either in a complex stomach or enlarged caecum. The monogastric digestive systems of primates allow only limited bacterial fermentation, and then only at the expense of considerable slowing of the rate of processing (Chivers and Hladik 1984, Waterman 1984); colobine monkeys are the only exception, having foregut fermentation. Many otherwise adequate sources of nutrients are therefore unavailable, and typical monkeys and apes eat a wide diversity of foods and must combine items of many types to build up a diet that gives adequate nutrition. This may necessitate eating dozens or hundreds of plant species (Nishida and Uehara 1983, Post 1982), and to do so may require searching over a large home range (Sigg and Stolba 1981, Tutin *et al.* 1983). Particularly when these food plants are asynchronously seasonal in tropical forests, foraging based on memory has been argued to present a major intellectual problem for primates (Mackinnon 1978, Milton 1981). However, if so it is a problem solved equally by monkeys and apes, and indeed monkeys have to date provided more convincing evidence of the use of spatial memory in planning an optimal foraging route than have apes (Garber 1988).[1]

Feeding procedures in monkeys

When we turn to the procedures used to process natural foods before ingestion, differences do become apparent between apes and other simian primates. In general, monkeys eat foods which require little processing, and the procedures they then use are of a rudimentary sort: 'pick up and put in mouth', or 'pull off and put in mouth'. Presumably for this reason, manual techniques of food preparation in monkeys have attracted little scientific interest. African baboons, *Papio anubis*, and vervet monkeys, *Cercopithecus aethiops*, however, have both been observed processing foods which need more significant pro-

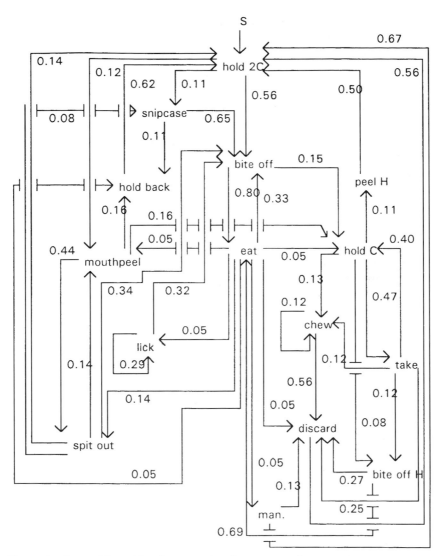

Figure 18.1. Sequential transitions between actions in a vervet monkey processing sugarcane (Harrison 1996). The complexity of this diagram has by no means reached asymptote. Processing begins at the top of the figure at the point marked S. Each term corresponds to an element of manual processing; these are defined in full in Harrison (1996). In some cases, an element may be employed in different ways. Where this is the case, the particular manner used is indicated by a suffix: C means using a cylindrical grip; 2C means using a two-handed, cylindrical grip; H means using only one hand, and so on.

cessing, and in each case a considerable repertoire of actions are employed. Baboons typically feed while travelling, and most of their manual processing consists of single actions: twisting off a leaf, yanking out a rooted tuft, stripping leaves off a branch in a single movement (Whiten 1988). Vervets, when feeding on larger items, sometimes employ many actions one after the other, and their actions are drawn from a large repertoire (Harrison 1996). However, the sequence of actions is ever-changing, so that it is not possible to see any organising principle behind the choice of what to do next. Some actions do tend to follow others, but the sequence varies on different occasions (Figure 18.1). Attempts to find a single, underlying process for an individual produce an appearance of considerable complexity, but it is unlikely that this reflects underlying complexity in the organising process. What is happening is that the food objects themselves are continually changing in size and structure, as a result of the monkey's attentions, and the changing problem elicits a continually different response.

In monkeys, then, a *large repertoire* of mechanical actions is available for food processing, a consequence of the dextrous five-fingered primate hand and the sensitive grip allowed by fingernails and pads (Napier 1961), and different actions are elicited by the sight of particular configurations of the visible food stimulus: one plant-problem, one action. Learning which action is most effective at dealing with each problem-state could easily be achieved by socially facilitated trial-and-error, with little to be gained from an ability to learn novel procedures by observation. The obvious model for any such observational learning would be the mother, yet the mother's larger and stronger hands enable a range of actions to be effective that the infant – if it were to copy them – would find useless. Their 'one problem, one action' style does not typically produce any sequential organisation in behaviour, and to date no monkey species has been found to show sophisticated organisation in learned feeding skills. Where a big food item changes during processing, different actions may be elicited one after the other, but in an unordered, *unsystematic* way. No doubt apes sometimes process food in this way – especially the novel cultivars often given them in zoos, which have been selectively bred to be easy for humans to process. But do great apes also use any more complex techniques?

Feeding procedures in great apes

The well-known tool using behaviours of the chimpanzee, *Pan troglodytes*, do constitute a very different type of food preparation. However, the *fact* of tool use may reflect only an ecological need, not a difference in cognitive ability.

Indeed, under captive or laboratory conditions, capuchin monkeys, *Cebus* spp., have been found to use objects as tools in a range of ways, matching chimpanzee uses: probes, chisels, sponges, knives, pestles, hammers, and so on (Parker and Gibson 1977, Visalberghi 1987, Westergaard and Fragaszy 1987, Westergaard and Suomi 1993). More significant for the issue of social learning is a series of analyses of what capuchin monkeys *understand* about tool use (Fragaszy and Visalberghi 1989, Visalberghi and Fragaszy 1990, Visalberghi and Limongelli 1994, Visalberghi and Trinca 1987). To give one example, a capuchin is presented with a fixed, horizontal transparent tube containing a peanut, and a piece of doweling rod. The monkey readily discovers how to use the rod in order to poke out the peanut. Then the rod is removed, various other items are substituted, and the monkey's behaviour is observed: which objects does it try out as potential replacement tools? Tested in this way, capuchins do *not* restrict their selection only to items that are rigid, thin enough to enter the tube, and long enough to emerge at the far end. Instead, they attempt to poke with small blocks, or with overly thick rods, even on one occasion with a flexible chain. These actions reveal failure of these monkeys to understand the mechanics of the task, although capuchins in other laboratories have given evidence of understanding how things work in other, perhaps simpler, tasks (for details see review by Anderson 1996). The errors made by tool-using capuchins also highlight the importance of the conspicuous *absence* of such errors in the development of tool use in wild chimpanzees. Infant chimpanzees simply do not attempt to probe for termites with rocks or logs, nor do they attempt to hammer open nuts with vines or leaves. Wild chimpanzees can also *anticipate* a future food processing need. At Gombe National Park, Tanzania, individuals sometimes make appropriate probing tools – by stripping a stem of leaves and tidying the tip – before moving to a termite mound, often over 100 m away and out of sight (Goodall 1986). At Taï Forest, Côte d'Ivoire, individuals sometimes carry appropriately selected hammer-stones up to 500 m to a site of nuts requiring cracking (Boesch and Boesch 1984). It seems that chimpanzees possess some mental specification (or 'representation') of an adequate tool for a particular purpose even when they lack the tool itself, and this mental specification can be activated in advance of the particular stimulus configuration that the task presents. Capuchins are able to learn that a certain object can be used as a tool, but it is not clear that they can develop a mental specification of the properties an object necessarily requires to do the job of tool.

The difference between monkey and ape tool use is even greater than these issues suggest. In a probe-tool experiment, the capuchin monkey is typically presented with a highly simplified world: an obvious hole, food visible within it, and a choice of only a few moveable items, one of which is often an adequate probe. The wild chimpanzee is 'presented' with a termite mound, initially without holes in it, in the African bush. (To be precise, a few dozen, rather inconspicuous, *Macrotermes* termite nests are found in a home range of 15–20 square miles of forest and bush. A day's walk at random would miss all of them.) Not only is there a large range of inappropriate objects, readily to hand but always ignored, but a mound might be prodded or struck as easily as probed. Even the hole has to be made, and this can only be done in certain parts of the mound in which termites are readying to emerge, at a certain time of year (the mud capping the emergence tubes is then softened by the termites). The edible termites are not visible at all. Probing must be done delicately, and sometimes the other hand is needed as a guide and support. Hasty poking destroys the tool; hasty pulling out dislodges any attached termites. There is, in short, a lot more to learn: termite fishing is an *elaborated* skill. Nor is it the only elaborated skill that wild chimpanzees show; ant-dipping for safari ants involves deft bimanual coordination and precise timing, since an error results in many painful bites from *Dorylus* ants, and nut-cracking involves using two tools in conjunction, as hammer and anvil, and takes years to perfect (Boesch and Boesch 1983, Sugiyama *et al.* 1993). Even under careful scrutiny, chimpanzee tool use does emerge as a skill which would benefit from enhanced social learning.

Until very recently, these insect-gathering and nut-cracking skills stood alone as elaborated techniques of food processing in great apes. Other species of ape readily use tools in captivity, but never in the complex, sophisticated forms that might require traditional transmission of information by social learning (McGrew 1989). In one Sumatran population, orangutans, *Pongo pygmaeus*, have recently been discovered to prepare stout chisel tools from live branches, removing twigs and leaves, and then carry them up into trees and manoeuvre them with the mouth or by hand to break into insect nests and to prompt the insects to exit (van Schaik *et al.* 1996). In the same population, individuals use smaller, thinner tools to clean the matrix of irritating hairs from inside part-open *Neesia* fruits, then push out the seeds to eat. Although nothing is yet known about the development of these skills, they appear comparable to chimpanzee insect fishing in their complexity. Nevertheless, most orangutans, all gorillas, *Gorilla gorilla*, and all pygmy chimpanzees, *Pan paniscus*, feed without using tools. Since it is entirely possible that this reflects their different ecologies – other sources of nutrition rendering tool use unnecessary – we must turn instead to *plant* feeding in order to compare these great ape species with monkeys. So far, techniques of plant feeding have only been systematically investigated in the mountain gorilla, *G.g.beringei*.

Figure 18.2. Adult female mountain gorilla processing a large stalk of *Peucedanum linderi*, showering her day-old infant with feeding remains. Photo R. W. Byrne.

Learning what to eat would seem to present little problem for mountain gorillas. The herbaceous vegetation in which they spend most of their day is generally low in secondary compounds (Waterman *et al.* 1983), and food plants are relatively abundant among the total vegetation (Watts 1984). Further, there is almost no seasonal or spatial variation in availability of gorilla foods; all year round, and throughout their home range, gorillas are close to prolific plant foods (Watts 1984). Learning which plants are edible is aided by the fact that infants are showered with remains of food plants from the first day of life (Figure 18.2), and their feeding is usually synchronised with the mother's (Watts 1985). However, these herbaceous plants are defended *physically*, in ways that make them unpleasant to eat without careful pre-processing. Leaves and stems are covered with spines, stings, or tiny hooks; soft pithy stems are encased in hard, woody sheaths (Byrne and Byrne 1991). Gorillas are efficient at dealing with these challenges, and their techniques are remarkably *complex*, in a technical sense.

As an example, consider preparation of the nettle *Laportea alatipes*, an

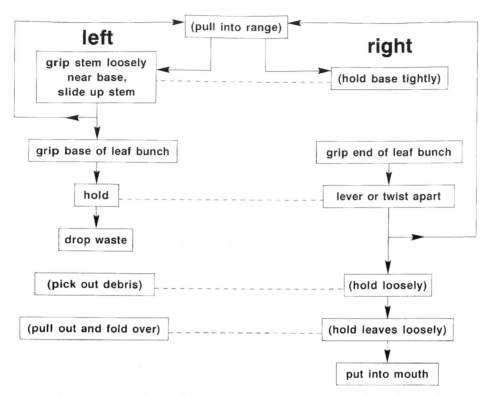

Figure 18.3. Flow diagram of the processing sequences used by a 'right-handed' mountain gorilla for preparing bundles of nettle leaves to eat. The process starts at the top and ends when a handful is put into the mouth. Actions are shown in square boxes; most actions are highly lateralised, and left and right sides of the figure show where there are significant hand preferences; horizontal dotted lines show when the actions of the two hands must be accurately coordinated.

important food plant which is rich in protein. The technique used by gorillas rapidly amasses a bundle of leaf-blades, without the stems or petioles which host the strongest stings, and folds the bundle (Figure 18.3) so that only a single leaf underside is exposed when the parcel is popped into the mouth (Byrne and Byrne 1993). This minimises the number of stings that contact the palm, fingers and especially the lips. Note that, for successful performance, multiple stages have to be sequenced correctly; several of these stages require bimanual coordination in which the two hands are used together in complementary roles. The gorilla's actions are very strongly lateralised (Byrne and Byrne 1991), a characteristic also of chimpanzee tool use (McGrew and Marchant 1996). And most importantly, some parts of the process can be repeated until a criterion is reached ('enough food for a handful'); this means

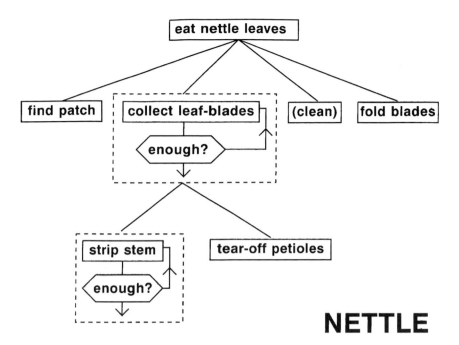

Figure 18.4. Hierarchical organisation of nettle leaf processing. This is the hypothetical program organisation of minimum hierarchical complexity needed to produce the actual behaviour observed in a mountain gorilla preparing nettle leaves to eat. The 'top goal' is placed at the top of the figure, indicating its control of the goals below it, and this may occur recursively, increasing hierarchy 'depth'.

that portions of the process are treated as subroutines. These subroutines can be single actions, or quite long sequences of actions. The ability to iterate subsections of program, and the fact that optional processes can be inserted at some points (usually for cleaning a part-processed handful), show that the apparent stages in the processing sequence are discrete entities, and reveal some of the structural organisation in the behaviour (Figure 18.4), which is *hierarchical* (Byrne and Russon 1998).

One possible explanation for such sophisticated complexity of design would be that nettle-eating skill is a unique, species-typical adaptation in gorillas: a skill that develops under tight genetic channelling, and consequently seen in almost every individual of the species. There are two reasons for rejecting this hypothesis. The gorillas of Karisoke, Rwanda, are the only ones known to use the nettle-processing technique – indeed, it would make no *sense* for most gorillas to be able to process nettle, which is a temperate plant only growing above 3000 m in Central Africa. Even at slightly lower altitudes in the Virunga Volcanoes, mountain gorillas encounter no *Laportea* nettles, and most gorillas live in very different, lowland habitats. Secondly, the techniques Karisoke gorillas use for other plants are quite different, yet show similar signs of design complexity (Byrne and Byrne 1991). For bedstraw, *Galium ruwenzoriense*, for instance, the problem is that the plant is covered with tiny hooks that enable the plant to clamber. The technique used by gorillas to deal with this problem involves the concertina-like folding of carefully picked out green stems, followed by rolling the bundle into a tight wad, allowing consumption by shearing bites of the molar teeth. In this way, the dead stems that form much of a large *Galium* mass are avoided, and the clinging hooks covering green stems and leaves are crushed into a mass of plant tissue. Although a very different process to nettle preparation, it also necessitates that the gorilla correctly sequences multiple stages, can interate parts of the process, and can coordinate the two hands in complementary roles. Bedstraw preparation is an elaborated and hierarchically organised skill, showing complex design well suited to the problem. For wild celery, *Peucedanum linderi*, where the problems are to eat soft pith inside very hard casing and to deal with long unwieldy stems, and for thistle *Carduus nyassanus*, where rigid spines are found on leaf-edges and stem-flanges, the same applies; elaborated techniques are used, but they differ in structure and elements. These four plants make up over 80% of a mountain gorilla's diet (Watts 1984); the remainder includes many foods that are not physically defended, and these are processed in much simpler ways.

Gorilla plant feeding is therefore as complex and apparently demanding of skill as the methods of insect-fishing and nutcracking in chimpanzees (McGrew and Marchant 1996). Examination of the technique used by ex-captive orangutans to eat a wild coconut palm, *Borassodendron borneensis*, found evidence of hierarchical organisation in the several stages of preparation (Russon, 1998).

A consensus may be emerging, that great apes forage with techniques that show *technical complexity* compared with those of monkeys (Byrne 1997), just as great apes may exhibit greater depth of understanding in social manipulations such as deception (Byrne and Whiten 1992) and communication (Gómez 1996; and see other chapters in Russon *et al.* 1996). Great apes possess the ability to build up, during learning, hierarchically structured routines, using one program as an embedded subroutine in another (Byrne and Russon 1998). While this is basic to human problem solving (Byrne 1977, Case 1985, Gibson 1990, Greenfield 1991, Miller *et al.* 1960, Newell *et al.* 1958), I know of no evidence that any monkey or nonprimate learns skills of this kind, and learning the structural organisation of processes is a prime candidate for learning from skilful practitioners.

Acquisition of complex feeding techniques in great apes

Evidence of active maternal interventions that might be directed at learning are rare, even in great apes[2]. Least rare are reports of the removal of *nonfoods* from the infant. This has been seen occasionally in chimpanzees (Goodall 1973, Nishida and Uehara 1983, Wrangham 1977) and gorillas, where mothers have been noted to remove brightly coloured flowers and dung (Fossey 1979), poisonous *Solanum* fruit (P Sicotte pers. commun.) and non-food *Crassocephalum* and *Senecio* leaves (personal observations) from infants which put the items in their mouths. Mother chimpanzees sometimes pluck all poisonous leaves within an infant's reach and drop them out of reach, after seeing a leaf in the infant's grasp and removing it (Hiraiwa-Hasegawa 1986). Some instances have occurred with leaves of *Ficus exasperata*, a species which chimpanzees eat from *some* individual trees and not others, suggesting that this species is polymorphic for concentration of a toxic secondary compound. If so, learning the distinction between 'good' and 'bad' individual trees would present a considerable challenge for an infant chimpanzee. However, leaf-removal only occurred in individual trees whose leaves adults avoid, so there was no evidence that the mother understood her infant's learning difficulty. Some observations from gorillas imply a greater ability to understand the infant's needs. Young gorillas must learn to avoid eating certain plant foods *before* they have developed the technical ability necessary to deal with physical defences of the plants, even though the mother is eating such foods daily. Mother gorillas have twice been noted to remove pieces of the hook-covered *Galium ruwenzoriense* plant from infants before they could ingest them:

> 21:9:89 Samvura (4 months old) picked up a piece of *Galium* and a leaf of *Crassocephalum duciiaprutii* (not a known gorilla food) and put them to mouth; Maggie, her mother, immediately reached over and took both items away, dropping them out of reach.
> 15:11:89 Samvura (6 months old) took up a small piece of *Galium*; Maggie, her mother, immediately took it away and threw it out of reach. She later did just the same when Samvura started chewing on a common *Senecio* (a nonfood, but often torn up in gorilla displays), but allowed him to chew on a discarded celery *Peucedanum* stalk.

Removal of *poisonous* items from infants can readily be explained as a protective action, based on no more than an association of danger with certain plants. This cannot apply when the plant is in fact highly edible, as in these cases, and the records suggest that an adult gorilla can understand the difference between infant and adult manual competence and the implications of lacking the competence to deal with a particular 'hard-to-process' plant.

Active teaching is rarest of all, reported only in the context of chimpanzee nut-cracking with tools (Boesch 1991). One mother, after her juvenile had placed a nut haphazardly on the anvil, 'took it in her hand, cleaned the anvil, and replaced the piece carefully in the correct position', whereupon the juvenile was successful. Another mother, after her juvenile had failed in using an irregular-shaped hammer held inefficiently, 'in a very deliberate manner, slowly rotated the hammer into the best position . . . it took her a full minute to perform this simple rotation'. The mother then cracked 10 nuts, sharing most of the kernels. When her daughter resumed attempts, she was more successful and crucially 'always maintained the hammer in the same position as her mother did'. No deliberate instruction has been observed during the longer-term studies at East African chimpanzee sites. Instruction may be specific to nut-cracking: young chimpanzees require more years to master nut-cracking than any other tool-using technique (Boesch and Boesch 1990), so this may be the hardest tool-using technique to learn.

Other chimpanzee tool-use techniques are acquired more readily, and even the most complex techniques of gorillas are acquired rapidly, reaching adult efficiency by the time of full weaning, as measured by processing speed (Byrne and Byrne 1991). For these skills, there is no evidence of teaching, but some evidence that infants learn procedures by observation. In general, this applies at the level of the techniques' organisational structure (sequence, hierarchical embedding of routines, and bimanual coordination) rather than the detail of motor actions. Indeed, there is great variation across individuals in all the details of the precise action used, the type of grip and behavioural laterality (Byrne and Byrne 1993). Given this pronounced idiosyncratic variation, it is striking that each Karisoke gorilla acquires a very similarly organised procedure for dealing with each problem. Since there are many other possible ways of processing the plants, this standardisation suggests that the organisation of processing is imitated, while the motor detail is clearly not; observational learning of this kind has been labelled 'program-level imitation' (Byrne 1993, 1995). Unfortunately, the consistency of technique is *so* great that there is little room for individual differences in style which might have enabled tracing copied routines from mother to offspring. Data from captive gorillas and chimpanzees (Custance *et al.* 1995, Hayes and Hayes 1952, Patterson and Linden 1981), and ex-captive orangutans (Russon and Galdikas 1993, 1995), strongly suggest that all great apes can imitate novel actions. It is therefore possible that the 'cultural' differences in chimpanzee tool-use techniques result from the role of apprenticeship leaning in skilled techniques (McGrew 1992, Parker 1996).

Although the idiosyncratic variation among gorillas, in the precise actions they use, implies that these details are learnt by individual experience, one observation suggests that imitation may occasionally play a part even at this level. One female, Picasso, born in a group living at lower altitudes than those at which *Laportea* nettles are a major food, immigrated into the study population. Yet five years later she could still not perform the folding of the prepared bundle of leaf-blades which minimises contact of stings with lips; she ate rather few nettles, perhaps as a consequence. As an adult, Picasso would not have been permitted to feed near enough to other adults to see the fine detail of their food processing. Intriguingly, her juvenile also lacked this behaviour, which was found in every one of the 36 other adults and juveniles. Under controlled conditions, evidence of imitation of this level of detail is controversial (Byrne and Tomasello 1995), although it is thought to underlie some experimental results (Tomasello *et al.* 1993b, Whiten *et al.* 1996).

Conclusions

Examination of the details of feeding techniques points to a difference between monkeys and apes in the complexity and sophistication of their naturally acquired, manual feeding procedures. Further, the difference is one that is likely a consequence of different mental mechanisms for organising and learning novel manual procedures.

Although monkeys use many distinct actions to cope with a wide range of foods, the actions are applied in an unsystematic, stimulus-driven way. When monkeys do learn techniques that superficially appear more elaborate, such as tool-use, their response to altered circumstances often suggests no underlying mental representation of the skill. Organisational complexity is in the eye of the beholder, and the beholder is human. On present evidence, learning by individual exploration – often, no doubt, aided by rather general social mechanisms, such as stimulus enhancement (Galef 1988, Spence 1937) and response facilitation (Byrne 1994) – remains a plausible explanation of the observations.

In various great apes, in contrast, there is evidence for greater sophistication. Tool-using chimpanzees show an understanding of objects and structures based on mental representation (remembering *that*, not just remembering *how*), and the ability to evoke these representations in anticipation to guide future behaviour, for example the specification of what is needed to make an efficient tool. Gorillas and orangutans build up novel, hierarchical organisations of behaviour (Byrne and Russon 1998), and give some evidence that they can learn the overall organisation of novel behaviour from others (Byrne and Byrne 1993, Russon 1996). These abilities also seem to be based on mental representation, in this case a representation of structured actions rather than of interacting objects. The consequences are that all great apes feed using methods that show technical complexity (Byrne 1997), and cognitively complex mechanisms of social learning, such as imitation, have a chance of real pay-off for efficient learning in these species. This evidently increases the range of food resources that great apes can exploit. Most strikingly, in the case of the mountain gorilla, it gives a freedom to occupy an ecological niche normally restricted to species with more specialised guts. In the Virunga Volcanoes, all the gorilla's food competitors are ruminating ungulates which rely on bacterial symbiosis to deal with problems for which gorillas use manual skill.

Several cognitively complex mechanisms of social learning are apparently available to great apes, but current evidence is too sparse for precision on when each is employed. Teaching is evidently little used, having been reported only in one species, for a task which is particularly hard to acquire. Program-level imitation (Byrne 1993, 1994), and in general the ability to organise actions into novel, hierarchical constructions (Byrne and Russon 1998), is capable of explaining other data on complex feeding techniques in apes. Together, they enable an individual to use the overall organisation of another's behaviour as a guide to construction of a novel, complex routine, the details of which may be fleshed out by individual experience. Although great apes are capable of imitative learning at a detailed, 'action-level' (Custance *et al.* 1995, Tomasello *et al.* 1993b), this seems to contribute little to the acquisition of natural feeding skills. However, individuals of species such as great apes, with the ability to understand ('mentally represent') the cause-and-effect of actions upon objects, may also be able to learn a good deal about how to deal with novel problems by attending to the object transformations in another's behaviour, rather than to the behaviour itself (Call and Tomasello 1995, Tomasello 1994, Tomasello *et al.* 1993a). This ability is perhaps best termed 'affordance learning by observation', to avoid confusion with the cognitively simple mechanism of goal emulation (see Byrne 1994, Whiten and Ham 1992).

The challenge for future research is to discover the precise contributions of associative learning, hierarchical construction, program-level imitation, and affordance learning, to the acquisition of the complex manual techniques of great apes.

Notes

1 But note that great apes' large body size and the inefficiency for long-distance travel of their means of locomotion – brachiation or clambering through trees, knuckle- or fist-walking on the

ground – may impose on apes a need for greater foraging efficiency than that of monkeys (Byrne 1997).

2 In some non-primate groups, maternal interventions are commoner, or even regular, in particular in cats and killer whales (Caro and Hauser 1992).

References

Anderson, J. R. (1996). Chimpanzees and capuchin monkeys: comparative cognition. In *Reaching into Thought. The Minds of the Great Apes*, ed. A. E. Russon, K. A. Bard and S. T. Parker, pp. 23–56. Cambridge: Cambridge University Press.

Boesch, C. (1991). Teaching among wild chimpanzees. *Anim. Behav.*, **41**, 530–2.

Boesch, C. and Boesch, H. (1983). Optimisation of nut-cracking with natural hammers by wild chimpanzees. *Behaviour*, **26**, 265–86.

Boesch, C. and Boesch, H. (1984). Mental map in wild chimpanzees: an analysis of hammer transports for nut cracking. *Primates*, **25**, 160–70.

Boesch, C. and Boesch, H. (1990). Tool use and tool making in wild chimpanzees. *Folia Primatol.*, **54**, 86–99.

Byrne, R. W. (1977). Planning meals: problem-solving on a real data-base. *Cognition*, **5**, 287–332.

Byrne, R. W. (1993). Hierarchical levels of imitation. Commentary on M Tomasello, A C Kruger and H H Ratner 'Cultural learning'. *Behav. Brain Sci.*, **16**, 516–17.

Byrne, R. W. (1994). The evolution of intelligence. In *Behaviour and Evolution*, ed. P. J. B. Slater and T. R. Halliday, pp. 223–65. Cambridge: Cambridge University Press.

Byrne, R. W. (1995). *The Thinking Ape: Evolutionary Origins of Intelligence*. Oxford: Oxford University Press.

Byrne, R. W. (1997). The technical intelligence hypothesis: an additional evolutionary stimulus to intelligence? In *Machiavellian Intelligence II: Extensions and Evaluations*, ed. A. Whiten and R. W. Byrne, pp. 289–311. Cambridge: Cambridge University Press.

Byrne, R. W. and Byrne, J. M. E. (1991). Hand preferences in the skilled gathering tasks of mountain gorillas (*Gorilla g. beringei*). *Cortex*, **27**, 521–46.

Byrne, R. W. and Byrne, J. M. E. (1993). Complex leaf-gathering skills of mountain gorillas (*Gorilla g. beringei*): variability and standarization. *Am. J. Primatol.*, **31**, 241–61.

Byrne, R. W. and Russon, A. (1998). Learning by imitation: a hierarchical approach. *Behav. Brain Sci.*, **21**(5), 667–721.

Byrne, R. W. and Tomasello, M. (1995). Do rats ape? *Anim. Behav.*, **50**, 1417–20.

Byrne, R. W. and Whiten, A. W. (1992). Cognitive evolution in primates: evidence from tactical deception. *Man*, **27**, 609–27.

Call, J. and Tomasello, M. (1995). The use of social information in the problem-solving of orangutans (*Pongo pygmaeus*) and human children (*Homo sapiens*). *J. Comp. Psychol.*, **109**, 308–20.

Caro, T. M. and Hauser, M. D. (1992). Is there teaching in non-human animals? *Q. Rev. Biol.*, **67**, 151–74.

Case, R. (1985). *Intellectual Development: Birth to Adulthood*. New York: Academic Press.

Chivers, D. J. and Hladik, C. M. (1984). Diet and gut morphology in primates. In *Food Acquisition and Processing in Primates*, ed. D. J. Chivers, B. A. Wood and A. Bilsborough, pp. 213–30. New York and London: Plenum Press.

Custance, D. M., Whiten, A. and Bard, K. A. (1995). Can young chimpanzees (*Pan troglodytes*) imitate arbitrary actions? Hayes and Hayes (1952) revisited. *Behaviour*, **132**, 11–12.

Fossey, D. (1979). Development of the mountain gorilla (*Gorilla gorilla beringei*): the first thirty-six months. In *The Great Apes*, ed. D. A. Hamburg and E. R. McCown, pp. 138–84. Menlo Park, CA: Benjamin/Cummings.

Fragaszy, D. M. and Visalberghi, E. (1989). Social influences on the aquisition and use of tools in tufted capuchin monkeys (*Cebus apella*). *J. Comp. Psychol.*, **103**, 159–70.

Galef, B. G. (1988). Imitation in

animals: history, definitions, and interpretation of data from the psychological laboratory. In *Social Learning: Psychological and Biological Perspectives*, ed. T. Zentall and B. G. Galef, Jr, pp. 3–28. Hillsdale, NJ: Erlbaum.

Garber, P. (1988). Foraging decisions during nectar feeding by tamarin monkeys (*Saguinus mystax* and *Saguinus fuscicollis*, Callitrichidae, Primates) in Amazonian Peru. *Biotropica*, **20**, 100–6.

Gibson, K. R. (1990). New perspectives on instincts and intelligence: brain size and the emergence of hierarchical mental construction skills. In *'Language' and Intelligence in Monkeys and Apes*, ed. S. T. Parker and K. R. Gibson, pp. 97–128. Cambridge: Cambridge University Press.

Gómez, J. C. (1996). Ostensive behaviour in great apes: the role of eye contact. In *Reaching into Thought. The Minds of the Great Apes*, ed. A. E. Russon, K. A. Bard and S. T. Parker, pp. 131–51. Cambridge: Cambridge University Press.

Goodall, J. v. L. (1973). Cultural elements in a chimpanzee community. In *Precultural Primate Behaviour*, ed. E. W. Menzel, pp. 144–84. Basel: Karger.

Goodall, J. (1986). *The Chimpanzees of Gombe: Patterns of Behavior*. Cambridge, MA: Harvard University Press.

Goodman, M., Bailey, W. J., Hayasaka, K., Stanhope, M. J., Slightom, J. and Czelusniak, J. (1994). Molecular evidence on primate phylogeny from DNA sequences. *Am. J. Phys. Anthropol.*, **94**, 3–24.

Greenfield, P. (1991). Language, tools and the brain: the ontogeny and phylogeny of hierarchically organized sequential behaviour. *Behav. Brain Sci.*, **14**, 531–95.

Guinet, C. and Bouvier, J. (1995). Development of intentional stranding hunting techniques in killer whale (*Orcinus orca*) calves at Crozet Archipelago. *Can. J. Zool.*, **73**, 27–33.

Harrison, K. (1996) Skills used in food processing by vervet monkeys, *Cercopithecus aethiops*. PhD thesis, University of St Andrews.

Hayes, K. J. and Hayes, C. (1952). Imitation in a home-raised chimpanzee. *J. Comp. Physiol. Psychol.*, **45**, 450–9.

Heyes, C. M. (1993). Imitation, culture and cognition. *Anim. Behav.*, **46**, 999–1010.

Hiraiwa-Hasegawa, M. (1986). A note on the ontogeny of feeding. In *The Chimpanzees of the Mahale Mountains*, ed. T. Nishida, pp. 277–83. Tokyo: University of Tokyo Press.

Kim, H.-S. and Takenaka, O. (1996). A comparison of TSPY genes from Y chromosomal DNA of the great apes and humans: sequence, evolution, and phylogeny. *Am. J. Phys. Anthropol.*, **100**, 301–9.

Mackinnon, J. (1978). *The Ape Within Us*. London: Collins.

McGrew, W. C. (1989). Why is ape tool use so confusing? In *Comparative Socioecology: The Behavioural Ecology of Humans and Other Mammals*, ed. V. Standen and R. A. Foley, pp. 457–72. Oxford: Blackwell Scientific Publications.

McGrew, W. C. (1992). *Chimpanzee Material Culture: Implications for Human Evolution*. Cambridge: Cambridge University Press.

McGrew, W. C. and Marchant, L. F. (1996). On which side of the apes? Ethological study of laterality of hand use. In *Great Ape Societies*, ed. W. C. McGrew, L. F. Marchant and T. Nishida, pp. 255–72. Cambridge: Cambridge University Press.

Miles, H. L., Mitchell, R. W. and Harper, S. E. (1996). Simon says: the development of imitation in an enculturated orangutan. In *Reaching into Thought. The Minds of the Great Apes*, ed. A. E. Russon, K. A. Bard and S. T. Parker, pp. 278–99. Cambridge: Cambridge University Press.

Miller, G. A., Galanter, E. and Pribram, K. (1960). *Plans and the Structure of Behaviour*. New York: Holt, Rinehart and Winston.

Milton, K. (1981). Distribution patterns of tropical plant foods as a stimulus to primate mental development. *Am. Anthropol.*, **83**, 534–48.

Napier, J. R. (1961). Prehensility and opposability in the hands of primates. *Symp. Zool. Soc. Lond.*, **5**, 115–32.

Newell, A., Shaw, J. C. and Simon, H. A. (1958). Elements of a theory of human problem solving. *Psychol. Rev.*, **65**, 151–66.

Nishida, T. and Uehara, S. (1983). Natural diet of chimpanzees (*Pan troglodytes schweinfur-*

thii): long-term record for the Mahale Mountains, Tanzania. *Afr. Stud. Monogr.*, **3**, 109–30.

Parker, S. T. (1996). Apprenticeship in tool-mediated extractive foraging: the origins of imitation, teaching and self-awareness in great apes. In *Reaching into Thought*, ed. A. Russon, K. Bard and S. T. Parker, pp. 348–70. Cambridge: Cambridge University Press.

Parker, S. T. and Gibson, K. R. (1977). Object manipulation, tool use, and sensorimotor intelligence as feeding adaptations in early hominids. *J. Hum. Evol.*, **6**, 623–41.

Patterson, F. and Linden, E. (1981). *The Education of Koko.* New York: Holt, Rinehart, and Linden.

Post, D. G. (1982). Feeding behaviour of yellow baboons in the Amboseli National Park, Kenya. *Int. J. Primatol.*, **3**, 403–30.

Russon, A. E. (1996). Imitation in everyday use: matching and rehearsal in the spontaneous imitation of rehabilitant orangutans (*Pongo pygmaeus*). In *Reaching into Thought: The Minds of the Great Apes*, ed. A. E. Russon, K. A. Bard and S. T. Parker, pp. 152–76. Cambridge, UK: Cambridge University Press.

Russon, A. (1998). The nature and evolution of intelligence in orangutans (*Pongo pygmaeus*). *Primates*, **3**, 485–503.

Russon, A. E., Bard, K. A. and Parker, S. T. (ed.) (1996). *Reaching into Thought: The Minds of the Great Apes.* Cam-

bridge: Cambridge University Press.

Russon, A. E. and Galdikas, B. M. F. (1993). Imitation in free-ranging rehabilitant orangutans. *J. Comp. Psychol.*, **107**, 147–61.

Russon, A. E. and Galdikas, B. M. F. (1995). Constraints on great ape imitation: model and action selectivity in rehabilitant orangutan (*Pongo pymaeus*) imitation. *J. Comp. Psychol.*, **109**, 5–17.

Sigg, H. and Stolba, A. (1981). Home range and daily march in a hamadryas baboon troop. *Folia Primatol.*, **36**, 40–75.

Spence, K. W. (1937). Experimental studies of learning and higher mental processes in infra-human primates. *Psychol. Bull.*, **34**, 806–50.

Sugiyama, Y., Fushimi, T., Sakura, O. and Matsuzawa, T. (1993). Hand preference and tool use in wild chimpanzees. *Primates*, **34**, 151–9.

Tomasello, M. (1994). The question of chimpanzee culture. In *Chimpanzee Cultures*, ed. R. W. Wrangham, W. C. McGrew, F. B. M. d. Waal and P. G. Heltne, pp. 301–17. Cambridge, MA: Harvard University Press.

Tomasello, M. (1996). Do apes ape? In *Social Learning in Animals: The Roots of Culture*, ed. C. M. Heyes and B. G. Galef, pp. 319–46. San Diego: Academic Press.

Tomasello, M., Kruger, A. C. and Ratner, H. H. (1993a). Cultural learning. *Behav. Brain Sci.*, **16**, 495–552.

Tomasello, M., Savage-Rumbaugh, E. S. and Kruger, A. C.

(1993b). Imitative learning of actions on objects by children, chimpanzees, and enculturated chimpanzees. *Child Devel.*, **64**, 1688–705.

Tutin, C. E. G., McGrew, W. C. and Baldwin, P. J. (1983). Social organisation of savanna-dwelling chimpanzees, *Pan troglodytes verus*, at Mt. Assirik, Senegal. *Primates*, **24**, 154–73.

van Schaik, C. P., Fox, E. A. and Sitompul, A. F. (1996). Manufacture and use of tools in wild Sumatran orangutans: implications for human evolution. *Naturwissenschaften*, **83**, 186–8.

Visalberghi, E. (1987). Acquisition of nut-cracking behaviour by two capuchin monkeys (*Cebus apella*). *Folia Primatol.*, **49**, 168–81.

Visalberghi, E. and Fragaszy, D. M. (1990). Do monkeys ape? In *'Language' and Intelligence in Monkeys and Apes*, ed. S. T. Parker and K. R. Gibson, pp. 247–73. Cambridge: Cambridge University Press.

Visalberghi, E. and Limongelli, L. (1994). Lack of comprehension of cause-effect relationships in tool-using capuchin monkeys (*Cebus apella*). *J. Comp. Psychol.*, **103**, 15–20.

Visalberghi, E. and Trinca, L. (1987). Tool use in capuchin monkeys: distinguishing between performing and understanding. *Primates*, **30**, 511–21.

Waddell, P. J. and Penny, D. (1996). Evolutionary trees of apes and humans from DNA sequences. In *Handbook of Symbolic Evolution*, ed. A. J. Lock and C. R. Peters, pp. 53–73. Oxford: Clarendon Press.

Waterman, P. G. (1984). Food acquisition and processing as a function of plant chemistry. In *Food Acquisition and Processing in Primates*, ed. D. J. Chivers, B. A. Wood and A. Bilsborough, pp. 177–211. New York and London: Plenum Press.

Waterman, P. G., Choo, G. M., Vedder, A. L. and Watts, D. (1983). Digestibility, digestion-inhibitators and nutrients and herbaceous foliage and green stems from an African montane flora and comparison with other tropical flora. *Oecologia*, **60**, 244–9.

Watts, D. P. (1984). Composition and variability of mountain gorilla diets in the central Virungas. *Am. J. Primatol.*, **7**, 323–56.

Watts, D. P. (1985). Observations on the ontogeny of feeding behaviour in mountain gorillas (*Gorilla gorilla beringei*). *Am. J. Primatol.*, **8**, 1–10.

Westergaard, C. G. and Fragaszy, D. (1987). The manufacture and use of tools by capuchin monkeys (*Cebus apella*). *J. Comp. Psychol.*, **101**, 159–68.

Westergaard, G. C. and Suomi, S. J. (1993). Use of a tool-set by capuchin monkeys (*Cebus apella*). *Primates*, **34**, 459–62.

Whiten, A. (1988). Acquisition of foraging techniques in infant olive baboons. *Int. J. Primatol.*, **8**, 469.

Whiten, A. and Ham, R. (1992). On the nature and evolution of imitation in the animal kingdom: reappraisal of a century of research. In *Advances in the Study of Behavior*, ed. P. J. B. Slater, J. S. Rosenblatt, C. Beer and M. Milinski, pp. 239–83. San Diego: Academic Press.

Whiten, A., Custance, D. M., Gomez, J-C., Teixidor, P. and Bard, K. A. (1996). Imitative learning of artificial fruit processing in children (*Homo sapiens*) and chimpanzees (*Pan troglodytes*). *J. Comp. Psychol.*, **110**, 3–14.

Wrangham, R. W. (1977). Feeding behaviour of chimpanzees in Gombe National Park, Tanzania. In *Primate Ecology*, ed. T. H. Clutton-Brock, pp. 503–38. New York: Academic Press.

19

Social transmission of facts and skills in the human species: neural mechanisms

Kathleen R. Gibson

The species *Homo sapiens* inhabits six continents and has experienced repeated population explosions that have caused the extinction of other species and that now threaten the ecology of the entire planet. Several factors contribute to human population explosions and reproductive success. First, humans are extremely adept at accumulating and sharing factual information pertaining to foraging and mating opportunities and to environmental hazards such as weather, predators and hostile human groups. Second, humans are the most technologically competent species on this planet. Technology permits individual humans to harvest far more food than they alone can consume, and, thus, provides the nutritional basis for human population explosions. It permits humans to inhabit temperate, arctic and desert habitats and to migrate across large bodies of water to places such as Australia and the South Pacific. Technology further facilitates human reproductive rates by enhancing human social cohesiveness and child-rearing strategies. It is, for example, only because we can make and use tools, that we can engage in group bonding activities, such as rituals, dance and sports, that use musical instruments, extracted pigments, costumes and athletic equipment. It is only because of carrying devices, such as slings and cradle-boards, that we can transport infants over long distances, and only because of our technical abilities to preprocess 'baby' foods that we can wean our infants at early ages in order to reproduce again.

This chapter reviews current understandings of human versus nonhuman primate brains in relationship to the mental and neurological capacities that facilitate the social transmission of factual information, tool-making and learned motor skills, such as tool-use, dance and ritualised motor behaviours. It suggests that human abilities to transmit facts and skills socially reflect the functioning of a number of neural areas, all of which are enlarged in humans in comparison with great apes and that the enlargement of the human brain necessitated changes in human developmental and social parameters which enhance human social learning opportunities.

Human versus nonhuman primate brains in relationship to learning and other mental skills

All monkeys, apes and humans have brains that, at least upon gross inspection, appear to be organised in a highly similar fashion. In particular, no neural structure is known to exist in humans which does not also exist in the brains of apes and monkeys. The only clearly demonstrated differences in neural organisation between human brains and those of other higher primates are quantitative in nature (Gibson 1996, Holloway 1996). The human brain, at about 1300 gm, is approximately three times the size of the average great ape brain, and it is many times the size of most monkey brains. In mammals, each unit increase in body size is accompanied by a 0.67 or 0.75 unit increase in brain size (Jerison 1973, Martin 1981). When viewed from the perspective of great ape versus human body sizes, the human brain enlargement is even more impressive. Indeed, measures that provide estimates of predicted brain size based on brain sizes of 'typical' mammals or insectivores with human body sizes suggest that relative to body size humans have the largest brain sizes of any mammals. One such measure is the Jerison's Encephalisation Quotient (EQ) (Jerison 1973).

Although humans exceed all other mammals in EQ and in similar measures of relative brain size, elephants and some cetaceans exceed humans in absolute brain size. For this reason, many investigators have assumed that EQ is a better predictor of intelligence and learning capacity than is absolute brain size. Few attempts, however, have been made to determine whether EQ actually correlates better with measures of learning or intelligence than does absolute brain size. Indeed, attempts to do so are thwarted by the widely divergent sensory and motor adaptations that characterise mammalian groups. For example, tests that rely primarily on olfactory capacity would be expected to favour canids and other highly olfactory species, while those that rely on echolocation capacity would be expected to favour bats and dolphins. The brains of widely divergent taxonomic groups, such as cetaceans, elephants, canids and primates, also differ profoundly in organisational parameters, making it difficult to distinguish between the behavioural effects of brain size versus those of brain organisation.

Nearly all monkeys, apes, and humans, however, have well-developed vision, tactile sense and manual dexterity (Clark 1960). These common sensorimotor adaptations render it possible to test different primate species on similar visually or tactilely based tasks, and, hence, to determine whether EQ or absolute brain size correlates with mental capacities in higher primates. Such studies indicate that absolute brain size, but not EQ, correlates with performance on learning tasks which measure mental flexibility (Rumbaugh *et al.* 1996, Beran *et al.* 1999). Unfortunately, no studies have attempted to determine whether EQ or absolute brain size correlates with any measures of social learning.

Comparisons of great ape and monkey brain sizes and behaviours, however, lend support to the view that absolute brain size is a better predictor of mental capacities, including some social learning capacities, than is relative brain size. Absolute brain size is larger in all great apes than in any monkey (data from Stephan *et al.* 1981). In contrast, Jerison's (1973) data indicate much overlap in EQs between monkey and great ape species. Hence, if EQ is the most appropriate measure of intelligence or learning skills, we should expect to find overlap between monkeys and great apes in mental abilities. If, on the other hand, absolute brain size is the better measure, we should expect to find a mental gap between monkeys and apes. Accumulating behavioural evidence indicates that such a gap does exist (Byrne 1995, chapter 18, this volume, Parker and Gibson 1990, Parker *et al.* 1994, Russon *et al.* 1996). Great apes perform better than do monkeys on tests of mirror recognition, numerical skills, theories of mind, imitation and tool use. Great apes, but not monkeys, have also managed to acquire minimal gestural and visually based 'language' skills (Gardner *et al.* 1989, Miles 1990, Savage-Rumbaugh and Lewin 1994).

As primate and human brains enlarged, some neural components enlarged proportionately more than others including the cerebellum, the neocortex, the basal ganglia, and the hippocampus and other portions of the limbic system (Stephan *et al.* 1981). Modern research suggests that each of these neural areas makes vital contributions to learning processes and suggests that the enlargement of each of them may have enhanced human learning and social learning skills.

The limbic system, for example, plays a critical role in the mediation of emotional and social behaviour in all mammals (MacLean 1982). Some components of this system, including the amygdala, also facilitate memory for emotionally laden events such as trauma (LeDoux 1994, Rogan *et al.* 1997). Other components function in the perception of pleasure and in the mediation of maternal behaviour. Damage to the amygdala produces monkeys who exhibit a syndrome of inept social behaviour called the Klüver–Bucy syndrome (Klüver and Bucy 1937). These monkeys aggress against the wrong animals, have intercourse with the wrong sex or species, and lose fear of snakes and laboratory personnel. Female monkeys with damage to the amygdala do not properly rear their young and may kill them. The limbic system, thus, contains the major neural structures mediating the learning of emotionally significant information and socially appropriate responses. The functional significance of limbic system enlargement in the human species is unclear. It is clear, however, that a well-functioning limbic system is critical to the abilities of humans and other mammals to attend to appropriate social stimuli and, hence, to the ability to learn from appropriate social sources.

Other enlarged neural circuits mediate declarative and procedural learning (Mishkin *et al.* 1984, Squire and Zola-Morgan 1988). Declarative (or explicit) learning refers to the learning of factual information that reaches consciousness. In humans, information acquired through declarative learning processes can be verbalised. Declarative learning is primarily mediated by a neural circuit involving the hippocampus and the frontal lobes of the neocortex. Procedural (or implicit) learning refers to the learning of routines and habits, which once mastered, are performed rapidly and automatically without conscious thought. New procedures are often acquired slowly through much repetition, but once acquired, they are rarely forgotten. Procedural learning is mediated by at least two circuits, one involving the basal ganglia and the premotor regions of the neocortex and the other involving the cerebellum and the neocortex.

Humans and animals can acquire many facts and procedures individually in the absence of social influences. Hence, neither declarative nor procedural learning can be considered forms of social learning. As discussed elsewhere in this volume, however, many mammals do transmit factual information or procedures through social channels, and humans routinely do so. In humans, socially shared knowledge serves as the foundation of culturally accumulated historical, scientific, mythological and other 'facts'. Socially shared procedures serve as the foundation of culturally specific tool-making, artistic and architectural techniques, speech, gesture, dance, ritual and sport.

Humans transmit factual information primarily via spoken, gestural and written languages. Hence, in humans, the declarative learning system can be said to have expanded into a declarative learning–language system. Procedural learning skills, however, are critical to this system, because learned procedures underlie all forms of language (Gibson and Jessee in press). Spoken languages, for example, require the performance of complex movements of the lips, tongue, palate and larynx automatically without conscious thought. Similarly, sign and written languages require the automatic, subconscious performance of complex manual movements. When communicating, we also automatically generate routine grammatical constructs, and, when writing, those who are highly literate spell many words automatically. Hence, one reason humans are adept at socially transmitting information is that they have excellent procedural learning skills.

In contrast to factual information, which in the human species is primarily transmitted via language, tool-making and other learned motor procedures are socially transmitted primarily via imitation and demonstration, with language generally playing only a supplementary role (Wynn 1993). Hence, in addition to possessing a declarative learning – language system for transmitting factual information, humans have what can be called a procedural learning – imitation – demonstration system for transmitting learned procedures.

To explain fully behaviours such as imitation, demonstration and language, which function primarily for the social transmission of skills or information, we must look beyond questions of simple procedural, declarative and emotional learning skills to the area in which, above all others, human abilities appear to exceed those of the apes: hierarchical mental constructional capacities. Specifically, humans exceed great apes in their abilities to hold multiple items of information in mind simultaneously, to construct or note mentally relationships between different items of information, and to embed constructed concepts or actions into still higher order constructs (Gibson 1983, 1988, 1990, 1996, Greenfield 1991, Reynolds 1983, 1993). These abilities allow humans to combine and recombine individual motor acts, sensory impressions or concepts in order to create novel, more complex motor or mental concepts.

One example of human mental construction is the construction of motor routines from smaller motor units. Such routines include complex gymnastic and dance sequences, ritualised gestural sequences, such as making the sign of the cross, the gestural routines essential to sign language and mime, and many forms of tool use such as typing, writing or playing a musical instrument. Other examples include the manufacture of items such as cordage, wooden shafts, stone points and scraped hides that can then be combined together to construct tools, such as stone-tipped spears, or shelters, such as teepees. Such constructions are flexible and hierarchical in that individual motor units or manufactured objects can serve as sub-units of varied higher constructs.

Increased mental constructional abilities impact on human social learning skills in fundamental ways. First, by providing flexibility and the ability to combine existing sensory, motor and conceptual schema into new composite constructs, mental construction allows humans to invent new tools, architecture, art, dances, gymnastic routines, games, sports, music and linguistic utterances. The combination of many objects, concepts or motor actions together into new constructs assures that many inventions require mental capacities far beyond those possessed by great apes and that many are so complex that two individuals working in isolation would be unlikely to chance upon them. Hence, social transmission is the most time and cost-effective mechanism for assuring that novel and useful constructs remain in a population.

Mental constructional processes also underlie several capacities essential for the social transmission of factual knowledge and procedural skills: imitation, language and the ability to understand the minds of others. For example, the imitation of visible motor routines, such as tool-using techniques or dance sequences, requires constructing mental relationships between the visual images of another individual's actions and the kinesthetic images of one's own body movements (Mitchell 1994). Similarly, the imitation of sounds requires constructing mental relationships between the sounds produced by another, the sounds produced by the self, and the kinesthetic image of the self's vocal tract. In addition to vocal imitation, human speech requires the construction of multiple simultaneous and sequential relationships between numerous fine motor movements of the tongue, lips, soft palate and larynx. To use the sounds of speech in a meaningful way requires constructing mental relationships between spoken words and objects, events or thoughts. Humans also regularly construct phrases from several words and then embed these phrases into sentences that are, in turn, embedded into longer discourses. Hence, human language is a constructive process in the sense that many diverse sounds, words and phrases are synthesised together to make larger communicative constructs. It is a flexible process in that individual words or phrases may be used as parts of many novel communicative constructs. It is a hierarchical process in that individual linguistic units are embedded within larger units that are, in turn, embedded in within still larger units. Although some great apes have mastered minimal aspects of gestural or other visual languages, all language-trained apes fall short of human children in their abilities to construct sentences and longer discourses, and several investigators have hypothesised that mental construction is the primary distinguishing feature between ape and human linguistic skills (Gibson 1983, 1988, 1990, Greenfield 1991, Reynolds 1983, 1993).

In order to communicate effectively, one must possess more than just language and imitative capacity, one must also comprehend the mental perspectives of others with whom one is communicating. This skill is especially critical for teaching and demonstration. One manifestation of perspective taking, often called theory of mind, develops in human children at about four years of age (Case 1985). A typical theory of mind test involves placing an object, such as a sweet, under or inside another object, such as a bag or cup, in full view of two children. One child then leaves the room, and the sweet is moved to a new location. Prior to about four years of age, the child who remains in the room assumes that the other child will expect the sweet to be in its new location, rather than in the location where the child initially observed it being placed. Case has argued that in human children the development of the mental skills needed to pass theory of mind tests rests on the emergence of sufficient information processing and mental constructional capacities to hold multiple items of information in mind simultaneously and to note the relationships between them. The philosopher Daniel Dennett explicitly argues for the existence of different hierarchical orders of theories of mind (Dennett 1988). A first order theory of mind might take the form of 'X believes that Y is hungry'. 'X wants Y to believe that X is hungry' would be a second order theory of mind. 'X wants Y to believe that X believes' would be a third order theory of mind. Hence, theory of mind, which is an essential component of effective human teaching and demonstration skills, depends on similar hierarchical mental constructional skills as those that serve as the foundation of other advanced human mental abilities.

Various considerations suggest that the mental constructional capacities which underlie complex human mental and motor routines, language, imitation and theory of mind relate to brain size and to increased neocortical information processing capacity (Gibson and Jessee in press). In particular, those areas of the neocortex that have experienced the greatest enlargement in human evolution, the frontal and parietal association areas, provide the working memory, planning and synthetic capacities essential for mental constructional behaviours. Lesions in these areas also often produce deficits in language, imitation and object use.

In summary, the human ability to transmit information and skills socially reflects the functioning of several neural regions that have enlarged in human evolution and that contribute diverse skills to the social learning process including capacities for procedural and declarative learning and for the mental construction essential for taking the perspective of others, imitation, language and tool-construction.

Relative brain size, dependence on technology and increased opportunities for social learning

As noted above, relative brain size correlates neither with learning capacity nor with performance on tests of mental ability. The relatively large human brain, however, indirectly contributes to human social learning capacities by demanding birthing, developmental, and rearing conditions that provide human infants and children with greater opportunities for the vertical, intergenerational transmission of information than those generally encountered by non-human primates.

At birth, the human brain has reached about 25% of its final adult size. In contrast, the neonatal brains of most monkeys are approximately 50% of final adult size and those of great apes are about 35 to 40% of final size (Portman 1967, Sacher and Staffeldt 1974). Despite the somewhat earlier birth of human neonates, the neonatal human head is sufficiently large in comparison with the maternal pelvic outlet that it sometimes presents problems during the birthing process that can result in death to the mother or brain damage to the infant. Hence, the early birth of human infants may be necessitated by the relative enlargement of the human brain (Leutenegger 1982). The extent to which the early birth may also directly contribute to human social learning processes is unclear, but it does provide human neonates with social contact with adults at an early highly plastic stage of central nervous system development (Gibson 1991).

At all stages of development, neural tissue is extremely metabolically demanding (Rumbaugh *et al.* 1996, Armstrong 1983, Leonard and Robertson 1992). In adult humans, the brain utilises 20% to 25% of all metabolic energy produced by the body. The metabolic demands of neural tissue may account, in part, for the strong correlations that exist between body size and brain size in all vertebrate classes and in all mammalian orders (Jerison 1973). If, for example, two species produce equal amounts of metabolic energy per unit of body tissue, the one with the larger body will be able to produce the energy needed to sustain a larger brain.

During prenatal, infantile and early childhood development, the metabolic demands of the human brain are even greater than in adults and may amount to approximately 50% of body's metabolic energy. Extreme nutrient deprivation during these periods of early growth can result in smaller brain sizes and lower IQs (Morgan and Gibson 1991). The extraordinary metabolic demands of growing human brains may be one factor that has selected for the prolonged period of human brain development (Gibson 1991, Foley and Lee 1991). Human brains, for example, are not fully myelinated until their late teenage years or later, whereas rhesus monkey brains are fully myelinated by about $3\frac{1}{2}$ years of age. By extending the growth period, growth-supporting nutrients can be obtained over a longer period of time.

In addition to being prolonged, human growth differs from that other primates in other critical parameters (Bogin 1997). Nonhuman primates generally manifest three periods of postnatal growth: infancy, which corresponds to the period of nursing and which usually ends with the eruption of the first molar, the juvenile period, which usually begins with weaning, and adulthood. During the juvenile period most primates continue to travel with their mothers or other group members, but juveniles provide most or all of their own food. In contrast, in most human hunting and gathering societies, children are weaned at about three years of age, but their first molars do not erupt until six or seven years of age. Hence, human children are weaned at an earlier developmental stage than other primates. The period between weaning and the eruption of the first molar is defined by students of human development as the period of childhood – a period that does not exist in nonhuman primates. During this period, human children require a diet that is very high in nutrient concentrations, but they are unable to provide any portion of that diet for themselves, and may have difficulties masticating tough foods, because their dentition is still immature. By age six or seven, human children have sufficient intellectual and emotional maturation that literate societies consider them ready for school, and many preliterate societies consider them ready for expanded duties such as care of younger children and help in the gathering of plant foods. Nonetheless, during this preadolescent period that corresponds to the juvenile stage in other primates, human juveniles remain dependent upon adult caretakers for most of their food. In humans, the juvenile period is followed by another period that is not formally recognised in nonhuman primates – an adolescent period of rapid growth. It is only during adolescence that males in most hunting and gathering societies become actively engaged in the hunting of big game. During this period, they usually travel with their fathers or other adult males until proving their own hunting proficiency.

Humans address these developmental challenges with both technological and social strategies. Since the human baby cannot cling or keep up with its parents on its own, all societies use slings, baskets, prams or other carrying devices to carry their young. Indeed, without such technology, the early human birth would scarcely be possible. Human adults, unlike apes, are also able to provision their children and juveniles, because human technology allows proficient hunters and gatherers regularly to acquire and transport more food than they can consume themselves. Thus, all hunting and gathering societies possess spears that permit hunting big game, digging sticks that facilitate the procurement of deeply buried, but nutrient dense, tubers and roots, and hammerstones for the cracking of nuts. Many also possess fishing weirs and techniques for felling birds. All manufacture containers that allow for the transport of larger quantities of food than can be carried in the hands alone. Human technology also allows humans to preprocess food by cooking or mashing it, thus rendering food more readily ingested by children who do not yet have their permanent teeth and creating the possibility of weaning children at an early developmental stage.

In addition to being aided by human technology, the metabolic demands of growing human brains are facilitated by a human social structure that differs profoundly from that of other primates. Whereas in most primate societies, the mother is the sole caretaker of the young, humans form long-term male–female pair bonds in which both parents assume responsibility for provisioning the young. Moreover, human females experience a definitive menopause and may survive for 20 to 30 years or more after losing their reproductive capacity. In hunting and gathering societies, postmenopausal women may help provide food and other care-taking activities for their grandchildren (Aiello 1996, Hawkes *et al.* 1997). Hence, human children and juveniles, unlike most juvenile primates, have three potential sources of food supply: their mothers, their fathers and their grandmothers. In hunting and gathering societies, the father and related males usually provide the meat of big game, while the mother and grandmother provide small game, vegetables and fruits.   These human growth and child-rearing patterns impact very significantly on human social learning potentials. The combination of early birth, long developmental trajectories, the joint care of the children by both parents and often by grandmothers as well, and the tendencies of human juveniles and adolescents to travel with adults assures that developing humans spend large amounts of time in proximity of more than one adult caretaker and have far more opportunity to acquire skills and information via vertical, intergenerational transmission than do developing great apes who are likely to acquire social information and skills primarily from their mothers (Shennan and Steele chapter 20, this volume). Hence, human growth and developmental patterns not only depend on human technological competence, child-rearing practices and social structure, they may also facilitate technological advancement by rendering the vertical transmission of skills and information more probable (see Shennan and Steele chapter 20, this volume for a discussion of the vertical transmission of crafts in human societies).

Natural selection and the enlargement of the human brain

Some scholars assume that the enlarged human brain represents an accidental by-product of the evolution of the prolonged human developmental trajectory; hence, brain enlargement has little functional significance (Gould 1977). This view ignores the implications of the metabolic demands of relatively large brains and the implications of the potentially fatal consequences to both mother and infant of a neonatal infant head that is too large for the birth canal. These factors would be expected to exert strong selective pressures against enlarged brains, unless overridden by other, stronger, selective forces in favour of them. All evidence indicates that the brains of human phylogenetic ancestors experienced dramatic enlargement from the period of about 2 500 000 to 250 000 years ago (Aiello 1996). Hence, strong selective pressures appear to have favoured the enhancement of abilities provided by brain size enlargement throughout a period of more than 2 000 000 years of human evolution.

The concept that the brain enlarged in response to selective pressures is supported by modern behavioural evidence indicating that both the declarative learning–language complex and the procedural learning–imitation–demonstration complex provide survival and reproductive advantages for modern hunter–gatherer populations. Hunter–gatherers routinely master and transmit to others, primarily by means of language, information about the anatomy, behaviour and availability of plant and animal foods and about the best ways of finding, securing and processing food. Those hunter–gatherers who migrate seasonally or who otherwise travel long distances in search of food or trade also gather and communicate information about shelters, food, water resources and weather conditions along travel routes. The communication of such information is a major factor in the human ability to exploit a far wider variety of food items than are generally exploited by ape populations and, often, to travel long distances over hazardous routes in order to obtain food.

Humans, of course, also transmit factual information about each other, i.e. they gossip. Humans live in larger groups than do apes (Aiello and Dunbar 1993, Dunbar 1992). Hence, gossip may be a form of social grooming that substitutes for the direct physical grooming possible in smaller ape and monkey societies (Dunbar 1996). Human societies are also characterised by a 'release from proximity' (Rodspeth *et al.* 1991) that permits them to maintain strong kin and social ties with individuals living in other groups. Human hunter–gatherers, for example, routinely seek marriage partners from outside their natal bands, frequently travel through terrain dominated by other human groups, and may engage in complex political alliances or warfare with other tribes or bands. Unlike apes, those humans who leave their natal home to reside elsewhere continue to maintain ties with parents, grandparents and other kin still living in the group in which they were born. Indeed, via language, humans may be aware of kin that they have never met (Quiatt and Reynolds 1993). Consequently, it would be nearly impossible for a single individual to acquire on his/her own all pertinent information about all other humans with whom he/she might need to interact. Gossip serves as a means of transmitting essential information about potential mates, distant kin and distant foes. Such knowledge is essential for the choice of mates and for understanding who to ask for help or shelter when travelling to distant areas in search of food or water.

Socially transmitted procedural skills, like socially transmitted facts, enhance both survival and reproduction in human societies. Humans in all hunter–gatherer cultures learn to make and use survival-enhancing tools such as foraging tools, weapons, containers for transporting food and infant carrying devices. In most societies, humans also learn to make clothing and shelter and to tend or make fire. In all societies, some individuals learn to make pigments, paints, musical instruments and costumes that can be used for artistic and ritual purposes. Finally, humans in all societies routinely master culturally approved styles of dance, music, sports, religious rituals and symbolic gestures. Rituals, dance and music serve as powerful social and emotional bonding mechanisms during feasts which occur when peoples who have separated are reunited – often on a seasonal basis in migratory hunter–gatherer groups (Rodspeth *et al.* 1991). Rituals also serve as group bonding mechanisms during initiation rites, weddings and funerals and during preparation for warfare or for dangerous hunts. Great apes do exhibit some ritualistic behaviour, especially when individuals who have separated reunite. Ritual, however, is much more integral part of human culture than it is of great ape culture. One can only speculate on why such bonding mechanisms are more important in humans than in other primates. However, the human tendencies to maintain social and kin ties with individuals they see only occasionally, to engage in group warfare and dangerous hunts, and to travel long distances through areas inhabited by others would appear to provide powerful reasons for humans to develop strong emotional bonds with other humans who may help them in time of need.

In sum, both the declarative learning–language complex and the procedural learning–imitation–demonstration complex contribute to human foraging and other survival-enhancing behaviours and also to human mating, bonding, child-rearing and other social behaviours. Consequently, it is not possible, simply by examining the behaviours of living peoples, to determine what behavioural context served as the primary selective force for the evolution of language, imitation and other abilities that enhance human capacities to transmit information and learned skills socially. What is clear, however, is that, the human technological niche was the underlying factor permitting the expansion of human societies, increased childhood dependency and the increased foraging capacity essential for the nourishment of enlarged human brains. Hence, technology provided the contexts that render the social transmission of skills and information valuable and essential to human societies.

Archaeological evidence indicates that the genus *Homo* has occupied a technological niche since its first appearance in the archaeological record about 1.9 million years ago (Harris and Capaldo 1993, Wolpoff 1996). Hence, conditions in which natural selection would be expected to have favoured increased social learning capacities could well have existed throughout this long period of human evolution (see Mithen chapter 21, this volume).

Conclusion

Human abilities to transmit information and skills socially reflect the functioning of a number of brain structures that have enlarged in human evolution, including the neocortex, cerebellum, basal ganglia and limbic system. These neurological structures provide our species with enhancements in two major learning complexes: (1) a declarative learning–language complex characterised by the mastery of factual knowledge and by the social transmission of such knowledge primarily via social, gestural, or written languages; (2) a procedural learning–imitation–demonstration complex characterised by the mastery of motor skills and other routine procedures and by the social transmission of procedures and skills primarily by imitation and demonstration. Enlarged human neural structures also provide our species with information processing and mental constructional capacities that permit individual humans to invent complex behavioural schemes that others would be unlikely to reinvent through trial and error and, hence, render the social transmission of skills and concepts cost-effective.

The brain size increases that mediated the evolution of human abilities were not accompanied by comparable increases in human body size. Hence, the human brain is extremely large in relationship to body size and places significant metabolic demands on the human body. Humans meet these metabolic demands with developmental and social processes that further enhance human social learning abilities by providing growing human children with greater opportunities for the vertical transmission of skills and facts than those encountered by great apes.

References

Aiello, L. (1996). Hominine preadaptations for language and cognition. In *Modelling the Early Human Mind*, ed. P. Mellars and K. Gibson, pp. 89–99. The McDonald Institute for Archaeological Research, University of Cambridge.

Aiello, L. C. (1998). The expensive tissue hypothesis and the evolution of the human adaptive niche: a study in comparative anatomy. In *Science in Archaeology: An Agenda for the Future*, ed. J. Bayley, pp. 25–36. London: English Heritage.

Aiello, L. and Dunbar, R. I. M. (1993). Neocortex size, group size, and the evolution of language. *Curr. Anthropol.*, **34**, 184–93.

Armstrong, E. (1983). Relative brain size in mammals. *Science*, **220**, 1302–4.

Beran, M. J., Gibson, K. R. and Rumbaugh, D. M. (1999). Predicted hominid performance on the transfer index: body size and cranial capacity as predictors of transfer ability. In *The Descent of Mind*, ed. M. Corballis and S. E. G. Lea, pp. 87–97. Oxford: Oxford University Press.

Bogin, B. (1997). Evolutionary hypotheses for human childhood. *Am. J. Phys. Anthropol.* supplement, **25**, 63–89.

Byrne, R. (1995). *The Thinking Ape*. Oxford: Oxford University Press.

Case, R. (1985). *Intellectual Development: Birth to Adulthood*. New York: Academic Press.

Clark, W. E. Le Gros. (1960). *The Antecedents of Man*. Chicago: Quadrangle Press.

Dennett, D. C. (1988). The intentional stance in theory and practice, In *Machiavellian Intelligence*, ed. R. Byrne and A. Whiten. pp. 180–201. Oxford: Oxford University Press.

Dunbar, R. I. M. (1992). Neocortex size as a constraint on group size in primates. *J. Hum. Evol.*, **22**, 469–93.

Dunbar, R. I. M. (1996). *Grooming, Gossip, and the Evolution of Language*. Cambridge, MA: Harvard University Press.

Foley, R. and Lee, P. (1991). Ecology and energetics of encephalization in hominid evolution. *Phil. Trans. R. Soc. Lond. B*, **334**, 223–32.

Gardner, R. A., Gardner, B. T. and Van Cantfort, T. E. (eds.)

(1989). *Teaching Sign Language to Chimpanzees*. Albany: State University of New York Press.

Gibson, K. R. (1983). Comparative neurobehavioral ontogeny: the constructionist perspective in the evolution of language, object manipulation and the brain. In *Glossogenetics*, ed. E. de Grolier. pp. 52–82. New York: Harwood Academic.

Gibson, K. R. (1988). Brain size and the evolution of language. In *The Genesis of Language: A Different Judgement of Evidence*, ed. M. Landsberg, pp. 149–72. Berlin: Mouton de Gruyter.

Gibson, K. R. (1990). New perspectives on instincts and intelligence: brain size and the emergence of hierarchical mental constructional skills. In *'Language' and Intelligence in Monkeys and Apes: Comparative Developmental Perspectives*, ed. S. T. Parker and K. R. Gibson, pp. 97–128. Cambridge: Cambridge University Press.

Gibson, K. R. (1991). Myelination and behavioral development: a comparative perspective on questions of neoteny, altriciality and intelligence. In *Brain Maturation and Cognitive Development: Comparative and Cross-cultural Perspectives*, ed. K. R. Gibson and A. C. Petersen, pp. 29–63. Hawthorne, NY: Aldine de Gruyter.

Gibson, K. R. (1996). The ontogeny and evolution of the brain, cognition and language. In *Handbook of Symbolic Evolution*, ed. A. Lock and C. Peters, pp. 407–32. Oxford: Oxford University Press.

Gibson, K. R. and Jessee, S. (in press). Language evolution and expansions of multiple neurological processing areas. In *The Evolution of Language: Assessing the Evidence from the Non-Human Primates*, ed. B. King. Santa Fe: School for American Research.

Gould, S. J. (1977). *Ontogeny and Phylogeny*. Cambridge, MA: Harvard University Press.

Greenfield, P. M. (1991). Language, tools and the brain: the development and evolution of hierarchically organized sequential behavior. *Behav. Brain Sci.*, **14**, 531–95.

Harris, J. W. K. and Capaldo, S. D. (1993). The earliest stone tools: their implications for an understanding of the activities and behaviour of late Pliocene hominids. In *The Use of Tools by Human and Non-human Primates*, ed. A. Berthelet and J. Chavaillon, pp. 196–200. Oxford: Clarendon Press.

Hawkes, K., O'Connell, J. F. and Blurton Jones, N. G. (1997). Hadza women's time allocation, offspring provisioning, and the evolution of postmenopausal lifespans. *Curr. Anthropol.*, **38**(4), 551–77.

Holloway, R. (1996). Evolution of the human brain. In *Handbook of Symbolic Evolution*. ed. A. Lock and C. Peters, pp. 74–116. Oxford: Oxford University Press.

Jerison, H. J. (1973). *Evolution of the Brain and Intelligence*. New York: Academic Press.

Klüver, H. and Bucy, P. (1937). 'Psychic blindness' and other symptoms following bilateral temporal lobectomy in Rhesus monkeys. *Am. J. Physiol. 246 (Regulatory Integrative Comparative Physiology)*, **119**, 352–3.

LeDoux, J. E. (1994). Emotion, memory and the brain. *Sci. Am.*, 50–7.

Leonard, W. R. and Robertson, M. L. (1992). Nutritional requirements and human evolution. *Am. J. Hum. Biol.*, **4**, 179–95.

Leutenegger, W. (1982). Encephalization and obstetrics in primates with particular reference to human evolution. In *Primate Brain Evolution: Methods and Concepts*, ed. E. Armstrong and D. Falk, pp. 85–95. New York: Plenum Press.

MacLean, P. D. (1982). On the origin and progressive evolution of the triune brain. In *Primate Brain Evolution: Methods and Concepts*, ed. E. Armstrong and D. Falk, pp. 291–316. New York: Plenum Press.

Martin, R. D. (1981). Relative brain size and its relation to the rate of metabolism in mammals. *Nature*, **393**, 57–60.

Miles, H. L. W. (1990). The cognitive foundations for reference in a signing orangutan. In *'Language' and Intelligence in Monkeys and Apes: Comparative Developmental Perspectives*, ed. S. T. Parker and K. R. Gibson, pp. 511–39. Cambridge: Cambridge University Press.

Mishkin, M., Malamut, B. and Bachavelier, J. (1984). Memories and habits: two neural systems. In *Neurobiology of Learning and Memory*, ed. G. Lynch,

J. L. McGaugh and N. M. Weinberger, pp. 65–77. New York: Guilford Press.

Mitchell, R. (1994). The evolution of primate cognition: simulation, self-knowledge, and knowledge of other's minds. In *Hominid Culture in Primate Perspective*, ed. D. Quiatt and J. Itani, pp. 177–232. Niwot, CO: University Press of Colorado.

Morgan, B. and Gibson, K. R. (1991). Nutritional and environmental interactions in brain development. In *Brain Maturation and Cognitive Development: Comparative and Cross-cultural Perspectives*, ed. K. R. Gibson and A. C. Petersen, pp. 91–106. Hawthorne, NY: Aldine de Gruyter.

Parker, S. T. and Gibson, K. R. (eds.) (1990). *'Language and Intelligence in Monkeys and Apes: Comparative Developmental Perspectives*. Cambridge: Cambridge University Press.

Parker, S. T., Mitchell, R. W. and Boccia, M. L. (eds.) (1994). *Self-awareness in Animals and Humans: Developmental Perspectives*. Cambridge: Cambridge University Press.

Quiatt, D. and Reynolds, V. (1993). *Primate Behaviour: Information, Social Knowledge, and the Evolution of Human Culture*. Cambridge: Cambridge University Press.

Portman, A. (1967). *Zoologie aus vier Jahrzehnten*. Munich: Piper and Verlag.

Reynolds, P. C. (1983). Ape constructional ability and the origin of linguistic structure. In *Glossogenetics: The Origin and*

Evolution of Language, ed. E. de Grolier, pp. 185–200. New York: Harwood Academic Publishers.

Reynolds, P. C. (1993). The complementation theory of language and tool use. In *Tools, Language and Cognition in Human Evolution*, ed. K. R. Gibson and T. Ingold, pp. 407–45. Cambridge: Cambridge University Press.

Rodspeth, L., Wrangham, R. W., Harrigan, A. M. and Smuts, B. (1991). The human community as a primate society. *Curr. Anthropol.*, **32**, 221–49.

Rogan, T., Staubli, U. V. and LeDoux, J. E. (1997). Fear conditioning induces associative long-term potentiation in the amygdala. *Nature*, **390**, 604–7.

Rumbaugh, D. S., Savage-Rumbaugh, E. S. and Washburn, D. A. (1996). Toward a new outlook on primate learning and behavior: complex learning and emergent processes in comparative perspective. *Jap. Psychol. Res.*, **38**, 113–25.

Russon, A., Bard, K. and Parker, S. T. (eds.) (1996). *Reaching Into Thought: The Minds of Great Apes*. Cambridge: Cambridge University Press.

Sacher, G. A. and Staffeldt, E. F. (1974). Relation of gestation time to brain weight for placental mammals: implications for the theory of vertebrate growth. *Am. Nat.*, **108**, 593–615.

Savage-Rumbaugh, S. and Lewin, R. (1994). *Kanzi: The Ape at the Brink of the Human Mind*. New York: John Wiley and Sons.

Squire, L. R. and Zola-Morgan, S.

(1988). Memory: brain systems and behavior. *Trends Neurosci.*, **11**, 170–5.

Stephan, H., Frahm, H. and Baron, G. (1981). New and revised data on volumes of brain structures in insectivores and primates. *Folia Primatol.*, **35**, 1–39.

Wolpoff, M. (1996). *Human Evolution.* New York: McGraw Hill.

Wynn, T. (1993). Layers of thinking in tool behavior. In *Tools, Language and Cognition in Human Evolution.* ed. K. R. Gibson and T. Ingold, pp. 388–406. Cambridge: Cambridge University Press.

20

Cultural learning in hominids: a behavioural ecological approach

Stephen J. Shennan and James Steele

A model of the costs and benefits of social transmission of craft skills

Humans, of all mammals, display the greatest dependence on social learning for the acquisition of skills and of social and environmental knowledge. We are capable not only of transmitting traditions by imitation and teaching, but also of storing socially learned information culturally in linguistically encoded forms such as stories, and in durable external symbolic information stores such as artistic representations and writing (Donald 1991). It is simpler, when analysing such processes, to restrict the analysis to only a part of this range. Our focus here will be on the evolution of human craft skills, and we shall focus in particular on stone tool making skills. This is not just a reflection of the constraints of the archaeological record of hominid behaviour (which is, to be sure, inordinately biased toward craft skills in durable media, notably stone tools). Transmission of craft skills is also an appropriate area in which to look for continuities and discontinuities with the social learning of other mammals, since it takes place largely through nonlinguistic channels, by social facilitation, imitation and active teaching, and seems to involve a process more of implicit learning through training of motor habits than of linguistic transmission of abstract plans.

To analyse hominid social learning in a comparative framework, we must answer three questions. The first question is ecological: what kinds of factors might have led to selective pressures in favour of social learning? The second question is social: who are the parties in an episode of social information transfer, and how is their participation influenced by other properties of their relationship? The third question is cognitive: what is the mechanism of social learning (see Gibson, chapter 19, this volume)? It is the first two questions that we intend to address here. The thrust of our argument is that the social transmission of craft skills represents a form of parental investment in offspring, and that we have to consider the costs and benefits to both parties if we are to understand how it evolved.

Our primary data are stone tools, and so we must ask how much they can tell us about the overall complexity and the learning costs of social learning processes in a given social group. Given the preservation bias of the archaeological record towards stone tools, it is highly relevant here to note that the closest analogy available from animal social learning studies – the stone tool use for nut-cracking of our nearest living relatives, the chimpanzees – has been variously described as 'probably the most demanding manipulatory technique yet known to be performed by wild chimpanzees' (Boesch 1993: 174), and as perhaps 'the most complex tool use by wild chimpanzees because a set of two detached objects is needed' (Matsuzawa 1996: 199). Acquisition of this technique at the level of proficiency seen in adult chimpanzees may take almost ten years, with successful use of stone tools by juveniles beginning at the age of three years, and a possible 'critical period' for learning the skill in the age range of three to five years (Matsuzawa 1996: 201). While acknowledging the difficulty of discriminating between different social learning processes in uncontrolled field observation settings, Boesch proposes that the acquisition of nut-cracking behaviours by infant Taï Forest chimpanzees entails processes of true imitation guided by frequent maternal intervention – processes which represent the highest order of social learning seen in wild chimpanzees (Boesch 1993: 179–80). By extension, we might expect that hominid stone tool related behavioural traditions which demonstrably evolved in complexity and efficiency within the span of existence of a given hominid species, and the implementation of which involved many complex, serially ordered procedures that would not readily have been rediscovered in the trial-and-error phase of a socially facilitated learning episode, were also the product of true imitative learning, most likely aided by active teaching, and that they also indicate the highest order of social learning process which occurred in the social groups of their makers and users.

If learning such technological traditions was difficult and time-consuming, then we must ask how the costs were balanced by compensating benefits. It is now well established that social learning will be beneficial when individual learning is costly and/or difficult, and when environments are reasonably slow to change, so that the experience of those transmitting the behaviour in question is still relevant to those who are learning it (cf. Sibly, chapter 4, this volume). However, this generalised statement does not take the economic calculus far enough. We need to specify the overall benefits and costs of particular episodes of social learning both to the learner and the demonstrator or model. To the demonstrator or model the costs will include time spent in demonstrating or otherwise intervening (where this is intentional and targeted at a specific learner), and the potential costs of transmitting the behaviour to another in the social group if that means diluting any associated frequency-dependent advantages. To the learner, these costs include time budgeted for acquiring the behaviour, and costs associated with the risk that the behaviour may prove maladaptive. It may be that these costs will be affected in contrasting ways by the level of difficulty of the novel behaviour, since while the complexity of the behaviour modelled will correlate with the time taken to acquire it by a learner, more complex behaviours may (by the same token) be the hardest to 'steal' by a potential competitor in an initially undetected episode of copying. Overall, if a behaviour is either difficult to acquire, or gives frequency-dependent advantages, we propose that transaction costs would be minimised for both demonstrator and learner if the trait is transmitted when the learner is at a developmental stage where (s)he is most receptive to such learning.

As for benefits, they must outweigh the costs for both the model and the learner. This issue is discussed by Sibly (chapter 4, this volume), who points out that for a given benefit and a given cost, learners are selected to demand more help than transmitters are selected to supply (although it seems likely that in the case of the complex skills discussed in this paper the time-costs of learning are far greater than those of transmitting). However, as Sibly (chapter 4, this volume) also points out, any evaluation of the fitness benefits and costs to the parties concerned must take into account their genetic relatedness. Two-way benefits can certainly be seen to occur in cases of social transmission from parents to offspring. Boesch (1993) points out that in the Taï chimpanzee community, sharing nuts (a rich nutrient source) with their offspring commits mothers for the first eight years of the youngster's life. The efforts of the mothers to accelerate the rate of acquisition of nut-cracking behaviours in their young may reflect the importance of this transfer of knowledge in freeing up maternal investment for further offspring, in addition to the benefit of improving the survival chances of that particular offspring, for example if the mother dies young. Accelerated social transfer of this skill is thus beneficial to both parties.

Figure 20.1 illustrates a hypothetical example, in which age-specific learning rates for skills accelerate to reach a plateau, and then decline. Let us assume that time-taken-to-learn is the principal cost of social transmission for both parties involved, and let us predict that for individuals of a given species there will be an absolute ceiling for the maximum cost that can be sustained in a social transmission episode. We might then expect that ceteris paribus, the more difficult the behaviour is to learn, the more important it will be both that it should be acquired at the lowest point on the age-specific cost curve, and that the attempt should be made to shift the whole cost curve downwards by switching to a more efficient social learning mechanism (or by simplifying the behaviour to reach the same goal).

For social transfer of information to be adaptive the net benefits for both

Figure 20.1. Time taken to learn a skill (a proxy for units of fitness) as a function of learner's age at onset of acquisition, for five hypothetical skills of differing degrees of difficulty.

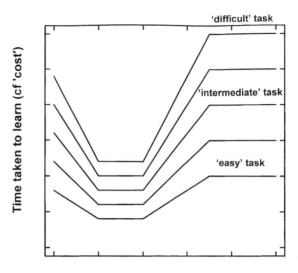

Age at onset of acquisition

parties must also outweigh the costs (Figure 20.2). Let us suppose that in a given case, the net benefits decline linearly with the age of acquisition of the behaviour by the learner (since, for example, there will be a net reduction in lifetime maternal investment available to all possible offspring, for every year that one offspring continues to monopolise that resource) (Figure 20.3). Integrating the data from the curve (Figure 20.3) describing net benefits from social transmission of a skill (here, specific to the age of the *learner*) with the data from the curve describing net costs (Figure 20.1), we can derive a set of values for net costs and net benefits to both parties of social transmission of any skill, specific to the age at onset of learning it, for a given social learning mechanism (Figure 20.4). The optimal age for onset of acquisition of a behaviour will be that point of inflection of the curve at which the rate of decline of age-specific learning costs (in units of fitness) begins to be exceeded by the rate of decline in net fitness benefits of acquisition.

This model is very schematic: obviously, variation in any of the parameters affecting the calculation of costs and benefits will also affect estimation of viability of transmission and of optimal age at onset of acquisition. There is, in particular, enormous scope for work on the effects of age on ease of acquisition of complex motor skills, if we are to understand the relative importance of gradual, cumulative maturation and of possible 'critical periods'. The curve in our hypothetical example is only one possibility among many – and even there, we have not given specific age values to the abscissa. However, we think that

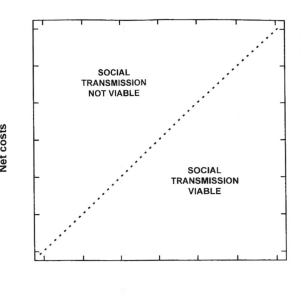

Figure 20.2. Viability of social transmission as a function of relative costs and benefits (measured in units of fitness).

Net costs

Net benefits

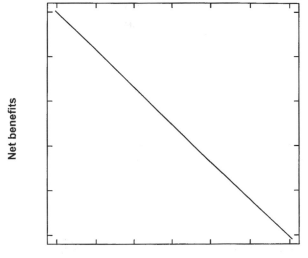

Figure 20.3. Hypothetical relationship between benefits of social transmission of a skill (in units of fitness), and learner's age at onset of acquisition (in years).

Net benefits

Age at onset of acquisition

Figure 20.4. Learner's age-specific cost and benefit curves (in units of fitness) for five hypothetical skills of differing degrees of difficulty. The optimum learner's age for onset of acquisition of a skill is that point on the curve which is below, and furthest from the dashed line (the line where costs equal benefits). Here, this point – (B/C)$_{max}$ – is shown for the 'difficult' skill (cf. Figure 20.1).

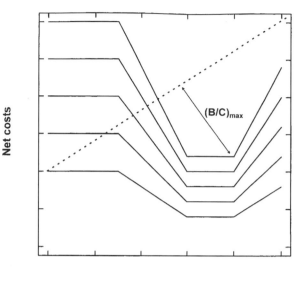

the general model captures some of the issues that need to be considered in a behavioural ecological account of the transmission of complex motor skills of the type involved in such crafts as the production of stone tools. These could not be passed on horizontally(from and to individuals within the same generation) to any great extent because they must be transferred from experienced to inexperienced individuals and they take a long time to learn. Nor is it likely that they would be transmitted vertically (from one generation to the next) outside the parent–offspring relationship, since there would be little benefit if any to the transmitter to outweigh the costs (exceptions would be socially learned behaviours whose increased frequency in a group increases the average fitness of all members of the group, as opposed to simply increasing the differential fitness (*vis-à-vis* their fellows) of those who have learned them.) (again, cf. Sibly, chapter 4, this volume). But not only would we expect such behaviour to be transmitted primarily via the parent–offspring channel. If we assume that the adaptive nature of such transmission is conditional on benefits outweighing the costs for both demonstrator and learner, then we might also expect such transmission of the more costly skills to involve primarily types of behaviour which enhance the ability of the learner to gain independent access to key feeding resources.

Social learning and Palaeolithic knapping traditions

Estimating benefits of Palaeolithic knapping traditions

According to our model, the costs of transmitting complex craft traditions must be compensated for by a matching level of benefits. By analogy with chimpanzee nut-cracking behaviours we have suggested that such benefits would most likely relate to the ability of the learner to gain independent access to key feeding resources (cf. Parker and Gibson 1979, Gibson 1990). The functions of early stone tools appear to conform to these predictions. Function can be inferred from the context in which stone tools are found, from traces of wear on their edges which may be indicative of use, and from experimental use of replica tools in different tasks to assess their range of efficiency. While it is certain that these early stone tools served many other functions which we have yet to identify, it is also clear that they were used to gain access to the key feeding resources of animal carcasses. To take the case of handaxes, tools which appear to involve some of the most complex manufacturing skills seen in the surviving culture of Middle Pleistocene hominids, we find them archaeologically associated with carcass remains (as at Boxgrove, in southern England, dating to c. 450 kyr BP – Pitts and Roberts 1997); use wear traces on their edges have been identified as consistent with meat processing (Keeley 1977); and experimental use of replicas in butchery tasks suggests that they are ergonomically efficient in extended butchering episodes (Jones 1981).

Estimating general teaching and learning costs of Palaeolithic knapping traditions

We cannot directly observe social transfer of skills and the mechanisms by which it took place two million years ago, nor indeed in much later hominid societies, but we can try to quantify the complexity of the sequence of actions required to produce stone tools of the different types found in the archaeological record, and associated with this or that fossil hominid species. This can be done experimentally, by replicating the tools, and measuring the difficulty involved in terms of motor skill and cognitive complexity; such experiments can be reinforced by rebuilding the stone cores from the discarded waste flakes which often survive on Palaeolithic sites, and reconstructing the decision processes and motor skills of the original knapper (see Figure 20.5). Analysis of shape and size variability in assemblages of Palaeolithic artefacts can give us complementary information about the degree to which these decision processes were constrained by a narrow range of values for the tool's design parameters.

Such experimental work has been used to estimate that the earliest pebble

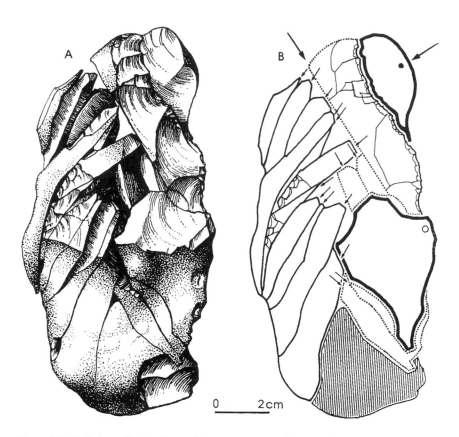

Figure 20.5. Refitting utilised and waste flakes to reconstruct the original stone core tells us about the decision processes of the ancient tool makers as they knapped their tools – the figure shows such a refitted core, from a site occupied by modern humans in northern France towards the end of the last glacial (from Karlin *et al.* 1992: 1112, Figure 3).

tool 'chopper' of the Oldowan industry can be made by a modern human adult with but a few minutes of practice (Pelegrin 1991; see Figure 20.6 and Mithen, chapter 21) of course, the learning costs will have been much more severe for the less encephalised and less dextrous hominids, either early *Homo* or robust Australopithecines, who originally made these tools. The developed handaxe industries of the Middle Pleistocene, associated with hominids of a more encephalised grade, show far greater standardisation of form, and a more complex knapping sequence involving two distinct types of flaking instrument (hard and soft hammers). Their manufacture seems to demand an order of magnitude more 'know-how' (Pelegrin 1993). Finally, to acquire full competence in production of the highly standardised blades of the Upper Palaeolithic Magdalenian industry, the product of modern human hunter–gatherers in late

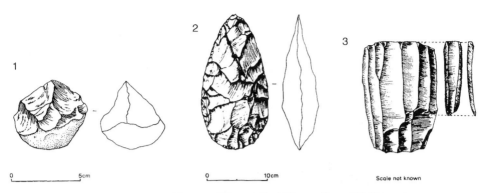

Figure 20.6. Stone tool types: (1) an Oldowan pebble tool, Koobi Fora, *c.* 2 myr BP; (2) an Acheulian biface or handaxe, Bed II, Olduvai Gorge, *c.* 1.5 myr BP; (3) an Upper Palaeolithic core and blade, Europe, *c.* 0.04 myr BP. After Schick and Toth (1993: 113, 233, 296).

glacial northern Europe about 15 000 years ago, would have taken hundreds of hours of training (Pelegrin 1993). Thus, through time there would have been an increasing investment in the acquisition of craft skills by both learners and teachers.

Inferring transmission mechanisms of Palaeolithic knapping traditions

Finding empirical evidence of transmission mechanisms for hominid craft traditions in the archaeological record is somewhat harder than inferring the functions of the stone tools which were produced. Indeed, we can cite only a single example. A detailed reconstruction of social learning of stone tool making skills has been proposed by archaeologists excavating two human hunter–gatherer camps in France, each dating from much more recent times – the late last glacial about 15 000 years ago. At Pincevent and at Etiolles, the various scatters of stone tool making debris around the hearths indicate the activities of different individuals in a small group, individuals with very different levels of competence (Pigeot 1990, Ploux 1991, Karlin *et al.* 1992, 1993). Fitting back together the various pieces of debris from their actions has enabled archaeologists reconstructing their decision sequences and motor skills to detect these varying skill levels (cf. Figure 20.5), and has shown that while the highly skilled knappers occupied places closest to the hearth, the less skilled knappers and the novices sat further back from it. These sites are interpreted as the remains of camps, each occupied by a group of the size of a family unit. In one case at Pincevent, a scatter of debris from an inexperienced knapper is mixed with that from a much more skilled knapper's 'demonstration' episode: a small core was entirely reduced to produce stone blades, which the archaeologist describes as a genuine educational demonstration, or 'here's how it's done' (Karlin *et al.* 1992: 1113).

We infer two key points for our argument from the situation described. First, that we have clear evidence of active teaching, as we would expect. Second, we know from ethnography that small hunter–gatherer groups are made up of close kin, so it is reasonable to infer that the teachers and learners in this case were closely related.

Modern ethnographic data on craft learning and the appearance of institutions of apprenticeship

However, until we can develop more widely applicable archaeological methods for estimating social transmission routes, the supposition that social transmission of hominid craft skills in the Pleistocene was largely from parent to offspring during childhood must remain just that – a supposition. Nevertheless, modern ethnographic information provides considerable support for this model. Table 20.1 reproduces summary information from Hewlett and Cavalli-Sforza's examination of cultural transmission among Aka foragers in the tropical forest of central Africa (1986).

It is apparent that the great majority of skills are learned from the parents (in fact, the same sex parent in most cases) and that the learning process is largely complete by age 10 and almost entirely so by age 15 years (Hewlett and Cavalli-Sforza's definitions of childhood and adolescence respectively).

Table 20.2 summarises a range of ethnographic information concerning the learning of craft skills. In all cases the transmission is mainly if not entirely vertical rather than horizontal. In the vast majority of cases where the information is available the transmission is from parent to offspring and specifically to same sex offspring. In virtually all cases where information is available the skill concerned takes quite a long time to learn and is transmitted in the course of childhood. Where master–apprentice transmission occurs it is similar to the parent–offspring case except that the individuals concerned are unrelated. This usefully emphasises the fact that skills transmission is usually a form of parental investment. Where the transmitter is not a parent the issue of the cost of the training process to the teacher/master has to be explicitly addressed by the parties involved, as the apprenticeship literature shows (e.g. Coy 1989,

Table 20.1. Column 1: parental contribution to transmission of skill and knowledge among the Aka foragers of Central Africa. Columns 2–7: percentages of individuals who have acquired the skill by a given age; child, by age 10; adolescent, by age 15; adult, by age > 15

Skill	Adults: % with parent transmitter	♂ Child (% with trait)	♂ Adolescent (% with trait)	♂ Adult (% with trait)	♀ Child (% with trait)	♀ Adolescent (% with trait)	♀ Adult (% with trait)
Net hunt	84.5	75	96	99	54	59	59
Other hunt	70.7	28	63	66	25	25	25
Food gathering	89.3	99	100	98	90	90	90
Food preparation	76.3	38	54	47	94	94	100
Maintenance	86.5	48	75	87	78	80	83
Infant care	85.6	36	70	82	69	84	86
Mating	77.5	0	94	100	13	100	100
Sharing	83.9	100	100	90	100	100	90
Special skills	58.2	88	100	100	100	100	98
Dance or sing	54.9	100	94	86	91	94	80
All traits	80.7						

Data from Hewlett and Cavalli-Sforza (1986).

Table 20.2. Compilation of information from ethnographic sources relating to the social transmission of traditional craft skills

Skill transferred	Age; length of time to learn	Transmitter	Transmittee	Transmission type	Place	Social context	Source
Making arrowheads	? but from age 3; ?.	Father	Son	Vertical (related)	Yamana (Tierra del Fuego)	Hunter–gatherer	Gusinde (1961)
Making bows and arrows	? but from age 5; ?.	Grandfather	Grandson	Vertical (related)	Comanche	?Hunter–gatherer	Wallace and Hoebel (1952)
Variety of different skills	By late childhood (age 10)/ adolescence; ?.	Parent (Father Mother) (81%)	Offspring (son daughter)	Vertical (related)	Aka, Central African Republic	Hunter–gatherer	Hewlett and Cavalli-Sforza (1986)
Basic string-bag making	? but age 2–7; ?.	Mother (mainly)	Daughter	Vertical (related)	New Guinea	Agricultural 'tribal' society for own use	MacKenzie (1991)
Complex string-bag making	?; ?.	Older woman	Younger woman	Vertical (including unrelated)	New Guinea	Agricultural 'tribal' society for own and men's use	MacKenzie (1991)

Glass working	?; ?.	Father (76%); other kin (7%); nonkin (17%)	Offspring (son)	Vertical (related)	Maya	Modern peasant communities	Hayden and Cannon (1984)
Construction wood mfg	?; ?.	Father (60%); other kin (17%); nonkin (23%)	Offspring (son)	Vertical (related)	Maya	Modern peasant communities	Hayden and Cannon (1984)
Fibre working	?; ?.	Father (64%); other kin (15%); nonkin (21%)	Offspring (son)	Vertical (related)	Maya	Modern peasant communities	Hayden and Cannon (1984)
Bone working	?; ?.	Father (73%); other kin (8%); nonkin (19%)	Offspring (son)	Vertical (related)	Maya	Modern peasant communities	Hayden and Cannon (1984)
Hunting	?; ?.	Father (64%); other kin (13%); nonkin (23%)	Offspring (son)	Vertical (related)	Maya	Modern peasant communities	Hayden and Cannon (1984)
Pottery making	?; ?.	Mother (49%); other kin/in-laws (29%); nonkin (22%)	Offspring (daughter)	Vertical (related)	Maya	Modern peasant communities	Hayden and Cannon (1984)
Weaving	?; ?.	Mother (65%); other kin (21%); nonkin (14%)	Offspring (daughter)	Vertical (related)	Maya	Modern peasant communities	Hayden and Cannon (1984)

Table 20.2. (cont.)

Skill transferred	Age; length of time to learn	Transmitter	Transmittee	Transmission type	Place	Social context	Source
Basket making	?; ?.	Mother (50%); other kin (18%); nonkin (32%)	Offspring (daughter)	Vertical (related)	Maya	Modern peasant communities	Hayden and Cannon (1984)
Sewing	?; ?.	Mother (82%); other kin (7%); nonkin (11%)	Offspring (daughter)	Vertical (related)	Maya	Modern peasant communities	Hayden and Cannon (1984)
Broom making	?; ?.	Father (85%); other kin (5%); nonkin (10%)	Offspring (son)	Vertical (related)	Maya	Modern peasant communities	Hayden & Cannon (1984)
Pot-making using hand and anvil	Age 35; 1 month.	Sister-in-law	Sister-in-law	Horizontal (related)	South Africa	Modern peasant society	Krause (1985: chapter 3)
Pottery-making using hand and anvil	?; ?.	Mother	Daughter	Vertical (related)	South Africa	Modern peasant society	Krause (1985: chapter 3)

Activity	Age/duration	Teacher	Learner	Transmission	Location	Society	Reference
Pottery-making (coil method)	As child; ?.	Mother	Daughter	Vertical (related)	South Africa	Modern peasant society	Krause (1985: chapter 4)
Pottery-making (coil method)	Age c.33; ?.	Mother	Daughter	Vertical (related)	South Africa	Modern peasant society	Krause (1985: chapter 5)
Specialist pottery making	During childhood; ?.	Mainly parent (sometimes older sibling)	Offspring	Vertical (related)	Andean South America	Modern peasant society	Hosler (1996)
Pottery decoration	Age 5–15 years; c. 10 years.	Mother (usually)	Daughter	Vertical (related)	Shipibo-Conibo, Peru	'Tribal'/peasant society, not strongly specialist	DeBoer (1990)
Complex bead knapping (full-time)	?; > 7 years.	Master	Apprentice	Vertical (unrelated)	India	Specialist craft in society with complex division of labour	Roux et al. (1995)
Basic bead knapping (full-time)	?; > 2 years.	Master	Apprentice	Vertical (unrelated)	India	Specialist craft in society with complex division of labour	Roux et al. (1995)

Table 20.2. (*cont.*)

Skill transferred	Age; length of time to learn	Transmitter	Transmittee	Transmission type	Place	Social context	Source
Complex bead knapping (ancient)	?; ?.	?	?	?	India	Harappan civilisation	Roux *et al.* (1995)
Specialist pottery production	?; ?.	Father	Son	Vertical (related)	India	Specialist craft in society with complex division of labour	Mahias (1993)
Wheel-thrown pots	?; up to 10 years.	Master	Apprentice	?	India	Craft in society with complex division of labour	Roux (1990)
Coil-built pots	?; up to 1 year.	?	?	?	India	Craft in society with complex division of labour	Roux (1990)

Wheel-made pottery	?; ?10 years.	?Master	?Apprentice	Vertical (unrelated)	India	Craft in society with complex division of labour	Roux and Corbetta (1990), in van der Leeuw and Papousek (1992)
Pottery mould making	?; ?.	Father	Son	Vertical (related)	Mexico	Craft in society with complex division of labour	van der Leeuw and Papousek (1992)
Blacksmithing and other traditional crafts	As child; ?.	Father	Son	Vertical (related)	Yoruba Nigeria	Craft lineages in a traditional urban chiefly society	Lloyd (1953)
Modern crafts	Age 15–20; ?.	Master	Apprentice	Vertical (unrelated)	Yoruba Nigeria	Modern crafts in traditional urban chiefly society	Lloyd (1953)

Lloyd 1953). Such apprenticeships arise when the costs and benefits in the social economy shift so that it is perceived to be worthwhile to learn a new skill rather than imitate one's parent, for example with an expanding division of labour (Goody 1989). The point is well made by the last two entries in Table 20.2, concerning craft traditions in Yoruba towns in the early 1950s; the traditional crafts were passed on from fathers to sons whereas the new skills appearing because of modernisation required the setting up of master–apprentice relations (Lloyd 1953).

The one definite exception in the table to the general rule of childhood learning and vertical transmission – pottery making using hand and anvil by a woman in South Africa – is in fact a confirmation of the model because the learning process was a relatively simple one which only took a month.

Interestingly enough, the table shows that this pattern of vertical, especially parent to same sex offspring, transmission of complex skills is prevalent in agrarian societies, where there is at least some evidence that parental influence and vertical transmission may have been of decreasing importance in other areas of life as a result of increasing peer group interaction among children. A study by Draper and Cashdan (1988) looked at the impact of the adoption of a sedentary way of life on the interactions between parents and children among !Kung foragers in the Kalahari desert. They found that children were increasingly involved in interactions with peers rather than adults because the new structure of work in sedentary communities was more demanding and time-consuming for adults and kept parents out of the village for longer periods. Peer interaction of this kind provides increasing opportunities for horizontal transmission of information and practices among children, a very different mode of transmission (cf. Cavalli-Sforza and Feldman 1981, Harris 1995), but apparently not one relevant to the transmission of craft skills.

Summary and implications of the model for processes of cultural evolution

Our model has proposed that the acquisition of craft skills involved benefits and costs both to the learners and to those transmitting them. The costs to the latter involved the time spent modifying their behaviour in order to make learning easier and would most probably have increased as the complexity of the skills required increased, although this may have been compensated, at least to some extent, by increasing encephalisation. The most obvious context in which the balance of fitness benefits in relation to the costs of transmitting skills is favourable is where the experienced and inexperienced individuals are closely related. Given the time taken to learn craft skills these individuals are likely to be of different generations and, of course, the best cost–benefit relation will be if they are parent and offspring. Benefits to both sides will be greatest if individuals start learning as young as possible. There is some evidence for this model of parent–offspring transmission in the learning of skills by chimpanzees and also in the ethnographic literature on the learning of craft skills, although very little direct archaeological evidence as yet.

However, if we follow the chimpanzee model, we would expect the default early hominid pattern to have been primarily mother–infant transmission of tool-making and using behaviours (see also Maestripieri 1995). The modern human pattern known ethnographically, on the other hand, is one of transmission from parent to same sex offspring. This shift links the evolution of craft skills and technology to hominid social organisation not in the way discussed by Mithen (1994) but in connection with the appearance of male parental investment in offspring, especially at the post-weaning stage. If, as seems likely, increased paternal investment emerged as a hominid strategy at a particular point in human evolution, then we would expect it to have been associated with an increase in father–male offspring transmission (Takahasi and Aoki 1995), which could have had an effect on the stone tool record and might have been linked to the emergence of greater differentiation of activities between males and females, and even a division of labour. Before this time all such skills would have been transmitted along the female line, to both male and female offspring.

These issues have recently been discussed by Aiello (in press) in the context of the evolution of diet and brain size. She suggests that the marked increase in brain size which occurred after 0.5 million years ago would have put increased stress on both adult females and post-weaning juveniles, and that the resulting increased infant mortality would have provided a selective pressure in favour of increased paternal investment. She also notes that the first definite evidence for big-game hunting occurs at this time and suggests that this might be an indication of a separate male sphere of activity, which would have involved the transfer of skills from fathers to sons. In the light of our model these ideas point to a new and important set of questions with which to return to the examination of the lithic archaeological record.

A further consequence arises from the fact that our model points to parent–offspring transmission as the strongly predominant route for the transfer of skills. This implies that patterns in skills transmission would have been closely linked with patterns in genetic transmission and the spread of innovations would have been associated with the spread of individuals belonging to specific genetic lineages. Such a suggestion is supported at the global scale by a recent analysis of the association between specific lithic 'industries' and specific hominid species as early hominids gradually colonised large parts of the world (Larick and Ciochon 1996).

More generally, the link postulated between cultural and genetic transmission means that the survival and success of particular techniques and skills would have resulted from three processes: the success of particular hominid lineages whose possession of a particular set of skills had given some selective advantage; an incidental link between the presence of particular skills in a particular lineage and the reproductive success of that lineage for some other selective reason; and finally, the processes of cultural drift and stochastic extinction of specific local lineages always associated with small populations.

Acknowledgements

We are grateful to the following colleagues for useful discussions of some of the ideas developed in this paper: Leslie Aiello, Clive Gamble, Paul Graves-Brown, Rob Hosfield and Tim Sluckin. Kathryn Knowles kindly redrew some of the figures. An earlier version of this paper was read at a University College, London Anthropology Seminar, and we are grateful to members of the audience for their comments and suggestions. The comments of the editors on an earlier draft contributed greatly to its clarification and simplification during revision.

References

Boesch, C. (1993). Aspects of transmission of tool-use in wild chimpanzees. In *Tools, Language and Cognition in Human Evolution*, ed. K. R. Gibson and T. Ingold, pp. 171–83. Cambridge: Cambridge University Press.

Cavalli-Sforza, L. L. and Feldman, M. W. (1981). *Cultural Transmission and Evolution*. Princeton: Princeton University Press.

Coy, M. W. (ed.) (1989). *Apprenticeship*. New York: SUNY Press.

DeBoer, W. (1990). Interaction, imitation and communication as expressed in style: the Ucayali experience. In *The Uses of Style in Archaeology*, ed. M. Conkey and C. Hastorf, pp. 82–104. Cambridge: Cambridge University Press.

Donald, M. (1991). *Origins of the Modern Mind*. Cambridge, MA: Harvard University Press.

Draper, P. and E. Cashdan (1988). Technological change and child behavior among the !Kung. *Ethnology*, **27**, 339–65.

Gibson, K. R. (1990) Cognition, brain size and the extraction of embedded food resources. In *Primate Ontogeny, Cognition and Social Behaviour*, ed. J. G. Else and P. C. Lee, pp. 93–103. Cambridge: Cambridge University Press.

Goody, E. (1989). Learning, apprenticeship and the division of labor. In *Apprenticeship*, ed. M. W. Coy, pp. 233–94. New York: SUNY Press.

Gusinde, M. (1961). *The Yamana:*

The Life and Thought of the Water Nomads of Cape Horn. New Haven, NJ: Human Relations Area Files.

Harris, J. R. (1995). Where is the child's environment? A group socialization theory of development. *Psychol. Rev.*, **102**, 458–89.

Hayden, B. and Cannon, A. (1984). Interaction inferences in archaeology and learning frameworks of the Maya. *J. Anthropol. Archaeol.*, **3**, 325–67.

Hewlett, B. S. and Cavalli-Sforza, L. L. (1986). Cultural transmission among Aka pygmies. *Am. Anthropol.*, **88**, 922–34.

Hosler, D. (1996). Technical choices, social categories and meaning among the Andean potters of Las Animas. *J. Mat. Cult.*, **1**, 63–92.

Jones, P. R. (1981). Experimental butchery with modern stone tools and its relevance for Palaeolithic archaeology. *World Archaeol.*, **12**, 153–65.

Karlin, C., Pigeot, N. and Ploux, S. (1992). L'ethnologie préhistorique. *La Recherche*, **23**, 1106–16.

Karlin, C., Ploux, S., Bodu, P. and Pigeot, N. (1993). Some socioeconomic aspects of the knapping process among groups of hunter–gatherers in the Paris Basin region. In *The Use of Tools by Human and Non-human Primates*, ed. A. Berthelet and J. Chavaillon, pp. 318–37. Oxford: Clarendon Press.

Keeley, L. (1977). The functions of Paleolithic stone tools. *Sci. Am.*, **237**, 108–26.

Krause, R. (1985). *The Clay Sleeps.* Alabama: University of Alabama Press.

Larick, R. and Ciochon, R. L. (1996). The African emergence and early Asian dispersals of the genus Homo. *Am. Sci.*, **84**, 538–51.

Leeuw, S. E. van der and Papousek, D. A. (1992). Tradition and innovation. In *Ethnoarchaeologie: justification, problemes, limites (XIIe Rencontres Internationales d'Archeologie et d'Histoire d'Antibes)*, pp. 135–58. Juan-les-Pins: Editions APDCA.

Lloyd, P. (1953). Craft organisation in Yoruba towns. *Africa*, **23**, 30–44.

MacKenzie, M. A. (1991). *Androgynous Objects: String Bags and Gender in Central New Guinea.* Chur: Harwood Academic Publishers.

Maestripieri, D. (1995). Maternal encouragement in nonhuman primates and the question of animal teaching. *Hum. Nat.*, **6**, 361–78.

Mahias, M. C. (1993). Pottery techniques in India. In *Technological Choices*, ed. P. Lemonnier, pp. 157–80. London: Routledge.

Matsuzawa, T. (1996). Chimpanzee intelligence in nature and in captivity: isomorphism of symbol use and tool use. In *Great Ape Societies*, ed. W. C. McGrew, L. F. Marchant and T. Nishida, pp. 196–209. Cambridge: Cambridge University Press.

Mithen, S. (1994). Technology and society during the Middle Pleistocene. *Cambr. Archaeol. J.*, **4**, 3–33.

Parker, S. T. and Gibson, K. R. (1979). A developmental model for the evolution of language and intelligence in early hominids. *Behav. Brain Sci.*, **2**, 367–407.

Pelegrin, J. (1991). Les savoir-faire: une très longue histoire. *Terrain*, **16**, 106–13.

Pelegrin, J. (1993). A framework for analysing prehistoric stone tool manufacture and a tentative application to some early stone industries. In *The Use of Tools by Human and Non-human Primates*, ed. A. Berthelet and J. Chavaillon, pp. 302–14. Oxford: Clarendon Press.

Pigeot, N. (1990). Technical and social actors: flintknapping specialists and apprentices at Magdalenian Etiolles. *Archaeol. Rev. Cambr.*, **9**, 126–41.

Pitts, M. and Roberts, M. (1997) *Fairweather Eden*. London: Century.

Ploux, S. (1991). Technologie, technicité, techniciens: methode de determination d'auteurs et comportements techniques individuels. In *25 Ans d'Etudes Technologiques en Préhistoire*, pp. 201–14. Juan-les-Pins: Editions APCA.

Roux, V. (1990). The psychological analysis of technical activities: a contribution to the study of craft specialisation. *Archaeol. Rev. Cambr.*, **9**, 142–53.

Roux, V. and Corbetta, D. (1990). Wheel throwing technique and craft specialisation. In *The Potter's Wheel: Craft Specialisation and Technical Competence*, ed. V. Roux and D. Corbetta, pp. 3–92. New Delhi: Oxford and IBH Publishing Co.

Roux, V., Bril, B. and Dietrich, G. (1995). Skills and learning difficulties involved in stone knapping: the case of stone-bead knapping in Khambat, India. *World Archaeol.*, **27**, 63–87.

Schick, K. D. and Toth, N. (1993). *Making Silent Stones Speak.* London: Weidenfeld and Nicholson.

Takahasi, K. and Aoki, K. (1995). Two-locus haploid and diploid models for the coevolution of cultural transmission and paternal care. *Am. Nat.*, **146**, 651–84.

Wallace, E. and Hoebel, E. A. (1952). *The Comanches: Lords of the High Plains.* Norman: University of Oklahoma Press.

21

Imitation and cultural change: a view from the Stone Age, with specific reference to the manufacture of handaxes

Steven Mithen

The last decade has seen a substantial quantity of literature addressing the issue of social learning with a particular focus on the relationship between social learning and cultural behaviour (e.g. Tomasello *et al.* 1993, Boyd and Richerson 1996, Heyes and Galef 1996). One of the major themes within this research has been the comparison of social learning between apes and humans with a view to understanding the cause of the immense differences in the extent of cultural behaviour between the two groups. There has, however, been limited reference in this literature to the nature of social leaning of pre-modern humans, as reconstructed from the fossil and archaeological records. In this chapter I will argue that such evidence is critical to our understanding of the relationship between social learning and cultural behaviour. To do so I will consider the behaviour of hominids that lived between 1.5 million and 250 000 years ago. This period is traditionally described as the Lower Palaeolithic. It is one during which handaxes are a dominant artefact of the archaeological record. Indeed it is the evidence from handaxes that I wish to draw upon in considering how social learning and cultural behaviour are related.

I will use this evidence to challenge the view put forward by Boyd and Richerson (1996) regarding why cultural evolution is so rare among animals, even though cultural traditions seem widespread. They attribute this to the absence of imitation within the repertoire of social learning, a view also endorsed by Tomasello *et al.* (1993). The evidence from the Lower Palaeolithic challenges this explanation. When we consider the tool making activities of pre-modern humans we find strong evidence that (1) imitation was present, and (2) cultural evolution was absent. To present this evidence, I must first briefly introduce what species of hominids we are dealing with and the nature of their stone tool technology.

Early humans and the Acheulian industry: a brief introduction

The fossil record for human evolution is notoriously difficult to decipher in terms of which fossils belong to which species, how many species existed, and the nature of their phylogenetic relationships (Johanson and Edgar 1996). Trying to attribute artefacts from the archaeological record to particular species poses even greater challenges. During the period of interest we are dealing with at least three different species of *Homo*, all of which appear to have manufactured – in certain times and places – those artefacts which are described as handaxes. These species are *Homo ergaster*, the term now used for many African fossils which had once been described as *Homo erectus* (Johanson and Edgar 1996), and best represented by the 1.5 million year old skeleton of the Nariokotome boy (Walker and Leakey 1993); *Homo erectus* itself, which is most likely a descendant of *Homo ergaster* and may be restricted to Asia, surviving up until just 30 000 years ago (Swisher *et al.* 1996); and *Homo heidelbergensis*, another likely descendant of *Homo ergaster* in Africa and which appears to have been the first type of human to colonise Europe, with the earliest remains found at Atapuerca dating to 780 000 years old (Carbonell *et al.* 1995; although some attribute at least one of these early specimens to another species, *Homo antecessor*, Bermúdez de Castro *et al.* 1997).

While these species differed with regard to their specific morphology they shared several important characteristics, notably an efficient bipedal gait and a relatively large brain size, especially after 500 000 years ago. In addition, with regard to their behaviour they were all reliant on stone tools, and most probably those made of wood, such as the recently discovered spears from Schöningen (Thieme 1997). They lived by hunting, gatherering and scavenging, with sites such as Boxgrove providing particularly detailed pictures of their subsistence activities (Roberts 1986). Similarly, they all appear to have lacked certain traits that are common to modern humans, such as burials and art objects (Mithen 1996). Due to these similarities, these species are often grouped together and referred to as Early Humans, as they appear to represent a grade of *Homo* different to both the earliest hominids, such as *Homo habilis*, and anatomically modern humans.

There are many archaeological sites associated with Early Humans found in Africa, Asia and Europe. Most of these are collections of stone artefacts, often in highly disturbed contexts, and frequently having been secondarily redeposited, such as in river gravels. In only rare circumstances are faunal remains found, or artefacts located *in situ* – such as at Boxgrove. One type of stone artefacts that is pervasive within this archaeological record of Early Humans is the handaxe, first appearing in East Africa 1.4 million years (Asfaw *et al.* 1992) ago, and still being made by some of the last Neanderthals in Europe just 50 000 years ago in the Mousterian of Acheulian Tradition (Mellars 1996). Handaxes were made from a range of raw materials, including basalt, chert and flint, by the bifacial knapping of either a nodule or a large flake (Inzian *et al.* 1992). They are particularly characterised by an imposed, symmetrical form usually resulting in either an ovate or a pear shaped artefact.

Assemblages which have a substantial number of handaxes are referred to as part of the Acheulian industry, although this term has limited meaning in terms of past behaviour. At many Acheulian sites, handaxes are found in vast numbers, scattered randomly over old occupation surfaces, such as at Gadeb and Olorgesailie in Africa, or Boxgrove in England, or within river gravels, such as in the Thames gravels at Swanscombe in Kent or the gravels of the Somme in France (for reviews of the Acheulian see Roe 1981, Wymer 1983, and Schick and Toth 1993). Quite why handaxes were made in such numbers, and why they were predominantly made in such a highly symmetrical fashion remain unanswered questions. My own view is that these imply that handaxes were not simply functional artefacts with regard to interacting with the natural environment, but were partly items for display (Kohn and Mithen in press).

Handaxes were not the only artefacts made by Early Humans. At many sites, handaxes are lacking, or extremely rare, and instead a rather simpler technology consisting of cores and flakes is found, not dissimilar to the Oldowan industry of East Africa, which is dated as early as 2.5 million years ago. Some particularly important sites where this core/flake technology predominates are found in S. E. England, such as at Clacton and Barnham, for which the term Clactonian industry was once used (Wymer 1974; for a recent review see Roberts *et al.* 1995). Equivalent assemblages in France are referred to as the Tayacian. Quite why Early Humans sometimes made handaxes and sometimes relied on a core/flake technology is unclear, although this may be associated with prevailing environmental conditions and social organisation (Mithen 1994).

Social learning and cultural behaviour

The relevance of this archaeological record for our understanding of the relationship between social learning and cultural behaviour can be appreciated after a brief summary of some recent arguments concerning this issue, notably those which have made comparisons between the social learning abilities of humans and chimpanzees (e.g. Hayes and Hayes 1952, Tomasello *et al.* 1993, Whiten and Custance 1996, Tomasello 1996). Such comparisons are of considerable value due to the short phylogenetic distance, but massive behavioural gulf, between these species. Several features of chimpanzee behaviour, notably patterns of tool use, are under-determined by ecological factors and have been described as cultural traditions (Nishida 1987, Boesch and Boesch 1990, McGrew 1992, Wrangham *et al.* 1994). Such cultural traditions, however, differ from those of modern humans in several respects, most notably their lack of development through time (Tomasello 1996). It is only among humans that cultural traditions build upon each other, as in increasing technological complexity or the efficiency of design. Tomasello *et al.* (1993) have referred to this as the 'ratchet effect', and asked why it is absent among chimpanzee cultural traditions.

Boyd and Richerson (1996) asked the same question in the form of 'Why is cultural evolution so rare?' They noted that cultural traditions are present in many animals but only in one species, humans, is cumulative cultural change significant. This is a puzzle, they argue, since cumulative culture change appears to be such a powerful means of adaptation: it lies at the route of human success with regard to global colonisation and ever increasing population numbers. So if cumulative cultural change appears so highly adaptive, and if many animals possess cultural traditions, why is cultural evolution so rare?

The solution Boyd and Richerson propose is that cumulative cultural change is caused by observational learning, which they equate with true imitation (although one must note that imitation has proved notoriously difficult to define (see Galef 1988, Visalberghi and Fragaszy 1990, Russon 1996, Whiten 1996). Following Tomasello *et al.* (1993), Boyd and Richerson claim that imitation is only found within the human species. Other forms of social learning, such as local enhancement, may enable cultural traditions to develop but because these simply draw the attention of a novice to a particular form of behaviour, which must then be learnt from scratch, they are not sufficient for cumulative cultural change. For this to occur the novice must be able to build upon the novel behaviour by observational learning, rather than having to learn the behaviour from scratch. This is a similar argument to that forwarded by Visalberghi (1993), who argued that the co-occurence of tool use and imitation has a snowball effect on cultural behaviour.

The inadequacy of such arguments, however, can be immediately seen when we expand the number of species under consideration from two – the chimpanzee and modern human – to at least three, by including Early Humans. If we return to the Acheulian industry we can see strong evidence that although observational learning/imitation was present, cumulative cultural change – the ratchet or snowball effect – was absent.

The manufacture of handaxes: evidence for imitation

One of the most significant features of handaxes is that they are technically demanding to manufacture, especially those which possess a high degree of symmetry. This has become apparent from evidence from both experimental replication and the re-fitting of knapping debris from archaeological sites (e.g. Bergman and Roberts 1988, Wenban-Smith 1989). To make a fine symmetrical handaxe a range of different knapping actions is required (Pelegrin 1993, Schick and Toth 1993). At first, relatively large cortical flakes must be removed, requiring use of a dense stone, or hard hammer. When the approximate shape has been created, other types of flakes need to be removed, notably thinning flakes which travel across the surface of the artefact and are struck with use of an antler, bone or wooden hammer at quite different angles and with different degrees of force to those initial hard hammer removals. To remove these, preparatory flakes may need to be detached to create a striking platform. And throughout the manufacturing process, the edge of the artefact may need to be slightly ground to remove irregularities that might deflect the force of the strike. There is, therefore, a wide range of knapping actions that must be undertaken: hammerstone stones of appropriate density, shape and size must be acquired, prepared and selected; the artefact must be struck at precisely the correct angle and with the correct amount of force; flakes must be removed to prepare the way for later removals.

In light of this it seems a very strong possibility that the technical skills to manufacture handaxes must have been acquired by observational learning; individual trial and error learning from scratch – which may have sufficed for a core/flake technology – seems unlikely to have been adequate. Whether this observational learning involved teaching, especially that using spoken language, is unclear. There is certainly no evidence that this is the case. Such evidence might be spatially associated piles of knapping debris reflecting differing degrees of knapping skill deriving from an experienced and novice tool maker working side by side. We certainly find such evidence for modern humans, such as at the 11 000 year old sites of Etiolles in France (Pigeot 1990) and Trollesgrave in Denmark (Fisher 1990). But even on the best preserved Early Human sites such as Boxgrove there is no clear evidence for such social interaction in the transmission of technical skills.

Nevertheless, observational learning seems a necessity for acquiring the technical skills to manufacture handaxes. Indeed, although imitation may prove difficult to define, it seems the most appropriate term to use, as both the knapping actions and the final form of the objects, the goal, were copied. This is most evident from the remarkably high degrees of similarity in form between handaxes, whether one is comparing artefacts in assemblages from broad geographical areas (Wynn and Tierson 1990) or artefacts from within a single assemblage (e.g. Tyldesly 1986). Indeed, within many assemblages almost identical handaxes are found. This fact is particularly important. The nodules from which handaxes were made would have been highly variable. A unique sequence of decisions and knapping actions would have been required to 'extract' an artefact of a specific size and shape. Hence this close similarity in form provides further evidence for the high degrees of technical skill and observational learning employed by Early Humans.

The absence of cumulative cultural change

While observational learning/imitation appears to have been a central element in the social learning repertoire of Early Humans, cumulative cultural change – the 'ratchet' or 'snowball' effect – was not present. As I remarked above, handaxes are found in the archaeological record for more than 1 million years. During that time there is no significant change in their range of forms, their manufacturing techniques or the way they were used (Wynn 1995). The only technological change during this period was the introduction of the levallois technique – a means of removing flakes of predetermined sizes from cores (Inzian *et al.* 1992; and see Dibble and Bar-Yosef 1995 for detailed descriptions). But once present, the levallois technique did not lead to any cumulative cultural change. As several archaeologists have commented, it is the remarkable stability in technology, and culture in general, during the Middle Pleistocene which is in most need of explanation, not the presence of change (e.g. Isaac 1977, Binford 1989, Mithen 1996).

This absence of cumulative cultural change is evident within the detailed archaeological records of individual regions. In southern England, for instance, which provides one of the most studied and best understood archaeological sequences (Roebroeks 1996), there is no chronological trend in either artefact form, the level of technical skill or the refinement of artefacts (Briggs *et al.* 1985, Cook and Ashton 1991). The very earliest artefacts, such as those found at Boxgrove and High Lodge 500 000 years ago, are comparable in technical skill and form to those found at sites such as the Upper sequence at Hoxne, 200 000 years later. The simpler core/flake technology, as dominates at sites such as Clacton and Barnham, is interspersed throughout the chronological sequence with no evident pattern (Roberts *et al.* 1995).

Figure 21.1 illustrates this technological stasis by illustrating three handaxes from different stages of the British Quaternary, all made by a similar technique

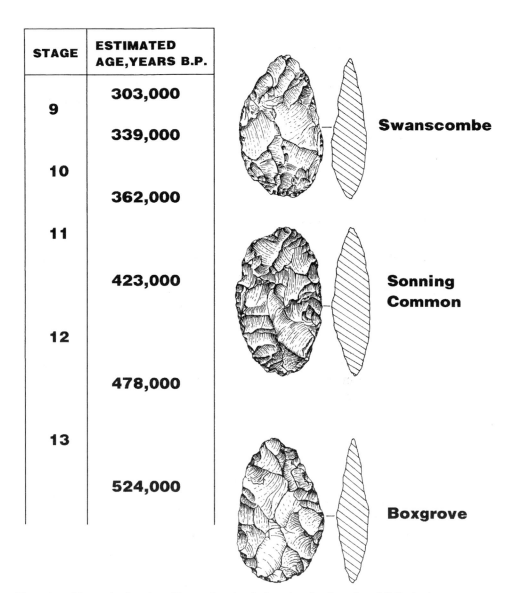

STAGE	ESTIMATED AGE, YEARS B.P.
9	303,000
10	339,000
	362,000
11	
12	423,000
	478,000
13	
	524,000

Swanscombe

Sonning Common

Boxgrove

Figure 21.1. Observation learning without cultural evolution : three handaxes from S.E. England made at different stages of the Quaternary showing high degrees of technological and typological similarity. Handaxes are *c.* 15 cm in length (after Wymer 1988).

and all with the same shape preference. As Wymer (1988) has described, in each of these handaxes one side of the cutting edge has been produced by the removal of a tranchet flake near the tip. What is significant is that these three handaxes span a chronological period of between at least 200 000 years between *c.* 50 000 and 300 000 years ago: that from Boxgrove is pre-Anglian in date (Oxygen Isotope stage 13), that from Sonning Common late Anglian (OI stage 12) and that from Swanscombe post-Hoxnian, (OI stage 11). There can be as little doubt that these handaxes required observational learning to manufacture, as there is that the 'ratchet' effect was absent.

This pattern found in S.E. England during the Lower Palaeolithic is repeated many times elsewhere in the Old World. For instance, between 250 000 and 30 000 years ago *Homo neanderthalensis* was present in Europe and the Near East (Stringer and Gamble 1993). This species has a brain size as large as that of modern humans and exhibited considerable technical skill in the production of artefacts such as levallois points. The level of skill required, and the highly similar morphologies of artefacts, again suggest that observational learning/ imitation would have been essential to the acquisition of the knowledge to make such stone tools. Yet such Mousterian technology also shows a remarkable stability through space and time (Kuhn 1993, Mellars 1996), with no evidence for the ratchet or snowball effect.

In summary, Tomesello *et al.* (1993: 508) are simply wrong to invoke a 'gradual increase in complexity' of 'hammer-like tools' during prehistory, as are Boyd and Richerson (1996: 80) when they claim that 'gradual change' is documented in the archaeological record. Such cumulative change is only a feature of the most recent prehistory, that after 50 000 years ago, after Early Humans with fully modern brain size and evident powers of imitation had been present for at least one million years.

Social learning and cultural behaviour

The evidence I have summarised indicates the dangers of relying on comparisons between chimpanzees and modern humans alone when trying to understand the relationship between social learning and cultural behaviour. As Boyd and Richerson (1996), Tomasello (1996; Tomasello *et al.* 1993) and Visalberghi (1993) have noted cumulative cultural change appears absent among apes. Observational learning/imitation is also either absent or only present in a very weak form. It is rash to conclude that there is necessarily a causal relationship between these before one has examined other species which are reliant on tools, and clearly show either cumulative cultural change or imitation/observational learning. Several species of Early Humans, such as *Homo heidelbergensis*, clearly exhibit the latter in their stone tool technology. Yet cumulative cultural change – the ratchet or snowball effect – remains absent. This only appears very late in human evolution, and appears associated with anatomically modern humans alone. Quite why, and when it appears, remains unclear. The potential for it may be associated with further cognitive evolution, what I have termed the emergence of cognitive fluidity (Mithen 1996), and its expression may be dependent upon particular social and ecological conditions that only arose in the later Pleistocene (Gibson 1996). What is evident, however, is that the data from the archaeological record regarding the thought and behaviour of pre-modern humans should not be ignored when trying to understand the relationship between social learning and cultural behaviour.

References

Asfaw, B., Beyene, Y., Suwa, G., Walter, R.C., White, T., Wolde-Gabriel, G. and Yemane, T. (1992). The earliest Achelean from Konso-Gardula. *Nature*, **360**, 732–5.

Bergman, C. A. and Roberts, M. B. (1988). Flaking technology at the Acheulian site of Boxgrove, West Sussex (England). *Revue Archéologique de Picardie*, **1–2**, 105–13.

Bermúdez de Castro, J. M., Arsuaga, J. L., Carbonell, E., Rosas, A., Martinez, I. and Mosquera, M. (1997). A hominid from the Lower Pleistocene of Atapuerca: possible ancestor to Neanderthals and Modern Humans. *Science*, **276**, 1392–5.

Binford, L. R. (1989). Isolating the transition to cultural adaptations: an organizational approach. In *The Emergence of Modern Humans: Biocultural Adaptations in the Later Pleistocene*, ed. E. Trinkaus, pp. 18–41. Cambridge: Cambridge University Press.

Boesch, C. and Boesch, H. (1990). Tool-use and tool-making in wild chimpanzees. *Folia Primatol.*, **54**, 86–99.

Boyd, R. and Richerson, P. (1996). Why culture is common, but cultural evolution is rare. In *Evolution of Social Behaviour Patterns in Primates and Man*, ed. W. G. Runciman, J. Maynard-Smith and R. I. M. Dunbar, pp. 77–93. Oxford: Oxford University Press.

Briggs, D. J., Coope, G. R. and Gilbertson, D. D. (1985). *The Chronology and Environmental Framework of Early Man in the Upper Thames Valley* (BAR 137). Oxford: British Archaeological Reports.

Carbonell, E., Bermúdez de Castro, J. M., Arsuaga, J. C., Diez, J. C., Rosas, A., Cuenca-Bercós, G., Sala, R. Mosquera, M. and Rodriguez, X. P. (1995). Lower Pleistocene hominids and artefacts from Atapuerca-TD6 (Spain). *Science*, **269**, 826–30.

Cook. J. and Ashton, N. (1991). High Lodge, Mildenhall. *Curr. Archaeol.*, **123**, 133–8.

Dibble, H. and Bar-Yosef, O. (eds.) (1995). *The Definition and Interpretation of Levallois Technology*. Madison: Prehistory Press, Monographs in World Archaeology No. 23.

Fisher, A. (1990). On being a pupil of a flint knapper 11 000 years ago. A preliminary analysis of settlement organization and flint technology based on conjoined flint artefacts from the Trollesgave site. In *The Big Puzzle: International Symposium on Refitting Stone Artefacts*, ed. E. Cziesala, S. Eickhoff, N. Arts and D. Winter, pp. 447–64. Bonn: Holos.

Galef, B. G. (1988). Imitation in animals: history, definition and interpretation of data from the psychological laboratory. In *Social Learning: A Comparative Approach*. eds. T. R. Zentall and B. G. Galef, pp. 77–100. Hillsdale, NJ: Erlbaum Press.

Galef, B. G. (1990). Tradition in animals: field observations and laboratory analysis. In *Methods, Inferences, Interpretations and Explanations in the*

Study of Behaviour, ed. M. Bekoff and D. Jamieson, pp. 74–95. Boulder, CO: Westview House.

Gibson, K. R. (1996). The biocultural human brain, seasonal migrations and the emergence of the Upper Palaeolithic. In *Modelling the Early Human Mind*, ed. K. R. Gibson and P. Mellars, pp. 33–46. Cambridge: The McDonald Institute.

Gowlett, J. (1984). Mental abilities of early man: a look at some hard evidence. In *Hominid Evolution and Community Ecology*, ed. R. Foley, pp. 167–92. London: Academic Press.

Hayes, K. J. and Hayes, C. (1952). Imitation in a home raised chimpanzee. *J. Comp. Physiol. Psychol.*, **45**, 450–9.

Heyes, C. and Galef, D. G (ed.) (1996). *Social Learning in Animals: The Roots of Culture*. San Diego: Academic Press.

Inizan, M-L., Roche, H. and Tixier, J. (1992). *Technology of Knapped Stone*. Paris: Cercle de Recherches et d'Etudes Préhistorique, CNRS.

Isaac, G. (1977). *Olorgesailie*. Chicago: University of Chicago Press.

Johanson, D. and Edgar, B. (1996). *From Lucy to Language*. London: Weidenfeld and Nicolson.

Jones, J. S., Martin, R. and Pilbeam, D. (eds.) (1992). *The Cambridge Encyclopedia of Human Evolution*. Cambridge: Cambridge University Press.

Kohn, M. and Mithen, S. (in press). Handaxes: products of sexual selection? *Antiquity*.

Kuhn, S. (1993). Mousterian technology as adaptive response. In *Hunting and Animal Exploitation in the Later Palaeolithic and Mesolithic of Eurasia*, ed. G. L. Peterkin, H. Bricker and P. Mellars, pp. 25–31. Archaeological Papers of the Amercian Anthropological Association, no. 4.

McGrew, W. (1992). *Chimpanzee Material Culture*. Cambridge: Cambridge University Press.

Mellars, P. (1996). *The Neanderthal Legacy*. Princton: Princton University Press.

Mithen, S. (1994). Technology and society during the Middle Pleistocene. *Cambr. Archaeol. J.*, **4**(1), 3–33.

Mithen, S. (1996). *The Prehistory of the Mind: A Search for the Origins of Art, Science and Religion*. Thames and Hudson.

Nishida, T. (1987). Local traditions and cultural transmission. In *Primate Societies*, ed. B. B. Smuts, R. W. Wrangham and T. T. Struhsaker, pp. 462–74. Chicago: Chicago University Press.

Pelegrin, J. (1993). A framework for analysing prehistoric stone tool manufacture and a tentative application to some early stone industries. In *The Use of Tools by Human and Non-human Primates*, ed. A Berthelet and J.Chavaillon, pp. 302–14. Oxford: Clarendon Press.

Pigeot, N. (1990). Technical and social actors: flint knapping specialists and apprentices at Magdalenian Etiolles. *Archaeol. Rev. Cambr.*, **9**, 126–41.

Roberts, M. B. (1986). Excavation of the lower palaeolithic site at Amey's Eartham Pit, Boxgrove, West Sussex: a preliminary report. *Proc. Prehist. Soc.*, **52**, 215–46.

Roberts, M., Gamble, C. S. and Bridgeland, D. R. (1995). The earliest occupation of Europe: The British Isles. In *The Earliest Occupation of Europe*, ed. W. Roebroeks and T. van Kolfschoten, pp. 165–91. Leiden: University of Leiden Press.

Roe, D. (1981). *The Lower and Middle Palaeolithic Periods in Britain*. London: Routledge and Kegan Paul.

Roebroeks, W. (1996). The English Palaeolithic record: absence of evidence, evidence of absence, and the first occupation of Europe. In *The English Palaeolithic Reviewed*, ed. C. S. Gamble and A. J. Dawson, pp. 57–62. Salisbury: Trust for Wessex Archaeology.

Russon, A. E. (1996). Imitation in everyday use: matching and rehearsal in the spontaneous imitation of rehabilitant orangutans. In *Reaching into Thought: The Minds of the Great Apes*, ed. A. Russon, K. Bard and S. T. Parker, pp. 152–76. Cambridge: Cambridge University.

Schick, K. and Toth, N. (1993). *Making Silent Stones Speak: Human Evolution and the Dawn of Technology*. New York: Simon and Schuster.

Stringer, C. and Gamble, C. (1993). *In Search of the Neanderthals*. London: Thames and Hudson.

Swisher, C. C. III, Rink, W. J., Antón, S. C., Schwarcz, H. P., Curtis, G. H., Suprijo, A. and Widasmoro. (1996). Latest

Homo erectus of Java: potential contemporaneity with *Homo sapiens* in Southest Asia. *Science*, **274**, 1870–4.

Thieme, H. (1997). Lower Palaeolithic hunting spears from Germany. *Nature*, **385**, 807–9.

Tomasello, M. (1996). Do apes ape? In *Social Learing in Animals*, ed. C. Heyes and D. G. Galef, pp. 291–318. San Diego: Academic Press.

Tomasello, M., Kruger, A. C. and Ratner, H. H. (1993). Cultural learning. *Behav. Brain Sci.*, **16**, 495–552.

Tyldesley, J. (1986). *The Wolvercote Channel Handaxe Assemblage: A Comparative Study*. Oxford: British Archaeological Reports, British Series 152.

Visalberghi, E. (1993). Capuchin monkeys: a window into tool use in apes and humans. In *Tools, Language and Cognition in Human Evolution*, ed. K. R. Gibson and T. Ingold, pp. 138–50. Cambridge: Cambridge University Press.

Visalberghi, E. and Fragaszy, D. M. (1990). Do monkeys ape? In *'Language' and Intelligence in Monkeys and Apes: Comparative Developmental Perspectives*, ed. S. T. Parker and K. R. Gibson, pp. 247–73. Cambridge: Cambridge University Press.

Walker, A. and Leakey, R. (eds.) (1993). *The Nariokotome Homo erectus Skeleton*. Berlin: Springer-Verlag.

Wenban-Smith, F. F. (1989). The use of canonical variates for determination of biface manufacturing techniques at Boxgrove Lower Palaeolithic site and the behavioural implications of this technology. *J. Archaeol. Sci.*, **16**, 17–26.

Whiten, A. (1993). Human enculturation, chimpanzee enculturation and the nature of imitation. *Behav. Brain Sci.*, **16**, 538–9.

Whiten, A. (1996). Imitation, pretense, and mindreading: secondary representation in comparative primatology and developmental psychology. In *Reaching into Thought: The Minds of the Great Apes*, ed. A. Russon, K. Bard and S. T. Parker, pp. 300–24. Cambridge: Cambridge University Press.

Whiten, A. and Custance, D. (1996). Studies of imitation in chimpanzees and children. In *Social Learing in Animals*, ed. C. Heyes and D. G. Galef, pp. 291–318. London: Academic Press.

Wrangham, R., McGew, W., de Waal, F. and Heltne, P. (eds.) (1994). *Chimpanzee Cultures*. Cambridge, MA: Harvard University Press.

Wymer, J. (1974). Clactonian and Acheulian industries from Britain: their character and significance. *Proc. Geol. Assoc.*, **85**, 391–421.

Wymer, J. (1983). *The Palaeolithic Age*. London: Croom Helm.

Wymer, J. (1988). Palaeolithic archaeology and the British Quaternary sequence. *Q. Sci. Rev.*, **7**, 79–98.

Wynn, T. (1995). Handaxe enigmas. *World Archaeol.*, **27**, 10–23.

Wynn, T. and Tierson, F. (1990). Regional comparison of the shapes of later Acheulean handaxes. *Am. Anthropol.*, **92**, 73–84.

Concluding remarks

Social learning and behavioural strategies among mammals

Hilary O. Box and Kathleen R. Gibson

The material presented in this volume complements ongoing interests in the field of social learning (see, for example, Heyes and Galef 1996). In some cases it also differs substantively from existing material but provides additional not contradictory perspectives. In these last pages we highlight some points of special interest that include directions for further study.

A central objective of the volume is to discuss opportunities for, and advantages of, the social transfer of information in the natural lifestyles of a wide diversity of mammalian taxa, mostly by studies in nature but also with reference to exemplary studies in captivity. Socially mediated behaviours are part of the adaptive strategies of different taxa; they are part of an evolved complex that has advantages for the success and fitness of individuals across a wide diversity of species, in different functional contexts. As yet, however, given the diversity of the mammals, as among their characteristics and life-styles, we have information for relatively few species. Hence, it is relevant to increase the comparative database in order to understand better the phenomena involved and their biological significance, and to redress the relative imbalance between functional approaches in the area – especially in the field – and the more frequent studies of mechanisms of social learning – in captivity.

We have new information on previously unstudied taxa in social learning contexts, as with marsupial macropodids, on relatively unfamiliar species in this domain, as with naked mole-rats, together with additional perspectives for relatively well known species, as among cats and rats. We include comparative studies of closely related species and of distantly related taxa living in the same environment. Studies of closely related species, as with the chapters on felids and canids by Kitchener (chapter 14) and Nel (chapter 15), can help us determine how species who resemble each other in their anatomical, sensory, motor and mental capacities adapt to the demands of different environments. In contrast, studies of unrelated taxa living in the same area (see Klein's paper on caribou, muskoxen and arctic hares, chapter 7) can help elucidate how species with significantly different biological propensities meet similar environmental challenges. Discussions for particular groups are given in the context of their taxonomic diversity, the diversity of their habitats, their social systems and the food that they eat. Opportunities for, and examples of, socially mediated behaviour are discussed, and environmental conditions are described in which these are potentially advantageous.

The material is of intrinsic interest for understanding behavioural strategies among different taxa. There is also a number of more general points. For example, in many mammals the mother is a predominant source of information from which young individuals may learn about the opportunities and hazards of their environments. She is not only a particularly salient referent in many species, but in some cases she is also the only experienced adult around. However, although there are notable exceptions, as with domestic cats and cheetahs (chapter 14), these social systems have not created a great deal of interest in studies of social learning.

Contributions in this volume emphasise different points in this regard. Hence, well controlled experiments by Hudson and her co-workers (chapter 8) show that in an extreme social system with minimal contact between mother and offspring, there are functionally important opportunities to acquire information at different stages in preweaning development. These data add to the growing body of data that a diversity of mammals, including humans, acquires dietary preferences via a variety of early developmental opportunities as presented in milk. In all mammals lactation not only frees a developing infant from foraging independently, but provides opportunities to acquire information (see Galef and Sherry, 1973, for example) that may influence subsequent dietary choices at a time when trial and error would expose individuals to a variety of hazards. In other cases that involve 'solitary' wild mammals – as with many of the cats (chapter 14) and the bears (chapter 13) – close bonds develop with nursing and protection, and mothers contribute to the survival and fitness of their offspring, as well to their own fitness, by presenting the offspring with a variety of opportunities to develop into competent individuals. Information about the identification and distribution of foods and how to deal with them are cases in point. It is also important to emphasise these cases because so much attention is paid to species that live in 'socially complex groups'. Social complexity is, in any case, difficult to define.

In a different domain, species of carnivores are of interest because, in many cases, they need to acquire skills in dealing with their food that herbivores do not. Canids, felids and killer whales (chapter 16), for example, learn to capture prey. In some instances, this may be a dangerous undertaking, either because of the potential ability of prey to injure their predators, or because the prey capture techniques used may be potentially harmful – as in the beaching behaviours of killer whales. These factors, coupled with the ability of prey animals to elude their predators, place a premium on skilful predator behaviours, some of which can only be acquired after much practice. Adults of some species, such as felids, bat-eared foxes, killer whales and some dolphins, adapt to these circumstances by providing practice opportunities for their young. In some instances they may also actively demonstrate appropriate prey capture techniques. These examples have sometimes created the impression that social learning is functionally more important for carnivores than for herbivores. However, plant materials are often defended by thorns, spines and hard outer shells or rinds; they may also have to be selected among numerous species – as to avoid toxins for example. Byrne (chapter 18) provides a detailed discussion of the manipulative skills of young gorillas in eating nettles. It is important also, of course, that these animals are great apes, a group for which Gibson (1996) and Parker (1996) have emphasised an adaptive complex of propensities that include a variety of mental abilities not found in monkeys, for instance. It is certainly relevant to consider the extents to which social learning may play a role in the development of plant processing techniques in other mammals. This has not been a popular topic for investigation. Other herbivores, such as giraffes, feed on plant foods that are defended with thorns and spines and may possess prehensile tongues (giraffes) or prehensile lips (black rhinoceros) to aid in plant ingestion. Elephants (chapter 6) possess a highly manipulative organ – the trunk – and may use combined trunk and tusk movements in their foraging endeavours. The extents to which young ungulates who possess highly manipulative tongues, lips or trunks learn appropriate food processing techniques by socially mediated behaviour is largely unknown. Once again, it is a question of emphasis. Carnivores certainly present interesting challenges for studies of social learning in comparative terms, but it is also important to consider challenges presented by a variety of herbivorous lifestyles. Klein's discussion of caribou (chapter 7) for example, emphasises the functional importance of young animals acquiring information about the microhabitats of lichen rich foods. Further, an excellent study of African buffalo by Prins (1996) considers the accessibility of grass material as food. Critically, it appears that 'different patches of even the same food species at the same time can be different from the herbivore's point of view . . . . This undoubtedly is also the case in other ecosystems and for other herbivores. The problem for the herbivore is then to discriminate between these patches, which to the human observer look alike' (p. 259). In his study area, for example, grazing grounds are not equivalent either in the availability of protein or in the condition of the sward. He suggests that buffalo utilising grass benefit from sociality both by sharing information about food patches, and by keeping the sward in optimal conditions for grazing. Shared information with immediate advantages for individuals, together with potential future advantages in terms of accumulated information, provide benefits of sociality at a potential cost of food competition. Conditions are provided then, for shaping sociality by natural selection, and the development of mental abilities 'to relate different unconnected pieces of information in new ways and to apply the results in an adaptive manner' (p. 259–60). Studies of herbivores raise new perspectives in behavioural ecology including those of social learning.

The term 'social learning' is very broad; it includes creating ecological opportunities in different ways by different mechanisms among a wide diversity of animals. We are at a point where we can appreciate social learning in the behavioural biology of a variety of mammalian taxa that include animals with very different life-history characteristics and lifestyles. A wide diversity of taxa is included in this volume. It is clearly inappropriate to assume that specific groups with specific characteristics are critical for social learning. For example, there are implicit assumptions that species that have large brains, live long lives in 'complex social groups' and develop complex skills as in obtaining their food have 'exclusive' opportunities to acquire information from conspecifics. However, fine grain comparative analyses of the influence of such characteristics is an important, but as yet imprecise, aspect of social learning studies. Moreover, given our general lack of information in this area for so many taxa, we should also consider, for example, the extents to which social learning plays significant roles in the behavioural strategies of animals that are relatively short-lived and spend much of their lives in relative social isolation. There is clearly a diversity of mammals that we may consider in this context. We have not even begun to consider possibilities. Social influences may be of short exposure for instance, but functionally important in the acquisition of information.

Information so far shows that different propensities and lifestyles among different taxa present different opportunities to learn different things in different ways (see also the final section of these concluding remarks). In this context, it is also relevant to consider the implications of the fact that much of the emphasis of research in social learning, both in nature and with experiments in captivity, has focused on foraging and feeding behaviour. Much of this volume also concerns foraging and feeding behaviour! This is where the information is! Feeding conditions are often more easily quantified than many other behavioural strategies; they are also critical to the survival and fitness of individuals. Exclusive reliance upon specific behavioural strategies, however, is unwarranted on various grounds. We obviously need more information about social learning and behavioural strategies as in avoiding and dealing with predators, for example. Until we have such information for behaviour that is suggested by the biology of the species, it is difficult to say where social learning makes functional impacts in the lives of a diversity of animals. The acquisition of social information as among the simian primates for example, has been relatively neglected. This is an unwarranted omission for species that live in long-lasting social groups with relatively complex patterns of social interaction that are demonstrably central to their life strategies (Harcourt and de Waal 1992). There *are* methods that provide sound observations about the acquisition of social information in controlled conditions (see chapter 3, section 3 for instance). It is a question of interest to develop them. Moreover, concentrating exclusively on particular strategies of behaviour may give rise to misleading biases of interpretation. For example, among simian primates evidence shows that the donation of information to young animals by experienced adults in feeding situations is rare across species of monkeys. This contrasts with information for species of great apes. However, additional observations, as with some aspects of social and feeding behaviour, now also show evidence for donation of information among species of monkeys as well as in the great apes. There are functional implications. It remains the case, however, that great apes are substantively different from monkeys in both regards.

It is also the case that much interest in social learning understandably concerns the acquisition of information among young animals – as in this volume. For many species however, behavioural strategies change as individuals get older and adapt to changing social and ecological challenges. There is much to learn about this.

In a different context, studies of sex differences raise critical but undersubscribed issues in natural social learning among mammals. The asymmetry between the sexes with regard costs of reproduction leads to a variety of social, organisational consequences. Females in different species may invest differentially in male and female offspring depending upon their relative future reproductive success. Further, differences among adult males and females in their life histories and social experiences may be considered with regard to opportunities that they have for socially mediated learning when they are young. For example, in eastern grey kangaroos, red kangaroos and red-necked wallabies (see chapter 5) play fighting is more frequent among females with their male offspring at certain ages than with their female young. There are various hypotheses to be considered here, but it is relevant that the development of fighting abilities and assessment of competitors is more important to males because it influences their access to females. Further, studies of African elephants (chapter 6) specifically present data on the developmental trajectories of males and females from birth to dispersal (males) and first reproduction (females), and raise central questions that consider how calves learn social responses that are appropriate for their age and sex, and integrate them into simultaneous physical changes that occur with age. Hence, differences in the physical processes of growth and reproductive maturity have consequences for their opportunities to learn social and environmental skills that fulfil their different requirements. Moreover, although young males and females have similar social environments when they are very young, their behaviour alters those environments as they get older; their behaviour interacts with previous experiences of their mothers and with their social contacts with each sex. These are important perspectives for studying the interrelationships among physical, physiological and social development.

From a different standpoint, one of our main aims in this volume is to stimulate more interest in social learning among field biologists working with a wide diversity of taxa. Field studies provide the natural ecological and social contexts for social learning, but it is important to recognise and address methodological difficulties that are commonly encountered. Apart from the physical difficulties of actually making close observations of animals in many groups, the influences of social transfer of information are often difficult to specify and clearly demonstrate in nature. For example, one problem that is not infrequently mentioned is that individuals of a diversity of taxa that include red foxes (see chapter 15), cats (see chapter 14) and monkeys (Milton 1993) may develop a fully competent behavioural repertoire as in foraging and feeding, without the opportunities to acquire information from experienced individuals. It is clearly important to know the extents to which the acquisition of information is critical and necessary to individual survival and fitness. Nevertheless, social learning may provide individuals with opportunities to accelerate learning about their environments in ways that have functionally important implications. Field studies show that the demands of the environment put inexperienced individuals at severe disadvantages without information that is available from experienced individuals (see chapter 14, for example). Behavioural flexibility in response to local environmental conditions, which is such a feature of behavioural adaptations among mammals, emphasises the significance of such observations. Furthermore, it is important to develop perspectives and techniques to address both the constraints and the advantages that field conditions offer (see chapter 2). It is inappropriate both to apologise for a lack of experimental rigour, and to undermine the potential value of observations in many field studies. These situations involve different and, at best, complementary conditions to well controlled experiments. King's distinction between social information acquisition and social information donation (see chapter 2) is an excellent example of a method for use in the field that deserves wide trial and application. It not only circumvents methodological problems that arise from problems of adequately sampling behaviour in nature, but allows specific functional questions to be asked, and subsequently considered comparatively. For instance, there is a diversity of examples in this volume in which the acquisition of information among inexperienced individuals is facilitated by specific patterns of behaviour directed towards them by experienced animals. These individuals may release prey in the presence of young; they may use particular calls to attract them to the feeding situation. In other cases feeding may occur in contexts of demonstration and teaching. Functional questions concern the kinds of information that is provided by which classes of individuals under different environmental conditions. Relatedly, there are implications for theories of evolutionary biology (see chapter 4) and for comparisons of sensorimotor and mental capacities among different taxa.

A long-standing implicit assumption of social learning studies is that information is passed among individuals as units of information, as 'packages'. This volume presents two new and distinct yet complementary and interrelated perspectives in this regard. First, King's discussion (chapter 2) that interactions among individuals interface individual experiences and propensities, and hence transform the information that is socially transferred, is an important and realistic perspective for future research. Second, the development of behavioural strategies, including those that are socially mediated, involve interactions among biological systems within individuals (chapter 3). Moreover, interactive approaches to study social learning also serve to emphasise another perspective, namely, that social learning is socially *mediated* learning, and is our preferred term. It has been used extensively throughout the volume. It does get a little cumbersome to use it in every instance, however!

Interest in individual differences presents something of a paradox. On the one hand, an important feature of the mammals is that individuals of the same species behave differently in different environmental circumstances; that they demonstrate marked flexibility in behaviour. Alternatively, although there is a growing interest (Hayes and Jenkins 1997) this has been an undersubscribed area in behavioural biology. With regard to specific interests in social learning, it is important that not only differences among species, but differences among individuals in their propensities for social learning, will have consequences for fitness and survival (Caro and Hauser 1992). In this context, individual differences in temperament and social learning provide one such approach (chapter 3). Moreover, within species, social units of individuals may show differences in behaviour that influence their survival and fitness. Importantly, if selection acts upon social groups then it acts upon the results of social learning within groups (chapter 1). Issues in this regard that involve differences in behaviour among social units within populations have received little consideration. There is, as Rowell points out, a pervasive unpopularity of 'group selection' from various perspectives, for example. Alternatively, the points that Rowell raises are realistic and important for the future development of studies in social learning. Differences among individuals in information that they have previously acquired, as among animals that migrate between social units, as well as differences in responsiveness among individuals to ongoing environmental challenges for example, provide opportunities for others to acquire information that may be differentially advantageous in local conditions.

One unusual aspect of this volume is the inclusion of humans within a comparative mammalian framework. A second is its use of the hominid capacities. As we conclude this volume, it is appropriate to question whether these approaches contribute new insights. We believe that they do. Human cultural capacities exceed those of other animals in technological, artistic, musical and other domains that rely on the manufacture of objects. As Shennan and Steele report (chapter 20), human craft production is primarily transmitted in the vertical dimension from parents to young. Other authors have recently suggested that vertical transmission of information is relatively rare in animals and that a major difference between human societies and those of other animals is that only human societies have both vertical and horizontal information transmission (Laland *et al.* 1993). Collectively, the papers in this volume demonstrate, however, that the transmission of information in the vertical direction, especially from mothers to young, is actually an extremely common phenomenon in mammals. Indeed, all mammals potentially can transmit information in the vertical direction via substances in maternal milk, breath and excretory products. All mammals that live in social groups can also potentially transmit information horizontally. Consequently, the ability to transmit information both in vertical and horizontal dimensions does not appear to be unique to humans and is unlikely to be a primary reason that human cultural capacities exceed those of other animals.

Another recent suggestion has been that imitation is the critical mental capacity that allows human cultural innovations to build upon each other and, thus, to produce cumulative cultural change (Boyd and Richerson 1996). It is, however, now clear that great apes can imitate the end results of a series of object manipulation procedures (chapter 18), and the hominid archaeological record clearly indicates that human imitative abilities existed long prior to any evidence of cumulative human cultural change (chapter 21). Hence, imitative capacity alone is not sufficient to produce cumulative cultural change. Other abilities appear to be needed.

The increased information processing capacity provided by the enlarged human brain provides humans with the ability to create new concepts, new motor routines and new objects by combining and recombining previously constructed subunits into still higher order constructs (chapter 19). Gibson suggests that it is this mental constructional capacity that allows humans to subsume previous cultural innovations under still later innovations, and that mental constructional capacity also provides humans with imitative, teaching and linguistic skills. She also notes that humans have well-developed procedural learning skills, which, in combination with their mental constructional capacity, provides them with tool-using, dance and other motor capacities that are essential for human culture.

These human abilities reflect the interactive functioning of numerous areas of the brain that have enlarged in human evolution. Other species discussed in this volume, including elephants, cetaceans and great apes, also have brains that are much larger both in absolute terms and with respect to body size than the brains of most mammals (Jerison 1973). This raises questions of whether these animals might also possess mental constructional capacity. Much information summarised elsewhere indicates that great apes do possess rudimentary mental constructional capacity in communicative, technical and social domains (Gibson and Ingold 1993). As a result, great apes possess the rudiments of many behaviours that are pertinent to social learning and that were once thought to be uniquely human, including language, tool-making, self awareness, imitation and theory of mind (Gibson 1996, Parker 1996).

The ability to hold several concepts in mind simultaneously and to note relationships between them is the critical component of mental constructional capacity. Dolphin and killer whale behaviours described in this volume (chapters 16 and 17) indicate that these animals do possess some mental constructional capacity. For instance, their vocal imitative skills require the ability to note relationships between kinesthetic and auditory images. Their abilities to herd fish cooperatively demonstrate that they can simultaneously keep in mind and note relationships between the behaviours of the fish, the behaviours of their conspecifics and their own behaviour. Those dolphins that herd fish into fishermen's nets must keep in mind still additional concepts pertaining to the net and to the men. Whether or not elephants also possess mental constructional capacities is less clear. Lee and Moss (chapter 6), however, do note that they learn to coordinate movements of the trunk. This might suggest they have some capacity to construct overall trunk movements from movements of individual parts of the trunk. If hypotheses presented in this volume are correct, one would predict that other evidence of mental constructional capacity will eventually be found in elephants. One might look, for example, for

behaviours such as imitating the trunk movements of other animals, constructing varied relationships between movements of the trunk and other body
parts, and constructing relationships between objects.

Brains are metabolically expensive organs. Hence, most species with large
brains also have large bodies, and those species whose brains are unusually
large in comparison with their body size tend to have short guts (Aiello and
Wheeler 1995) and to consume foods that are quickly digested and high in
nutrients. Brain growth is particularly demanding of nutrients, and perhaps, in
part, for this reason, animals with large brains also tend to grow and mature
physically for at least several years, sometimes for a decade or more, after birth.

Animals who reach full maturity only after years of growth may also
demand considerable maternal investment. Hence, we find that the large-
brained elephants, great apes, cetaceans and humans usually give birth to one
infant at a time and, except in technologically advanced modern human
societies, exhibit birth intervals of three or more years. Great ape infants may
nurse for five or more years, and elephants for up to four years. Although
human infants in hunter–gatherer societies typically nurse for about three
years, children in all human societies are usually provisioned by adults until
they reach their teenage years. In most hunter–gatherer societies both parents
help provision the young, and grandparents sometimes help as well. Time
spent nursing or provisioning individual offspring detracts from potential
investments in others. Hence, large brains are not only metabolically expensive, but they incur considerable parental reproductive cost. Shennan and
Steele (chapter 20) postulate that active teaching would be advantageous in
these conditions, because it would permit offspring to achieve independence
more quickly, and, hence, potentially increase parental reproductive fitness. In
fact, as Shennan and Steele note, chimpanzees of the Taï Forest have been
observed demonstrating nut-cracking techniques to their young. The Heim-
lich-Borans (chapter 16) also describe teaching behaviours in killer whales and
dolphins. Hence, theoretical considerations derived from analyses of human
behaviour and developmental patterns may also have some predictive potential with regard to the behaviours of other large-brained animals.

Given the metabolic, social and reproductive costs of large brains, the
question is 'why have them?'. The traditional answer has been that large brains
allow learning. It is clear, however, that both small-brained mammals, such as
rats and rabbits, and large-brained animals, such as elephants, cetaceans, great
apes and humans, possess learning, including social learning, capacities.
Hence, an enlarged brain is not essential for learning *per se*. Rather, differences
in brain size relate to differing sensorimotor capacities and, hence, to different
learning styles. In most mammals, for example, the size of the neocortex

reflects the extent to which they possess well-developed manipulative organs, such as hands or prehensile tails, and the extent to which they possess expanded visual, auditory and tactile sensory capacities. In humans, apes, monkeys and possibly in other animals as well, neocortical size also reflects mental constructional capacity (Gibson 1990).

Hence, the keys to the reasons for enlarged brains relate first to the advantages provided by enhanced motor skills and enhanced visual, auditory and tactile perception and, second, to the advantages of mental constructional capacities. It is beyond the scope of this volume to address questions of comparative mammalian sensory and motor adaptations. Since mental con-structional abilities appear integral to human social learning capacities and perhaps also to the social learning capacities of other large-brained animals discussed in this volume, including cetaceans, great apes and elephants, it is pertinent to ask if any common ecological or other adaptive advantages characterise these animals and humans.

Given that elephants and gorillas are herbivorous and killer whales and dolphins are carnivorous, the enlarged brains of these animals cannot be accounted for by any simple dietary explanation. Humans, elephants and dolphins are all considered to live in complex societies, but many other mammals, including many ungulates, naked mole rats, and monkeys, also live in complex societies. It is not at all clear in what, if any, ways elephant, cetacean or great ape societies should be considered more complex than those of many other species or in what way the social structure of these groups would demand mental constructional capacity above and beyond that possessed by other species.

One factor, however, does appear to unite this group of large-brained animals. They all use foraging techniques that require procedural skills that are mastered only after several years of practice on the part of the young. Hence, the young cannot fully provision themselves until several years of age or later. In each case, these foraging techniques appear to open feeding niches and/or to increase foraging efficiency. Human and great ape foraging procedures make it possible for them to exploit many high nutrient foods that must be extracted from shells, bark or other matrices (Parker and Gibson 1977, 1979). Gorillas' mental constructional techniques permit them to exploit nettles and other foods that contain stings or spines (chapter 18). Humans can also hunt big game, because they possess the mental constructional capacities to manufac-ture hunting tools. Dolphin and killer whale mental constructional abilities permit the cooperative hunting of schools of fish. Their ability to learn complex procedures also allows them to beach sea lions and, in the case of some dolphins, to use tools. The manipulative elephant trunk allows them to acquire foods from both the ground and from the tree tops and to manipulate both very small foods and very large ones. These considerations suggest that the primary advantage of the enlarged brains for these species is that it opens foraging niches not available to animals with lesser abilities to learn varied procedural skills.

This emphasis on the foraging advantages of mental constructional capacity is not meant to discount the social advantages of such capacities. Mental constructional capacities can also be of value in terms of cooperation and for understanding the perspectives of others. The nutrient demands of enlarged brains are such, however, that even though mental construction also enhances 'social' intelligence, the enhancement of foraging capacity is a basic requirement without which the enlarged brains could not exist.

References

Aiello, L. C. and Wheeler, P. (1995). The expensive tissue hypothesis: the brain and the digestive system in human and primate evolution. *Curr. Anthropol.*, **36**, 199–221.

Boyd, R. and Richerson, P. (1996). Why culture is common, but cultural evolution is rare. In *Evolution of Social Behavior Patterns in Primates and Man*, ed. W. G. Runciman, J. Maynard Smith and R. I. M. Dunbar, pp. 77–93. Oxford: Oxford University Press.

Caro, T. M. and Hauser, M. D. (1992). Teaching in non-human animals. *Q. Rev. Biol.*, **67**, 151–74.

Galef, B. G., Jr and Sherry, D. F. (1973). Mother's milk: a medium for transmission of cues reflecting the flavor of mother's diet. *J. Comp. Physiol. Psychol.*, **83**, 374–8.

Gibson, K. R. (1990). New perspectives on instincts and intelligence: brain size and the emergence of hierarchical mental constructional skills. In *Language and Intelligence in Monkeys and Apes: Comparative Developmental Perspectives*, ed. S. T. Parker and K. R. Gibson, pp. 97–128. Cambridge: Cambridge University Press.

Gibson, K. R. (1996). The biocultural human brain, seasonal migrations, and the emergence of the Upper Paleolithic. In *Modelling the Early Human Mind*, ed. P. Mellars and K. Gibson, pp. 33–46. Cambridge: The McDonald Archaeological Institute.

Gibson, K. R. and Ingold, T. (1993). *Tools, Language, and Cognition in Human Evolution*. Cambridge: Cambridge University Press.

Harcourt, A. H. and de Waal, Frans F. B. M. (1992). *Coalitions and Alliances in Humans and other Animals*. Oxford: Oxford Scientific Publications.

Hayes, and Jenkins, (1997). Individual variation in mammals. *J. Mammal.*, **78**(2), 274–93.

Heyes, C. M. and Galef, B. G., Jr (eds.) (1996). *Social Learning in Animals: The Roots of Culture*. San Diego: Academic Press.

Jerison, H. J. (1973). *Evolution of the Brain and Intelligence*. New York: Academic Press.

Laland, K. N., Richerson, P. and Boyd, R. (1993). Animal social learning: toward a new theoretical approach. *Perspect. in Ethol.*, **10**, 249–77.

Milton, K. (1993). Diet and social organisation of a free-ranging spider monkey population: the development of species-typical behaviour in the absence of adults. In *Juvenile Primates Life*

History Development and Behaviour, ed. M. E. Pereira and L. A. Fairbanks, pp. 173–81. Oxford: Oxford University Press.

Parker, S. T. (1996). Apprenticeship in extractive foraging: the origins of imitation, teaching, and self-awareness in great apes. In *Reaching into Thought: The Minds of Great Apes*, ed. A. E. Russon, K. A. Bard and S. T. Parker, pp. 348–70. Cambridge: Cambridge University Press.

Parker, S. T. and Gibson, K. R. (1977). Object manipulation, tool use, and sensorimotor intelligence as feeding adaptations in cebus monkeys and great apes. *J. Hum. Evol.*, **6**, 435–49.

Parker, S. T. and Gibson, K. R. (1979). A developmental model for the evolution of language and intelligence in early hominids. *Behav. Brain Sci.*, **1**, 367–408.

Prins, H. H. T. (1996). *Ecology and Behaviour of the African Buffalo: Social Inequality and Decision Making.* London: Chapman and Hall.

Index